Quantum Information

Gregg Jaeger

Quantum Information
An Overview

Springer

Dr. Gregg Jaeger
Quantum Imaging Laboratory
Department of Electrical and Computer Engineering
Boston University
8 Saint Mary's Street
Boston, MA 02215
USA
and
Department of Natural Sciences
College of General Studies
871 Commonwealth Avenue
Boston, MA 02215
USA
jaeger@math.bu.edu

e-ISBN-13: 978-1-4419-2258-8
e-ISBN-10: 0-387-36944-9
e-ISBN-13: 978-0-387-36944-0

Printed on acid-free paper.

9 8 7 6 5 4 3 2 1

springer.com

To my daughter Alia and all those who have inspired, supported,
and/or prodded me

Foreword

In one word, this is a *responsible* book; the rest is commentary.

Around 1992 a few of us were led by Charles Bennett into a Garden of Eden of quantum information, communication, and computation. No sooner had we started exploring our surroundings and naming the birds and the beasts, than Peter Shor put an end to that apparent innocence by showing that factoring could be turned—by means of quantum hardware—into a polynomial task. Fast factoring meant business; *everybody* seemed to be awfully interested in factoring. Not that anyone had any use for factoring *per se*, but it seemed that all the world's secrets were protected by factor-keyed padlocks. Think of all the power and the glory (and something else) that you might get by acting as a consultant to big businesses and government agencies, helping them pick everyone else's locks and at the same time build unpickable ones (well, *nearly* unpickable) for themselves. And if one can get an exponential advantage in factoring, wouldn't an exponential advantage be lying around the corner for practically any other computational task? Quantum information "and all that" has indeed blossomed in a few years into a wonderful new chapter of *physics*, comparable in flavor and scope to thermodynamics. It has also turned into a veritable "industry"—producing papers, conferences, experiments, effects, devices—even proposals for quantum computer architectures. Dutifully, also entire books on the subject have been appearing with a certain regularity. Every time I see a new one, my first reaction tends to be, "Who ordered that?" meaning, What needs does this book fill? What market does it address?

I'm convinced that a *bona fide* academic book (as contrasted to a commercial book) is first and foremost a knowledge-structuring exercise, a taut "clothesline" (to use an image by Herbert Wilf) on which to neatly pin one's thoughts and find them still there in the morning. In this respect, the present work is no exception. But one doesn't have to go through all the labor of producing a real book just for that. A second, also quite honorable motive, is to let your colleagues know that you've been there yourself; that you've seen a few things that they may have missed; that from a certain angle you get a much better view; and so on; that, in other words, by your *meisterstück* you claim full membership in the guild. In the meantime your colleagues had of course been looking at what you were doing all along, and probably had already made you member *in pectore*. I have no doubts that this applies to the present case.

But in many, perhaps most, of the books I've seen I believe I detected in an obstinate bass line, some sort of rumbling or blurbing that has no words to it but I would be tempted to interpret like this: "Yes, this book of ours will be good for an advanced undergraduate course in Quantum Computation, Quantum Cryptography, or Entanglement Distillation (or any other permutation of a number of similar sexy terms). But the real reason you must want it, for yourself and for your students, is because we are nearing the moment when a *Quantium*—rumored to be able to do *all* computations exponentially faster—is to be commercially available as a drop-in replacement for the Pentium! You need this book because you cannot afford not to be yourself one of the very designers of the Quantium, or at least one of the first to design it in!" I may just be hearing voices. But Quantum Information, Communication, and Computation is too rich a conceptual discipline to need debasing with the subliminal lure of "universal exponential speedup." For the moment such a promise should be kept in the same class as "Energy so cheap it won't be worth metering;" it doesn't even have to be false to be irresponsible. This book steers clear of all that.

I recall the title of E.T. Jaynes's book on information theory, *Probability— The Logic of Physics*, and paraphrase it as "the logic of incomplete information," thus stressing that physics, even though central for motivation, is, from a conceptual viewpoint, merely incidental to information theory. An *incomplete description* is just that, namely, one that is not sufficiently detailed to identify a single individual: several individuals may fit it. The art of probability is nothing more than doing ordinary Aristotelian logic *in parallel* on all those "several" (as often as not 10^{24}) individuals, and in the end lumping the results into "bins" according to whatever traits are relevant to our question of the moment. Introducing (that is, making up) a *probability distribution* is in essence equivalent to doing some of that binning *before* putting the system through the "logic engine" rather than after. In fact, if this is done properly, the two approaches *commute*, and the second, of course, may save much computational effort.

Quantum behavior has confronted physics with many novelties. What matters here is that it has introduced formerly unsuspected ways for a description to be incomplete. Introduced where? into physics? My gut feeling is that in this business physics *per se* is largely irrelevant. (Think, for example, of how *entropy*, introduced for very good reasons by physicists, is in fact the fundamental quantitative parameter of any *probability distribution*—and thus, as we've seen—an essential aspect of *any* incomplete description.) Be that as it may, one of the duties of information theory is to acknowledge these new aspects of incompleteness whose prototype is found in physics and incorporate them by adding the necessary new modules to the underlying "logic engine" mentioned above. There is no need to attempt to "interpret" or "explain" quantum mechanics before setting about this task; on the contrary, having in hand the resulting formal "quantum-enhanced" information theory may in the end make it easier to address the interpretation problem itself.

Jaeger's book seems to me consistent with the above strategy. That is, it takes quantum mechanics as a *premise*. It doesn't waste time arguing about, or changing, the premise itself, and concentrates instead on developing an inference engine capable of handling *premises of that kind*. That's how we get *Quantum Information Theory: An Overview* rather than one more book on "Quantum information communication computation and all that."

Tommaso Toffoli

Preface

Quantum information science is a rapidly developing area of interdisciplinary investigation at the nexus of quantum mechanics and information theory. It now plays a significant role in a number of subdisciplines of physics, information technology, and engineering. A number of books on quantum information are available but are becoming outdated and/or differ significantly in approach from this book, or cover only particular aspects of the subject. Historically, lecture notes for the first general course on quantum information, given at Hewlett-Packard and edited by Lo, Popescu, and Spiller, were published in 1998 [289]. Physicists have also long benefited from the generously provided on-line lecture notes of Preskill [341]. At least one comprehensive monograph on quantum information science was published at the turn of the century, namely, the meticulous nearly 700-page book of Nielsen and Chuang [315]. More recent books by Pavičić [325] and Stenholm and Suominen [404] are noteworthy for their utility. A monograph detailing the mathematical foundations of quantum information theory by Hayashi, which originally appeared in Japanese in 2003, has just appeared in English [208]. The books of Benenti, Casati, and Strini [33] and Gruska [201] are valuable textbooks for teaching the subject. Quantum key distribution and quantum computing are currently the most exciting applications of quantum information science that are sufficiently well developed that a number of books are specifically dedicated to one or the other of them: in the case of quantum cryptography and communication, comprehensive collections have been edited by Alber *et al.* [7], Beth and Leuchs [59], Bouwmeester, Ekert, and Zeilinger [72], Braunstein *et al.* [79], and Sergienko [375]; in the case of quantum computation, the books of Brylinski and Chen [89], Hirvensalo [216], Kitaev, Shen, and Vyalyi [252], and Pittenger [335], and that edited by Lomonaco [287] focus on quantum algorithms and/or associated mathematics. A number of popular books on quantum computing have also been published, for example [83, 121].

In our information age, electronic access to primary sources is widely available, allowing one to locate the finest details of original investigations once one is well oriented with tools and references in hand. Therefore, now what one often needs most when approaching the subject of quantum information science is an overview that efficiently yet rigorously presents the fundamentals and that provides a detailed weblinked bibliography to take one further [1]. This book is intended to be such a handy reference for practitioners and students of quantum physics and computer science that also treats foundational aspects of quantum mechanics connected with quantum information science, including those associated with quantum measurement which plays an essential role in relating classical and quantum information. Most of the examples provided here are quantum-optical ones as a pragmatic matter, arising from the fact that interferometry is central to quantum information processing and the fact that interferometry has primarily progressed through optical physics. However, exciting innovations have been made by experimental groups working with a range of physical systems. Hopefully, workers in areas of experimental physics and engineering other than optics will soon provide comprehensive and detailed overviews of each of the experimental methods of manipulating quantum information. For the time being, discussions of various devices for quantum information processing can be found [19, 74, 407]. Particularly noteworthy are the books edited by Everitt [165] and Leggett *et al.* [274].

In the twentieth century, the formalism introduced in Dirac's *The Principles of Quantum Mechanics* and von Neumann's *Mathematische Grundlagen der Quantenmechanik* was brought to bear on a broad range of physical problems. Elements of this formalism and related mathematics are outlined in the appendices, together with standard quantum postulates. During the last two decades of the twentieth century, investigations of the foundational problems of quantum mechanics and the physics of computation were pivotal in giving rise to quantum information science as a subject in its own right, providing a conceptual basis for the development of quantum protocols and algorithms. In turn, the investigation of foundational problems has benefited from the work of those seeking solutions to central issues in quantum information science, such as those of communication complexity. Aspects of this important interplay have been addressed here.

It is my hope that, in addition to its serving as a practical tool for researchers and students, this book will assist those seeking to understand the subject to appreciate the many decades of work back to which the origins of this exciting, relatively new field can be traced. Although the aim in including this material is not to present a history of the exploration of foundations of quantum mechanics or its philosophical underpinnings, a number of pertinent such results from earlier decades of the twentieth century are included because they will likely prove important to future progress in both quantum mechanics and information theory. The discussion of early work is here often in the language of quantum information so as to facilitate access to earlier, foundational work in quantum mechanics by those approaching fundamental issues from a twenty-first century perspective.

Gregg Jaeger   Cambridge, MA, August 2006

Acknowledgments

I gratefully acknowledge Paul Busch, Silvia Carrasco, John M. Myers, Sandu Popescu, and Tommaso Toffoli for their very helpful suggestions during the preparation of this book, my wife Savina Andonova Jaeger for her personal support, my colleagues Millard Baublitz, Peter Busher, Gianni di Giuseppe, Bahaa Saleh, Alexander Sergienko, and Malvin C. Teich for their professional support, and my editors and the staff at Springer for their assistance. Last, but not least, I acknowledge Abner Shimony as an inspiration to me, as to many others seeking to better understand the subtleties of Nature.

Contents

1

Qubits

The differences between quantum information and classical information are due to the difference of a *qubit* in a quantum-physical system capable of storing it from a *bit* in a classical-physical system capable of storing it.[1] This difference arises primarily from the superposition principle of quantum mechanics; despite its being bivalent in the chosen computational basis, a qubit system can be in one of an *infinite number* of significant states, whereas a bit is capable of being in only one of two significant states.[2] A qubit system in general also must be considered as *at the same time potentially being in one* measurable state *and/or the other* opposite state rather than *actually being in just one* of the two available states as must necessarily be the case for a bit encoded in a classical physical system. Furthermore, unlike a classical state, a single unknown qubit-system state cannot generally be found by a

---

[1] Physical bits in traditional digital computers are realized in memory elements, metal-oxide semiconductor field-effect transistors, and electrical wires, all of which carry substantial charge relative to a single electrical quantum [179]. Classical information processors use such elements to store bits of information and perform operations on them, whereas quantum information processors operate on individual quanta. The term "qubit" was coined by Benjamin Schumacher, "...although Holevo's theorem gives an information-theoretic significance to [quantum entropy]... it does not provide an interpretation of [quantum entropy] in terms of classical information theory. We could not use [it], for example, to interpret the quantum entropy of some macrostate of a thermodynamic system as a measure of resources necessary to represent information about the system's quantum microstate... [Instead] this is accomplished by replacing the classical idea of a binary digit with a quantum two-state system... These quantum bits, or 'qubits,' are the fundamental units of quantum information." [367].

[2] See Postulate I in Sect. B.1. Paul Dirac noted the unique character of the superposition principle, "*the superposition principle that occurs in quantum mechanics is of an essentially different nature from any occurring in the classical theory,* as is shown by the fact that the quantum superposition principle demands indeterminacy in the results of observations in order to be capable of a sensible physical interpretation . . . analogies are likely to be misleading" [Dirac's emphasis] [136].

single measurement. Rather, an *ensemble* of systems must be measured to discover their unknown shared quantum state.[3] It is the nature of quantum potentiality that alternative possibilities for reaching a given quantum state at a given moment superpose, and so are capable of interfering with each other.

Quantum computing benefits from the quantum superposition principle as it pertains to the states of a number of qubits forming a compound quantum system. The space of possible states available to such multiple-qubit systems grows more rapidly than does the space of states available to multiple-bit systems; the number of parameters describing a quantum system that can be used to encode information for the purpose of computing grows *exponentially* in the number of qubits, whereas in a classical system it grows linearly in the number of bits. Thus, quantum computers can be viewed as complex quantum interferometric devices providing a unique sort of parallelism of computational states described by these parameters. This novel parallelism can be harnessed to make tractable some important computational tasks that are thought intractable under the constraints of computing realized in systems describable by classical physics. Any improvement in efficiency provided by quantum algorithms over classical algorithms resulting from the exploitation of this parallelism is known as *quantum speedup*. Quantum speedup and the features enabling it are discussed in Section 1.7 and Chapters 13 and 14.[4]

Although the properties of a qubit system are bivalent and can only be probabilistically predicted, a qubit system differs from a probabilistic classical system that randomly takes one of two computationally relevant values, again because the latter can only actually *be* in one of two states at any time irrespective of how it may be measured.[5] The probabilities of the outcomes of measurements of any classical system are due only to the *ignorance* of the measurer of the actual state of the system, rather than from a *fundamental* indeterminacy of properties as is the case for quantum systems. That a quantum bit is not reducible to some probabilistic bit becomes clear when seeking a straightforward ignorance interpretation of quantum probabilities.[6]

---

[3] Here, the only exceptions to this are situations in which precise information as to the two particular alternative orthogonal states in which a single qubit happens to have been prepared is possessed by the measuring agent and only these potential states are measured. It is precisely this character of individual qubits that provides the possibility of secure quantum key distribution. Here, the term *ensemble* is meant in the sense of statistical thermodynamics, where it refers to a set of identically prepared systems. See Postulates II and III in Sect. B.1.

[4] It is important to note that speedup depends on the assumption that the time required for arithmetic operations in quantum computing grows less than exponentially with the number of qubits involved. On entanglement here, see [78, 246].

[5] The units of classical information are also sometimes referred to as *c-bits* (*cf.* [100]). One might call a putative inherently probabilistic bit a *probabit*.

[6] An argument supporting this statement is given in the following section. The deeper philosophical aspects of the important differences between quantum systems encoding qubits and classical systems encoding bits have been well explored

Because a familiarity with the various mathematical representations of the qubit, which is the simplest nontrivial quantum system that can be considered in quantum mechanics, is essential for understanding quantum information, various representations of qubit states are reviewed in this chapter, as is their interferometric behavior which endows quantum computers with added computational power. The reader is reminded that quantum states are associated with a complex Hilbert vector space, $\mathcal{H}$, via a special class of linear operators acting in it, the *statistical operators*, ρ, constituting the quantum state-space.[7] In the case of the pure qubit states, the statistical operators are projectors onto one-dimensional subspaces and can be uniquely associated with points on the boundary of the Bloch ball, known as the Poincaré–Bloch sphere; pure states can be equally well represented by these same one-dimensional subspaces (or *rays*) $\{e^{i\phi}|\psi\rangle|\phi \in \mathbb{R}\}$ or the state-vectors $|\psi\rangle \in \mathcal{H}$ spanning them.[8] The remaining, essentially statistical states are *mixed* states that can be formed from these pure states and lie in the interior of the Bloch ball.[9] The set of statistical states available to a qubit system is concretely representable by the 2×2 complex Hermitian trace-one matrices $[\rho_{ij}] \in H(2)$. By contrast, for the full physical state description of a quantum system *in spacetime*, an infinite-dimensional spatial representation is required in which the state-vectors are referred to as *wavefunctions*. However, because quantum information theory is based on the behavior of qubits and has thus far overwhelmingly dealt with quantities with discrete eigenvalue spectra in the nonrelativistic regime, the state-vectors considered here are usually taken to lie within finite-dimensional Hilbert spaces constructed by taking the tensor product of multiple copies of *two-dimensional* complex Hilbert space; in quantum mechanics, these are traditionally associated with the spin subspaces of elementary particles; for example, see [299]. Unless otherwise stated, the Hilbert spaces considered here are only finite-dimensional *subspaces* of the larger full physical state-spaces of particles, the other subspaces of which are rarely taken into account in the study of quantum information processing.[10] For example, in many cases we consider the polarization states of photons as the systems of interest, without

---

and subtly articulated by Abner Shimony [381], Peter Mittelstaedt [304], Michael Redhead [348], Jeffrey Bub [90], and others. For the most part, space does not allow these to be adequately addressed in this book.

[7] The term "Hilbert space" (*Hilbertraum*) was itself first introduced by John von Neumann [443]; see Sect. A.3 for its definition.

[8] Here, Paul Dirac's notation, described in Appendix A, has been used.

[9] The possible qubit states are illustrated in Fig. 1.1, below. Note, however, that mixed states *cannot* be written as linear combinations of state *vectors* but only of statistical operators. The natural structure generalizing the Poincaré–Bloch sphere is the *convex set*, which may be used to study a variety of quantum systems; see Appendix A and [300]. The distinction between pure and mixed states itself is immediately addressed in detail in Sect. 1.1, below.

[10] A review of quantum information in the context of continuous-variables systems can be found in [81].

considering the corresponding photon wavefunctions; *cf.* spin as considered in Section 1.1 of [359]. This greatly simplifies the mathematics required to discuss quantum information without compromising essentials.

In quantum mechanics, operators play several roles: they may represent system states, physical quantities *or* transformations of states, including temporal evolution (although *not* time itself) and measurement processes. Measurable properties of quantum systems are traditionally referred to as *observables* and correspond to quantities represented by Hermitian linear operators on Hilbert space, the eigenvalues of which are their possible measurable values. Here, we merely refer to these as quantum *properties*. Similarly, although the question of the role of the percipient (or *observer*) in quantum mechanics is a deep and interesting one, space does not allow it to be taken up here in any detail. Observers and observations are referred to as *measurers* (or *agents*) and *measurements*, respectively, in order to avoid the impression that these are assumed to have unusual *physical* characteristics beyond those attributed to other physical objects or processes.[11] It is also vital here to recognize the distinctions between a physical system, its representation, and the information the system is capable of storing, particularly when metaphysical and epistemic considerations come into play, such as in the context of the statistical descriptions of microscopic phenomena discussed in this book, because the term *qubit* is used ambiguously in the quantum-information literature. In addition to referring to the unit of quantum information, this term is often used to refer to a system that can store it and sometimes to refer to the mathematical set representing possible quantum states of such systems. In this chapter and elsewhere, we focus on the ideal physical system capable of storing one qubit of information and refer to it simply as the *qubit*, in accordance with the most common usage in the physics literature.

Readers unfamiliar with the postulates of quantum mechanics and its mathematics are requested to refer when necessary to Appendix B, where the standard postulates of quantum mechanics, including the superposition principle, are briefly outlined, and Appendix A, where the notation and mathematics of quantum mechanics used in the sequel are summarized; the Dirac notation is primarily used here because of its great practicality. Throughout the text, examples and details of secondary importance are often provided in separate boxes. Those very familiar with the various representations of quantum states and their basic properties may wish to proceed directly to Section 1.4, where the quantum circuit formalism and basic quantum gates are discussed. This chapter ends with a preview of the basic requirements for quantum information processing tasks to be discussed in more detail later. In particular, complex quantum information processing requires quantum coherence of states in multi-qubit Hilbert spaces (*cf.* Postulate IV, Section B.1).

---

[11] For discussions of the role of the observer in quantum mechanics see, for example, the papers included in [32, 384, 453].

1.1 Quantum state purity

The *purity*, $\mathcal{P}$, of a quantum state specified by the statistical operator ρ is the trace of its square,

$$\mathcal{P}(\rho) = \operatorname{tr} \rho^2 , \tag{1.1}$$

where $\frac{1}{d} \leq \mathcal{P}(\rho) \leq 1$ and d is the dimension of the Hilbert space, $\mathcal{H}$, attributed to the system it describes. The quantum state is *pure* if $\mathcal{P}(\rho) = 1$, that is, if it spans a one-dimensional subspace of $\mathcal{H}$. One can then naturally define state *mixedness* as the complement of purity, $\mathcal{M}(\rho) \equiv 1 - \mathcal{P}(\rho)$. The purity and mixedness of a quantum state are invariant under transformations of the form $\rho \rightarrow U\rho\, U^\dagger$, where U is unitary, most importantly under the *dynamical mapping* $U(t, t_0) = e^{-\frac{i}{\hbar}H(t-t_0)}$, where H is the Hamiltonian operator, which can readily be seen upon recalling that the trace operation $\operatorname{tr}(\cdot)$ is cyclic.[12]

Pure states are those states that are *maximally specified* within quantum mechanics.[13] A quantum state is pure if and only if the statistical operator ρ is *idempotent*, that is,

$$\rho^2 = \rho , \tag{1.2}$$

providing a convenient test for maximal state purity. It is then also a *projector*, $P(|\psi_i\rangle)$, where $|\psi_i\rangle$ is the normalized vector representative of the corresponding one-dimensional subspace of its Hilbert space; projectors are outer products defined in Section A.5.[14] A quantum state is thus mixed if it is *not* a pure state, that is, if $\mathcal{P}(\rho) < 1$.

---

[12] *Unitary* linear operators, U, are those for which $U^\dagger U = UU^\dagger = \mathbb{I}$, where "$\dagger$" indicates Hermitian conjugation (see Sect. A.3). Here, the time-evolution prescribed by Postulate V of quantum mechanics (*cf.* Sect. B.1) has been given with a time-independent Hamiltonian. However, temporal evolution in quantum mechanics need not be so simple (*cf.* Sect. 2.1 of [359] and Ch. 5). The cyclic-invariance property of the trace is simply that $\operatorname{tr}(BA) = \operatorname{tr}(AB)$, which is independent of changes of basis.

[13] Pure states in quantum mechanics are often also called *coherent states*. By contrast with the classical case, coherent states and superpositions of such states are meaningful in quantum mechanics only when described by linear wave equations. Note that coherent states in quantum optics are related specific states different from the general notion of quantum coherent state referred to above. The term "coherent state," in this book refers to the former.

[14] Note that rays cannot be added, whereas vectors $|\psi_i\rangle$ can be, making the latter better for use in calculations involving pure states, where superpositions are formed by addition. A Hermitian operator P acting in a Hilbert space $\mathcal{H}$ is a *projector* if and only if $P^2 = P$. It follows immediately from this definition that $P^\perp \equiv \mathbb{I} - P$, where $\mathbb{I}$ is the identity operator, is also a projector. The projectors P and $P^\perp$ project onto orthogonal subspaces within $\mathcal{H}$, $\mathcal{H}_s$, and $\mathcal{H}_s^\perp$, respectively, thereby providing a decomposition of $\mathcal{H}$ as $\mathcal{H}_s \oplus \mathcal{H}_s^\perp$; two subspaces are said to be *orthogonal* if every vector in one is orthogonal to every vector in the other. In the case of a general state of a single qubit, one may write $\rho = p_1 P(|\psi\rangle) + p_2 P(|\psi^\perp\rangle)$, where the weights p_i are eigenvalues of its statistical operator.

In the Dirac notation, projectors are written

$$P(|\psi_i\rangle) \equiv |\psi_i\rangle\langle\psi_i| . \tag{1.3}$$

Consider a finite set, $\{P(|\psi_i\rangle)\}$, of projectors corresponding to distinct, orthogonal pure states $|\psi_i\rangle$. Any state ρ' that can be written

$$\rho' = \sum_i p_i P(|\psi_i\rangle) , \tag{1.4}$$

with $0 < p_i < 1$ and $\sum_i p_i = 1$, is then a normalized mixed state.

Consider the normalized sum

$$| \nearrow \rangle = \frac{1}{\sqrt{2}}(|0\rangle + |1\rangle) . \tag{1.5}$$

of two orthogonal pure state-vectors $|0\rangle \doteq (1\ 0)^{\mathrm{T}}$ and $|1\rangle \doteq (0\ 1)^{\mathrm{T}}$ of a qubit, the r.h.s.'s being given in the matrix representation and $(\cdots)^{\mathrm{T}}$ indicating matrix transposition, the l.h.s.'s being given in the Dirac notation. The superposition in Eq. 1.5 is a pure state, as can be immediately verified by taking its square modulus. The similar linear combination formed by subtraction rather than addition is written $| \searrow \rangle$; see Eq. 1.11 below. The corresponding projectors are $P(| \nearrow \rangle) = | \nearrow \rangle\langle \nearrow |$, $P(| \searrow \rangle) = | \searrow \rangle\langle \searrow |$.

By contrast to the case of state-vector addition, the normalized sum of a pair of *projectors*, for example, $P(|0\rangle)$ and $P(|1\rangle)$ corresponding to pure states $|0\rangle$ and $|1\rangle$, namely,

$$\rho_+ = \frac{1}{2}\big(P(|0\rangle) + P(|1\rangle)\big) , \tag{1.6}$$

is a *mixed* state that can also be written

$$\rho_+ = \frac{1}{2}\big(P(| \nearrow \rangle) + P(| \searrow \rangle)\big) . \tag{1.7}$$

Furthermore, the statistical operator corresponding to the normalized sum of $| \nearrow \rangle$ and $| \searrow \rangle$ is $P(|0\rangle) \neq \rho_+$. Again, the pure state $| \nearrow \rangle$ is the result of the *quantum superposition* of two state-vectors, whereas ρ_+ is the result of the nontrivial *mixing* of two distinct pure ensembles and, therefore, cannot be represented as a projector.

A quantum system is said to be in a (partially) *coherent* superposition of states $|a_i\rangle$ from a given orthonormal basis if and only if its density matrix—the representation of its statistical operator in matrix form (see [62])—is *not* diagonal in the A-representation, where A is the Hermitian operator of which the $|a_i\rangle$ are eigenvectors; it is said to be in a *completely coherent* superposition if, in addition, it is in a pure state.[15]

Quantum mixed states, unlike their classical analogues, do not arise merely from ignorance of states of the systems they describe. To see this, consider that an *ignorance interpretation of quantum mixtures* would hold that a system in the state $\rho = p_0 P(|0\rangle) + p_1 P(|1\rangle)$ could actually *be* in some pure state—either the one described by $P(|0\rangle)$ or the one described by $P(|1\rangle)$)—where real coefficients p_0 and p_1 could be understood as the probabilities, summing to one, of the system being in either one or the other pure state, as in the example of Eq. 1.6. These probabilities would then be understood as *epistemic* probabilities, in that they represent *best estimates* of the chances of the eigenvalues corresponding to the pure states to be observed. One of the peculiarities of quantum systems such as the qubits above, compared to classical systems, is the *nonuniqueness of the decomposition of a mixed state into pure states* illustrated in the above box.[16] The nonuniqueness of the decomposition of any mixed state would simply mean that an experimenter's ignorance is greater than expected; one couldn't say *which* are the pure states one should assign to any particular pair of probabilities that add to one. For composite systems, however, ensembles described by states ρ can be formed that are *pure* but whose *component system states are mixed*, as illustrated by Eqs. 1.6–1.7.[17] Such an interpretation of mixed states is, therefore, untenable.[18]

Another peculiarity of quantum systems relative to classical systems is that the maximal specification of a quantum state by preparation or measurement can precisely determine the values of only *half* its properties. A basic example illustrating this is that of a quantum particle: either its position or its momentum can be precisely specified but not *both*. In classical physics, in principle, both of these quantities of a system can be precisely specified, corresponding to its location at a point in "phase space." By contrast, quantum systems can be located only within finite areas of phase space. This can be understood by reference to the interpretation of the Heisenberg–Robertson uncertainty relation for momentum and position as describing the impossibility of simultaneous specification of momentum and position more precisely than that the product of their variances be less than half $\hbar$, the quantum of action; see Appendix B.2 and [92] for discussions of uncertainty relations. For a detailed discussion quantum phase space for discrete systems such as qubits, which is not explicitly used in this book but is of ongoing interest, see [182].

---

[15] See Eq. 2.21 and Sect. A.5 for details of operator representation.

[16] Both the weights and the projectors may differ for any two of the infinite number of allowed decompositions of a statistical operator. The state space is not a Choquet simplex, that is, not a space for which such a decomposition is unique.

[17] For a careful treatment of this question see, for example [228, 430]. Chapters 3, 6, and 7 below treat composite quantum systems in detail.

[18] It is valuable in this regard to consider the information measures described in Chapter 4. Radical interpretations taking all probabilities as ignorance probabilities can nonetheless be found; for example, see [100].

1.2 The representation of qubits

The pure states of the qubit can be represented by vectors in the two-dimensional complex Hilbert space, $\mathcal{H} = \mathbb{C}^2$. Any orthonormal basis for this space can be put in correspondence with two bit values, 0 and 1, in order to act as the single-qubit *computational basis*, sometimes also called the *recti-linear basis*, and written $\{|0\rangle, |1\rangle\}$:[19] the elements of the chosen basis may be identified with the finite (Galois) field of two elements, $x_i \in GF(2)$, by writing them as $|x_i\rangle$ with $x_i \in \{0, 1\}$.[20] The computational basis states, $\{|x_i\rangle\}$, for the qubit Hilbert space is often taken, as is done here, to correspond to the poles of the Poincaré–Bloch sphere; see Fig. 1.1, below.

The superposition principle implies that any (complex) linear combination of qubit basis states, such as $|0\rangle$ and $|1\rangle$, that is,

$$|\psi\rangle = a_0|0\rangle + a_1|1\rangle \tag{1.8}$$

with $a_i \in \mathbb{C}$ and $|a_0|^2 + |a_1|^2 = 1$, is *also* a physical state of the qubit and is, as we have seen, also a pure state. The scalar coefficients a_0 and a_1 are referred to as *quantum probability amplitudes* because their square magnitudes, $|a_0|^2$ and $|a_1|^2$, are the *probabilities* p_0 and p_1, respectively, of the qubit described by state $|\psi\rangle$ being found in these basis states $|0\rangle$ and $|1\rangle$, respectively, upon measurement.[21]

The vectors of the computational basis can be represented in matrix form as

$$|0\rangle \doteq \begin{pmatrix} 1 \\ 0 \end{pmatrix} , \tag{1.9}$$

$$|1\rangle \doteq \begin{pmatrix} 0 \\ 1 \end{pmatrix} . \tag{1.10}$$

Another commonly used basis is the *diagonal basis*, $\{|\nearrow\rangle, |\searrow\rangle\}$, sometimes also written $\{|+\rangle, |-\rangle\}$, given by

$$|\nearrow\rangle \equiv \frac{1}{\sqrt{2}}(|0\rangle + |1\rangle) \text{ and } |\searrow\rangle \equiv \frac{1}{\sqrt{2}}(|0\rangle - |1\rangle) , \tag{1.11}$$

---

[19] This basis is generally taken either to be the z-axis of the traditional quantum mechanical description of spin-$\frac{1}{2}$ systems or to be the x-axis, as is typically the case in the representation of polarization states of light. Here, we follow the former convention, and identify $|0\rangle$ with the horizontal polarization state $|H\rangle$.

[20] $GF(2)$ is the Galois field of integers modulo 2. For the definition and properties of the Galois field $GF(N)$ and its relationship to the integers mod p, $\mathbb{Z}_p$, see Sect. A.1. Here we have $N = p^n$ with $p = 2$ and $n = 1$. Galois fields of higher values of n appear later.

[21] This relationship is given by the Born rule; see Sect. B.1 and [68]. A similar statement holds for components of quantum states in *any* basis of the Hilbert space of any finite-dimensional quantum system.

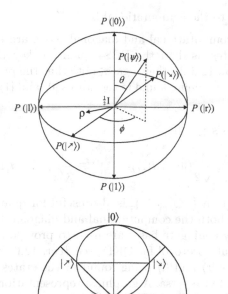

Fig. 1.1. a: Top figure. Statistical operators represented in the unit Bloch ball, a real-valued representation of the space of qubit states via the expectation values, S_i, of Pauli operators σ_i, $i = 1, 2, 3$; see Eqs. 1.19–1.22. Orthogonal quantum states are antipodal in this representation; the *conjugate* bases correspond to orthogonal axes. The pure qubit states, $P\big(|\psi(\theta, \phi)\rangle\big)$, lie on the periphery, known as the *Poincaré–Bloch sphere* [337]. The mixed qubit states, $\rho(r, \theta, \phi)$, lie in the interior and are weighted convex combinations of pure states. The maximally mixed state, $\frac{1}{2}\mathbb{I}$, lies at the center of the ball, being an evenly weighted linear combination of *any* two orthogonal pure states (*cf.* Eqs. 1.6–7). In the Poincaré presentation often used in polarization optics, the sphere is rotated counter-clockwise about the diagonal-basis axis by 90° with respect to the one here. **b**: Bottom figure. Pure states of the computational and diagonal bases jointly represented both in a Poincaré great circle, where orthogonal states are represented as antipodal, and in a single-qubit Hilbert-space semicircle, where orthogonal states are represented by orthogonal directions and the endpoints are identified with each other. The full circle in this figure is the great circle in the Poincaré–Bloch sphere that intersects both the computational and diagonal-basis axes. Note that the angle subtended by a pair of directions in Hilbert space is *half* the corresponding angle in the Poincaré–Bloch sphere (*cf.* Eq. 1.14), which corresponds to the fact that the special unitary group $SU(2)$ acting on the vectors in the complex representation is the universal (double) covering group of the special orthogonal group of rotations $SO(3)$ of the vectors of the real representation. Stereographic and sinusoidal projections of the Poincaré–Bloch sphere are also sometimes used; for an example of the rarer, latter case, see Section 3.1.1 of [33].

which is conjugate to the computational basis.[22]

Together, the computational and diagonal bases are used to provide the pairs of signal states used in the BB84 quantum key distribution (QKD) protocol; see Section 12.3. In that regard, note that the probabilities of qubits in states $|\nearrow\rangle$ and $|\searrow\rangle$ being found in the states $|0\rangle$ and $|1\rangle$ are $(\sqrt{1/2})^2 = 0.5$ and vice-versa.

The *circular basis* $\{|r\rangle, |l\rangle\}$,

$$|r\rangle \equiv \frac{1}{\sqrt{2}}(|0\rangle + i|1\rangle) , \qquad |l\rangle \equiv \frac{1}{\sqrt{2}}(|0\rangle - i|1\rangle) , \qquad (1.12)$$

sometimes also written $\{|\circlearrowright\rangle, |\circlearrowleft\rangle\}$, is also useful for quantum cryptography, being conjugate to both the computational and diagonal bases.[23] All three of the above mutually conjugate bases are used to provide three pairs of signal states in the six-state protocol for QKD (see Sect. 12.6); the probabilities of qubits in the states $|r\rangle$ and $|l\rangle$ being found in the states $|0\rangle$, $|1\rangle$, $|\nearrow\rangle$, and $|\searrow\rangle$ are all 0.5, and vice-versa. A graphical representation of the above three sets of basis vectors is shown in Fig. 1.1.[24]

Yet another basis, the *Breidbart basis*, is the "intermediate basis"

$$\left\{ \cos\frac{\pi}{8}|0\rangle + \sin\frac{\pi}{8}|1\rangle , -\sin\frac{\pi}{8}|0\rangle + \cos\frac{\pi}{8}|1\rangle) \right\} , \qquad (1.13)$$

which lies on the same great circle as the circular and rectilinear bases. It is used in QKD and for eavesdropping; see Section 12.5.

---

[22] Two bases are *conjugate* if the corresponding pairs of antipodal points of the Poincaré–Bloch sphere are 90° apart from each other [455]; see Fig. 1.1.a.

[23] In the convention where the computational and superposition bases lie on the equator of the Poincaré–Bloch sphere, which we do *not* follow here, these two states are identified with the poles; see the following footnote.

[24] Basis states are sometimes labeled on the Poincaré–Bloch sphere by state-vectors $|\psi_i\rangle$ rather than by the corresponding projectors $P(|\psi_i\rangle)$, which can be misleading because each line segment passing between antipodal points through the center of the Bloch ball corresponds to a set of real convex combinations of *projectors* $P(|\psi_i\rangle)$ rather than a complex linear combination of Hilbert-space vectors (*cf.* Sect. 1.1). In Fig. 1.1b, the correspondence between the complex and real representations is illustrated, allowing one to see the effect of vector addition on the Poincaré sphere. Note also that the Bloch ball is often presented differently, for example with the computational basis state $P(|r\rangle)$ at the "north pole," for example in the "Poincaré representation" of photon polarization, where the ball is rotated in θ by 90°, placing the chosen computational basis on the *equator*.

A *spinor* representation of the general pure state of a qubit is provided by

$$|\psi(\theta,\phi)\rangle = \cos(\theta/2)|0\rangle + e^{i\phi}\sin(\theta/2)|1\rangle \doteq \begin{pmatrix} \cos(\theta/2) \\ e^{i\phi}\sin(\theta/2) \end{pmatrix}, \qquad (1.14)$$

where $0 \leq \theta \leq \pi$ and $0 \leq \phi < 2\pi$; when $\theta = 0$ and π, ϕ is taken to be zero by convention (*cf.* Fig. 1.1). Thus ϕ is the relative phase between single-qubit computational-basis states. With this parametrization, the *general* qubit state is naturally visualized in the Bloch ball, the boundary of which is the Poincaré–Bloch sphere consisting entirely of the pure states, $|\psi(\theta,\phi)\rangle$. It is easy to see by inspection which pairs of values of the parameters θ and ϕ, corresponding to the altitudinal-complement angle and the azimuthal angle, respectively, provide the various states of the above bases. The Bloch vector associated with a pure state is $(\sin\theta\cos\phi, \sin\theta\sin\phi, \cos\theta)$, as described in the following section. The most general linear transformation of the qubit in the above representation is $(\theta,\phi) \rightarrow (\theta - \alpha, \phi - \beta)$, where $0 \leq \alpha \leq \pi$ and $0 \leq \beta < 2\pi$. This transformation is decomposable into two transformations, one with respect to θ and one with respect to ϕ, the former capable of being performed unitarily but the latter not.[25] The generic mixed state, ρ, lies in the interior of the Bloch ball, can be written as a convex combination of basis-element projectors corresponding to the pure-state bases described above (*cf.* Eq. 1.4), and can be most conveniently given in the Stokes-vector representation described in the following section.[26] The effect of a general operation on a qubit can be viewed as a (possibly stochastic) transformation within this ball; for illustrations of this in practical context, see [333]. The parametrization required to adequately describe mixed states is now discussed in detail.

1.3 Stokes parameters

The generic state of a qubit can also be specified by a real vector, most naturally one in Minkowski space $\mathbb{R}^4_{1,3}$, as well as by a convex combination $\rho \doteq \{p_i, P(|\psi_i\rangle)\}$ of projectors $P(|\psi_i\rangle)$ acting in the Hilbert space $\mathbb{C}^2$ as discussed above. A real description has most commonly been used to describe polarization via Stokes parameters in the restricted space $\mathbb{R}^3$ but can be used to describe *any* qubit and embedded in $\mathbb{R}^4_{1,3}$ [14, 15, 240]. The components of this four-vector, the four *Stokes parameters* S_μ, have the advantage of directly corresponding to empirical quantities, such as photon-counting rates

---

[25] This is particularly pertinent in regard to the performance of the universal-NOT operation [207].

[26] The position of a state ρ is often given by coordinates $(x, y, z) \equiv (\langle 0|\rho|1\rangle + \langle 1|\rho|0\rangle, \langle 1|\rho|1\rangle - \langle 0|\rho|0\rangle, i\langle 0|\rho|1\rangle - i\langle 1|\rho|0\rangle)$. We follow a *different* convention, provided just below Eq. 1.22, with respect to which this parametrization is rotated $90°$, where the position of ρ is given by Eqs. 1.19–22. See also the footnote above regarding the Poincaré representation, as well as the following section.

in selective measurements; see Chapter 8. Three of the dimensions ($\mu = 1, 2, 3$) associated with these parameters are conventionally taken to be those of the computational, diagonal, and circular bases, and correspond to orthogonal directions in the Poincaré–Bloch sphere; see Fig. 1.1. We now consider the relationship between the above state descriptions and this most general one.

The Stokes and density matrix descriptions are homomorphic:[27] the density matrix and the *Stokes four-vector*, S_μ, are related by

$$\rho = \frac{1}{2} \sum_{\mu=0}^{3} S_\mu \sigma_\mu, \tag{1.15}$$

where σ_μ ($\mu = 1, 2, 3$) are the *Pauli operators* which, together with the identity $\sigma_0 = \mathbb{I}_2$, are represented in the matrix space $H(2)$ by the *Pauli matrices*

$$\sigma_1 = \sigma_x = X \doteq \begin{pmatrix} 0 & 1 \\ 1 & 0 \end{pmatrix}, \qquad \sigma_2 = \sigma_y = Y \doteq \begin{pmatrix} 0 & -i \\ i & 0 \end{pmatrix},$$

$$\sigma_3 = \sigma_z = Z \doteq \begin{pmatrix} 1 & 0 \\ 0 & -1 \end{pmatrix}, \qquad \sigma_0 = \mathbb{I}_2 = I \doteq \begin{pmatrix} 1 & 0 \\ 0 & 1 \end{pmatrix},$$

where X, Y, Z are the quantum-logic-gate labels often used to specify the corresponding operations; see Section 1.4. The Pauli matrices form a basis for $H(2)$, which contains the qubit density matrices.[28] The nontrivial products of the four Pauli matrices—those between the σ_i for $i = 1, 2, 3$—are given by $\sigma_i \sigma_j = \delta_{ij} \sigma_0 + i \epsilon_{ijk} \sigma_k$, which defines their algebra.[29] Appropriately exponentiating the Pauli matrices provides the rotation operators, $R_i(\xi) = e^{-i\xi\sigma_i/2}$, for Stokes vectors about the corresponding directions i (*cf.* [359]); these rotations realize the group $SO(3)$.

The Stokes parameters S_μ ($\mu = 0, 1, 2, 3$) also allow one to directly visualize the qubit state geometrically in the Bloch ball via S_1, S_2, S_3. The Euclidean length of this three-vector (also known as the *Stokes vector*, or *Bloch vector*) is the radius $r = (S_1^2 + S_2^2 + S_3^2)^{1/2}$ of the sphere produced by rotations of this vector. With the matrix vector $\boldsymbol{\sigma} = (\sigma_1, \sigma_2, \sigma_3)$ and the three-vector $\mathbf{S} = (S_1, S_2, S_3)$, one has

$$\rho = \frac{1}{2}(S_0 \mathbb{I}_2 + S_1 \sigma_1 + S_2 \sigma_2 + S_3 \sigma_3) \tag{1.16}$$

$$\doteq \frac{1}{2} \begin{pmatrix} S_0 + S_3 & S_1 - iS_2 \\ S_1 + iS_2 & S_0 - S_3 \end{pmatrix}, \tag{1.17}$$

---

[27] For a discussion of the pertinent homomorphism, see [237].

[28] The qubit density matrices themselves are the positive-definite, trace-class elements of the set of 2×2 complex Hermitian matrices $H(2)$ of unit trace, that is, for which the total probability S_0 is unity, as prescribed by the Born rule and the well-definedness of quantum probabilities; see Appendix B. Density matrices are similarly defined for systems of countable dimension; see Sect. A.5 and [416].

[29] $\epsilon_{ijk} = 1$ for even permutations of 123, $= -1$ for odd permutations of 123, and $= 0$ otherwise.

known as the *Bloch-vector representation* of the statistical operator, in accord with Eq. 1.15. In optical situations, where $\mathbf{S}$ describes a polarization state of a photon, the degree of polarization is given by $P = r/S_0$, where S_0 is positive. For the qubit, when the state is normalized so that $S_0 = 1$, S_0 corresponds to total quantum probability. The density matrix of a single qubit is then of the form

$$\rho \doteq \begin{pmatrix} \rho_{00} & \rho_{01} \\ \rho_{10} & \rho_{11} \end{pmatrix}, \tag{1.18}$$

where $\rho_{00} + \rho_{11} = 1$, $\rho_{ii} = \rho_{ii}^*$ with $(i = 0, 1)$, and $\rho_{10} = \rho_{10}^*$, where * indicates complex conjugation.[30] One can write the Pauli matrices for $\mu = 1, 2, 3$ in terms of outer products of computational basis vectors, as follows.

$$\sigma_1 = |0\rangle\langle 1| + |1\rangle\langle 0| , \tag{1.19}$$
$$-i\sigma_2 = |0\rangle\langle 1| - |1\rangle\langle 0| , \tag{1.20}$$
$$\sigma_3 = P(|0\rangle) - P(|1\rangle) , \tag{1.21}$$

and $\sigma_0 = P(|0\rangle) + P(|1\rangle)$, which can be directly verified by inspecting their matrix representation given above. Using the above-mentioned homomorphism, the Stokes parameters are expressed in terms of the density matrix as

$$S_\mu = \mathrm{tr}(\rho\sigma_\mu) , \tag{1.22}$$

which are probabilities corresponding to ideal normalized counting rates of measurements in the standard eigenbases (see box below); in the standard normalized parametrization of Eq. 1.14, $S_0 = 1$, $S_1 = \sin\theta\cos\phi$, $S_2 = \sin\theta\sin\phi$, and $S_3 = \cos\theta$.

$$\boxed{\begin{aligned} &S_0 = \mathrm{tr}(\rho P(|0\rangle)) + \mathrm{tr}(\rho P(|1\rangle)), \ S_1 = \mathrm{tr}(\rho P(|\nearrow\rangle)) - \mathrm{tr}(\rho P(|\searrow\rangle)), \\ &S_2 = \mathrm{tr}(\rho P(|\mathfrak{l}\rangle)) - \mathrm{tr}(\rho P(|\mathbf{r}\rangle)), \ S_3 = \mathrm{tr}(\rho P(|0\rangle)) - \mathrm{tr}(\rho P(|1\rangle)). \end{aligned}}$$

The *four-vectors* formed by the individual Stokes parameters provide a basis in Minkowski space $\mathbb{R}^4_{1,3}$. The σ_μ are the generators of rotations and hyperbolic rotations in this space.[31] The proper, orthochronous Lorentz transformations $O_o(1,3)$ acting on the Stokes vector can be conveniently represented as products of six transformations $M_1, ..., M_6$, of which the following two, M_1 and M_4, are representative of the two basic types, ordinary and hyperbolic rotations, respectively.

$$M_1(\alpha) \doteq \begin{pmatrix} 1 & 0 & 0 & 0 \\ 0 & \cos\alpha & -\sin\alpha & 0 \\ 0 & \sin\alpha & \cos\alpha & 0 \\ 0 & 0 & 0 & 1 \end{pmatrix}, \quad M_4(\chi) \doteq \begin{pmatrix} \cosh\chi & \sinh\chi & 0 & \\ \sinh\chi & \cosh\chi & 0 & 0 \\ 0 & 0 & 1 & 0 \\ 0 & 0 & 0 & 1 \end{pmatrix}.$$

---

[30] Due to the constraints on density matrices, one can make use of the four convenient real parameters A, B, C and ϕ such that $\rho_{00} = A$, $\rho_{11} = B$, $\rho_{01} = Ce^{i\phi}$, and $\rho_{10} = Ce^{-i\phi}$, where $C \leq \sqrt{AB}$. For example, in Eq 1.6, $A = B = \frac{1}{2}$ and $C = 0$.

[31] For a discussion of the underlying mathematics, see [405].

Of the full set $\{M_i\}$ of six transformations, the first three are parameterized by α, β, γ, which are the angles of rotation about orthogonal directions in this real representation (when $i = 1, 2, 3$) leaving the zeroth component unchanged, and the second three are parameterized by χ, ω, ζ, which are the angles of *hyperbolic* rotation about the corresponding orthogonal directions in this real representation (when $i = 4, 5, 6$) that alter the zeroth component.[32]

The *Minkowskian (Lorentz-group invariant) length* associated with the transformation of the Stokes four-vector (S_0, S_1, S_2, S_3) under the Lorentz group is

$$S^2 = S_0^2 - S_1^2 - S_2^2 - S_3^2 , \tag{1.23}$$

which is familiar from its more well-known analogue in spacetime, the proper time.[33] The Euclidean length r and "degree of polarization" P are related to this invariant:

$$r^2 = S_0^2 - S^2 , \tag{1.24}$$

and

$$P^2 = \left(\frac{r}{S_0}\right)^2 = 1 - (S/S_0)^2 . \tag{1.25}$$

As we show in Chapter 7, the generalization of the Lorentz-group invariant length to multiple-qubit systems, in the product space formed from copies of $\mathbb{R}_{1,3}^4$, provides a measure of pure-state entanglement [240].

1.4 Single-qubit gates

The logic operations of quantum information processing can be carried out using *quantum gates*, which are unitary operations acting on quantum state-vectors. These operations realize, in the computational basis, the truth tables of the corresponding Boolean logic operations.[34] Single-qubit quantum gates are transformations on the vector spaces of individual qubits appropriately mapping the computational basis $\{|0\rangle, |1\rangle\}$ to itself. For example, just as the classical NOT gate takes the bit 0 to 1 and the bit 1 to 0, the quantum NOT gate takes the computational-basis vectors $|0\rangle$ to $|1\rangle$ and $|1\rangle$ to $|0\rangle$. The group of unitary transformations of the qubit state consists of operations described by four parameters. As we have just seen, the space of qubit states can be

---

[32] These parameters can be related to the effects of polarization-mode dispersion and polarization-dependent loss in optical fiber that can affect photons in practical applications such as QKD with polarization-based qubits [240].

[33] It is important in this regard to note that here the transformations of interest are qubit transformations, *not* spacetime transformations; the Stokes parameters are not the parameters of spacetime. For discussions of the effect of boosts on qubits in spacetime, see [116, 184, 331].

[34] These operations should be distinguished from those of traditional quantum logic, which is in a particular sense weaker than Boolean logic and in which, as a result, distributivity sometimes fails; see Sects. A.1 and A.7.

given as a three-dimensional real space (*cf.* Fig. 1.1) embedded in a larger one of four real dimensions. Unitary transformations are trace-preserving, and so *do not* change the norm of a state, that is, do not alter the value of the Stokes parameter S_0. A range of single-qubit gates are now described that are used in subsequent sections and chapters.

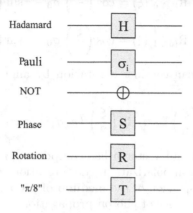

Hadamard	H
Pauli	σ_i
NOT	$\oplus$
Phase	S
Rotation	R
"$\pi/8$"	T

Fig. 1.2. Symbolic representations of single-qubit quantum logic gates.

The symbolic representations of a number of basic quantum logic gates are shown in Fig. 1.2. The effects, matrix representations, and interrelations of these and other gates are now described.

The *Hadamard gate*. This gate is one of the most significant quantum logic gates, because it can be used to enable the qubit interference vital to quantum computation, in which several qubits are transformed in parallel, typically involving tensor products of operators corresponding to this gate, that is, $H^{\otimes n}$ (*cf.* Figs. 14.1–2). It interchanges the computational and the diagonal bases: $|0\rangle \leftrightarrow |\nearrow\rangle$ and $|1\rangle \leftrightarrow |\searrow\rangle$. The Hadamard gate

$$H \doteq \frac{1}{\sqrt{2}} \begin{pmatrix} 1 & 1 \\ 1 & -1 \end{pmatrix}$$

induces a transformation equivalent to a rotation by the angle $\pi/4$ of the Poincaré–Bloch sphere about the y–axis, that is, the diagonal-basis axis, followed by a reflection through the x-y–plane, that is, the plane intersecting the equator.[35]

---

[35] The symbol "H" indicates the Hadamard transformation and is not to be confused with the symbols designating the space of Hermitian matrices $H(2)$, the Hamiltonian operator H, or Hilbert space $\mathcal{H}$. Hermitian matrices O (and operators) are those such that $O^\dagger = O$, where "$\dagger$" indicates Hermitian conjugation which, for matrices, corresponds to the operation of complex conjugation together with transposition. Note that $H^2 = \mathbb{I}$. One can write $H = \frac{1}{\sqrt{2}}(X + Z)$; see below.

The *rotation gates*. Each of these gates rotates a qubit state about a corresponding axis of the Poincaré–Bloch sphere by an angle ξ:

$$R_x(\xi) \equiv R_{(1,0,0)}(\xi) = \cos\left(\frac{\xi}{2}\right)\sigma_0 - i\sin\left(\frac{\xi}{2}\right)\sigma_1 , \qquad (1.26)$$

$$R_y(\xi) \equiv R_{(0,1,0)}(\xi) = \cos\left(\frac{\xi}{2}\right)\sigma_0 - i\sin\left(\frac{\xi}{2}\right)\sigma_2 , \qquad (1.27)$$

$$R_z(\xi) \equiv R_{(0,0,1)}(\xi) = \cos\left(\frac{\xi}{2}\right)\sigma_0 - i\sin\left(\frac{\xi}{2}\right)\sigma_3 . \qquad (1.28)$$

More generally, one can consider a rotation by angle ξ about an arbitrary direction **n**:

$$R_{\mathbf{n}}(\xi) = \cos\left(\frac{\xi}{2}\right)\mathbb{I} - i\,\sin\left(\frac{\xi}{2}\right)(n_x\sigma_1 + n_y\sigma_2 + n_z\sigma_3) , \qquad (1.29)$$

where the directional unit vector **n** has components n_x, n_y, n_z. For example, in the context of photon polarization, these rotations can result from photon polarization-mode dispersion of the medium of propagation, in which case ξ will depend on the distance of photon propagation.

The NOT (*bit-flip*) gate. This gate induces a change of the computational-basis value of the qubit, that is, takes $|0\rangle \leftrightarrow |1\rangle$:

$$\text{NOT} \doteq \begin{pmatrix} 0 & 1 \\ 1 & 0 \end{pmatrix} ,$$

performing a reflection through the x-y-plane, that is, that of the equator, which is identical to the Pauli matrix σ_1, that is, X, and which is accordingly often referred to as the "bit-flip operator." Note that $(\text{NOT})^\dagger = \text{NOT}$, and $(\text{NOT})^2 = \mathbb{I}_2$ (the identity) as one would expect from a logical-NOT operation.

The $\sqrt{\text{NOT}}$ gate. This gate,

$$\sqrt{\text{NOT}} \doteq \frac{1}{\sqrt{2}}\begin{pmatrix} 1 & i \\ i & 1 \end{pmatrix} ,$$

is so named because applying it twice is equivalent to applying the NOT gate once, up to an overall phase factor $(-i)$. The $\sqrt{\text{NOT}}$ gate is readily realized in beam optics by a beam-splitter acting on a spatial qubit; see Section 1.6.[36]

The *phase-flip gate*. This gate,

$$Z \doteq \begin{pmatrix} 1 & 0 \\ 0 & -1 \end{pmatrix} ,$$

induces a change of the phase angle ϕ of Eq. 1.14 by π and is identical to the Pauli matrix σ_3. It has the effect on computational-basis states that it takes

---

[36] Note that the Hadamard gate H is also somewhat loosely referred to as a "square-root of NOT." A beam-splitter must be supplemented to phase shifters in order to realize a Hadamard gate.

$|0\rangle \mapsto |1\rangle$, $|1\rangle \mapsto -|1\rangle$. It can also be viewed as a variant "NOT" gate when acting on the elements of the basis $\{|\nearrow\rangle, |\searrow\rangle\}$, because it interchanges them: $|\nearrow\rangle \leftrightarrow |\searrow\rangle$. The product iXZ is the Y (*bit+phase-flip*) gate identical to the Pauli matrix σ_2, which has the effect of inverting the qubit state-vector about the origin of the Poincaré–Bloch sphere, performing "universal state-vector inversion" [357].

The $\frac{\pi}{2}$-*phase gate*. This gate, also called the *i*-phase-shift gate, represented by

$$S \doteq \begin{pmatrix} 1 & 0 \\ 0 & i \end{pmatrix},$$

shifts the angle ϕ of Eq. 1.14 by $\frac{\pi}{2}$. It allows one to produce specific interferometric effects when implemented in conjunction with Hadamard gates. This gate is also the "square root" of the phase-flip gate, in that $S^2 = Z$.

The "$\frac{\pi}{8}$" *gate*. This gate, represented as

$$T \doteq \begin{pmatrix} 1 & 0 \\ 0 & e^{i\pi/4} \end{pmatrix} = e^{i\pi/8} \begin{pmatrix} e^{-i\pi/8} & 0 \\ 0 & e^{i\pi/8} \end{pmatrix}, \tag{1.30}$$

is commonly used in nuclear magnetic resonance simulations of quantum computing, and shifts the angle ϕ by $\frac{\pi}{4}$. Note that this gate is the "square root" of the $\frac{\pi}{2}$-phase gate, so that one has $T^4 = S^2 = Z$. It is sometimes also called "K."

The *phase rotation*. This gate,

$$P(\bar{\phi}) \doteq \begin{pmatrix} 1 & 0 \\ 0 & e^{i\bar{\phi}} \end{pmatrix} = e^{i\bar{\phi}/2} \begin{pmatrix} e^{-i\bar{\phi}/2} & 0 \\ 0 & e^{i\bar{\phi}/2} \end{pmatrix},$$

shifts the qubit phase by an angle $\bar{\phi}$, allowing, for example, the production of an interferogram when implemented together with $\sqrt{\text{NOT}}$ gates; for example, see Fig. 1.4. This operation thus rotates a state-vector by the angle $\bar{\phi}$ about the polar axis in the Poincaré–Bloch sphere.

A *generic* unitary quantum-logic operation on a single qubit can be represented as a combination of an overall phase shift and three rotations. In particular, one can represent the *general unitary gate* in terms of these gates as

$$U(\varsigma, \xi, \xi', \xi'') = e^{i\varsigma} R_z(\xi) R_y(\xi') R_z(\xi'') \tag{1.31}$$

(*cf.* Eqs. 1.26–28).

The above set of gates is formally similar to the set of Jones matrices of polarization optics.[37] Nonunitary transformations of qubit states have been thoroughly investigated as well, including those involved in the decoherence process; for example, see [207].

---

[37] For a thorough review of the practical creation, characterization, and manipulation of single qubits in optics, see [333].

1.5 The double-slit experiment

The simplest experimental situation in which uniquely quantum behavior is manifest is the *double-slit* (or *two-slit*) *experiment*, which has proven highly useful in illustrating the nature of quantum probability.[38] Consider the double-slit diaphragm and opaque-screen detector arrangement shown in Fig. 1.3. Take $a_1(x)$ to be the (complex) quantum probability amplitude corresponding (via retrodiction, *cf.* the introduction to Ch. 2) to the passage of a quantum system through one slit of a diaphragm toward the spatial point x on the measurement screen oriented perpendicularly to the direction of the initial beam. The corresponding probability density of later finding a particle at x upon measurement is then $p_1(x) = |a_1(x)|^2$. Similarly, let $a_2(x)$ be the amplitude corresponding to passage through a second slit and arrival at x; the corresponding probability density is $p_2(x) = |a_2(x)|^2$. The normalized quantum amplitude for the particle being found at x when *both* slits are passable, so that either slit might be entered on the way to the screen, is then

$$a_{12}(x) = \frac{1}{\sqrt{2}}\left(a_1(x) + a_2(x)\right) , \qquad (1.32)$$

according to the superposition principle.

The corresponding probability density of finding the particle at the point x on the collection screen is thus

$$p_{12}(x) = |a_{12}(x)|^2 \qquad (1.33)$$

$$= \frac{1}{2}|a_1(x) + a_2(x)|^2 . \qquad (1.34)$$

The probability density $p_{12}(x) \neq p_1(x) + p_2(x)$, being the squared modulus of the sum of amplitudes $a_1(x)$ and $a_2(x)$ which are complex numbers with nontrivial phases, exhibits *quantum interference* modulated by the phase difference between these amplitudes—most dramatically under the conditions of constructive and destructive interference—giving rise to locally "bright" (high probability) and locally "dark" (vanishing probability) regions in the pattern

---

[38] The two-slit experiment was first carried out with light by Fresnel and Young. For a discussion of double-slit experiments with electrons see, for example, [355]. Richard Feynman said that the interferometric behavior in this experiment, "In reality. . . contains the *only* mystery" of quantum mechanics and "We cannot make the mystery go away by 'explaining' how it works. We will just *tell* you how it works. In telling you how it works we will have told you about the basic peculiarities of all quantum mechanics" [169]. The mystery is the *nonclassical* nature of the quantum interference behavior described mathematically via quantum probability amplitudes and their addition. In practice, measurements will involve detection within a finite interval rather than a single point; the pointwise function described here is the probability *density*, from which the pertinent probability is obtainable by integration over the detection interval.

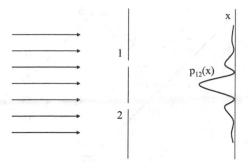

Fig. 1.3. The double-slit experiment, characterized by passage of a particle through slit 1 and/or slit 2 and detection at points x with single-particle interference on an opaque detection screen. Bright regions correspond to high values of the probability $p_{12}(x)$, dark regions to low probability values.

(interferogram) that results from the measurement of a collection of identically prepared particles; see Fig. 1.3. At these particular locations, quantum amplitudes add to and cancel out one another out maximally, respectively. Such an interference pattern is *not* observed when the measured systems are classical particles. For a more detailed discussion, see Ch. III.1 of [169].

Niels Bohr considered two versions of the double-slit experiment while exploring the nature of quantum interference [67]. In one, a rigid diaphragm with two slits and the ability to move in response to a collision is used, allowing the slit through which the particle passes to be determined by measurement of the recoil of the diaphragm, but no interference pattern is observed; an analogous apparatus for allowing path determination in two beams is shown in Fig. 1.4. In the other, the rigid diaphragm is fixed in place, as in Fig. 1.3; see Fig. 1.5 below for a two-beam analogue to that case. In this second version, an interference pattern in particle detections occurs, but the paths of the particles are not determinable [242]. Thus, there is complementarity between the distinguishability of the path of particles in the apparatus and the visibility of interference patterns formed by them at the detection screen.

Arrangements interpolating between the two extreme arrangements considered by Bohr have since been quantitatively investigated. One finds that measurements cannot be made that allow *a posteriori* precise determination of which slit any given particle of the ensemble passed through with high probability *without destroying* the interference pattern formed by the particles striking the measurement screen [373]. This was found to be expressible as a quantitative complementarity bound on particle path (*welcher weg*, or "which-way") determinability and the interference visibility [462]. It is helpful to consider discrete versions of the two-slit experiments of Bohr, which provide a "spatial qubit" corresponding to a pair of spatial paths, for example, those emerging from exit ports of a beam-splitter, that could be coherently recombined later, as in a Mach–Zehnder interferometer configuration. For our

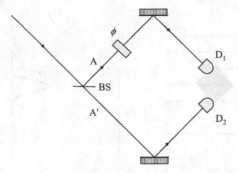

Fig. 1.4. An apparatus realizing a discrete two-beam experiment, in which detectors D_1 and D_2 are placed *before* the two orthogonal beams (A and A$'$) can merge. "BS" indicates a 50–50 beam-splitter and "ϕ" a variable phase-shifter.

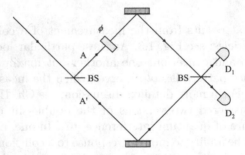

Fig. 1.5. Apparatus realizing a discrete two-beam experiment, in which detectors D_1 and D_2 are placed *after* the two orthogonal beams (A and A$'$) have merged. This apparatus has the advantage over the original two-slit apparatus of Fig. 1.3 that no intensity is lost from the original beam during quantum state preparation. "BS" indicates a 50–50 beam-splitter and "ϕ" a variable phase-shifter.

purposes, it is also convenient that this provides a (dual rail) realization of a qubit, as it makes use of *two* quantum field modes to represent a *single* qubit. The apparatus in Figs. 1.4–5 represent two such mutually exclusive experimental arrangements for a single qubit. The class of experiments interpolating between these is not shown here, but corresponds to a single wing of the apparatus shown in Fig. 3.2 in Chapter 3.

The problem of interest in each of these experiments is again that of understanding microscopic behavior given a particular *preparation*, $\bar{\mathcal{P}}$, of an ensemble of systems that emerges in two beams, A and A$'$, corresponding to quantum-field spatial modes, emerging from a beam-splitter, directed toward a given point.[39] Two distinct kinds of measurement apparatus can be placed away from the beam-splitter, allowing measurements of spatial path and interference to be made. In one apparatus, the features of the pattern of interference between beams A and A' that are allowed to merge, as in Fig. 1.5, enable measurement of the *visibility*, v, of the interference pattern arising given $\bar{\mathcal{P}}$; a variable phaseshifter introduces a phase shift similarly to the way that different path-lengths from diaphragm slits to a given point on the detection screen do in the double-slit experiment of Fig. 1.3. The other apparatus consists of detectors placed in beams A and A' before they reach a common point, and enables measurement of the *distinguishability*, D, of path, as shown in Fig. 1.4. Now, in order to consider individual systems without conceptual difficulty, let us consider the *prediction* of path rather than its retrodiction.

It may prove useful in this situation to introduce an ancillary quantum system to aid in the determination of the path a system may take. In such an *extended* class of experiments, there exist distinct ensemble preparations, $\bar{\mathcal{P}}$ and $\bar{\mathcal{P}}'$, both determining the *same* statistical operator ρ describing a *single* qubit but such that the resulting distinguishabilities are unequal, that is,

$$D(\bar{\mathcal{P}}) \neq D(\bar{\mathcal{P}}') \tag{1.35}$$

with

$$D(\bar{\mathcal{P}}) > P_D , \tag{1.36}$$

where P_D is any measure of path distinguishability that depends *only* on the statistical operator, ρ; one sees that path distinguishability is a function of *preparation* rather than of statistical operator alone [239].[40]

One must, therefore, consider *all* measurements (arrangements) that can be made consistently with the preparation $\bar{\mathcal{P}}$, not just two, and find a strategy for predicting for each system in the ensemble whether it will most likely be from beam A or from beam A', where a strategy may make use of knowledge of the preparation *as well as* the results of the measurements. The optimal

---

[39] General quantum state preparations are discussed in the following chapter and compared to quantum measurements.

[40] Such a measure P_D was proposed by Mandel [294]. Generically, preparations consist of the conditions leading to the arrival of a quantum system in an instrument; see Chapter 2. For example, preparations can involve ancillary systems correlated with the system of interest in such a way that a single reduced density matrix (about which, see Sect. 2.5) may result from two different preparations, which preparations potentially provide *additional* information about the state beyond that evident in the single-particle statistical operator. For example, the particle may be described by a fully mixed qubit state when uncorrelated with any other system *or* when part of a fully correlated composite system of two-qubits in a maximally entangled state, such as $|\Psi^-\rangle$; see Eq. 3.5, as well as Sect. 9.7. It is important to note, nonetheless, that two systems described by the same statistical operator are guaranteed to provide the same experimental results when *no additional information* of this kind is provided.

strategy, given $\bar{\mathcal{P}}$, is the one for which the probability of a correct prediction has the maximum value, p_{max}. The *path distinguishability*, $D(\bar{\mathcal{P}})$, for a given preparation $\bar{\mathcal{P}}$ can be taken to be the difference between this probability and the minimum probability of error, which can be written

$$D(\bar{\mathcal{P}}) = p_{max} - q_{min} = 2p_{max} - 1 , \qquad (1.37)$$

because the probability of an *incorrect* prediction having a minimum value is simply $q_{min} = 1 - p_{max}$ [239].

Those preparations in which measurements of possible ancillary systems that might interact with systems before they reach the beam-splitter can yield *no* information useful for predicting the path of a system of interest are called the *simple preparations*. Let us call the system initially under consideration "system I," propagating in beams A and/or A′, where "and/or" is used due to the presence of quantum-mechanical amplitude superposition, as discussed in the introduction to this chapter. A simple preparation where the ensemble is such that all its systems are described by the same pure quantum state $|\psi\rangle$ is called a *pure* simple preparation. For such preparations, one has a state

$$|\psi\rangle = a|A\rangle + a'|A'\rangle \qquad (1.38)$$

and the path distinguishability is given by

$$D(\bar{\mathcal{P}}) = |\,|a|^2 - |a'|^2\,| , \qquad (1.39)$$

as per Eq. 1.37, without loss of generality taking beam A to be that one most likely to be entered. For these preparations, there is a complementarity between path distinguishability D and visibility v given by an equality [239]:

$$D^2(\bar{\mathcal{P}}) + v^2(\bar{\mathcal{P}}) = 1 . \qquad (1.40)$$

For *mixed* simple preparations, where the systems of the ensemble are described by a mixed state ρ and distinguishabilities are similarly obtained from probabilities of the form $\rho = \mathrm{tr}(\rho O_i)$, where $O_1 = |A\rangle\langle A|$ and $O_2 = |A'\rangle\langle A'|$, the complementarity is instead expressed by an *inequality* [239]:

$$D^2(\bar{\mathcal{P}}) + v^2(\bar{\mathcal{P}}) \leq 1 . \qquad (1.41)$$

Now consider the broader class of preparations, the *correlated preparations*, consisting of situations where measurements of correlated ancillae *may* be useful for predicting the path. The least complex such case involves the use of a second system, "system II." Let $|\Theta\rangle \in \mathcal{H}_I \otimes \mathcal{H}_{II}$ be entangled, that is, *not* factorable into a state-vector in $\mathcal{H}_I$ and a state-vector in $\mathcal{H}_{II}$; for a discussion of entanglement, see Chapter 6. For such a *pure correlated* preparation, partially tracing out the variables associated with system II from the corresponding projector $P(|\Theta\rangle)$ provides the (reduced) state of system I.[41] An

---

[41] See Sect. 2.5 for a discussion of reduced statistical operators and the partial-trace operation.

ensemble formed by mixing *several* pure correlated cases, each with a distinct $|\Theta_k\rangle$ in $\mathcal{H}_I \otimes \mathcal{H}_{II}$ with respective proportion w_k, is the case of *mixed correlated* preparations. The first, strong complementarity relation above holds when D is distinguishability in the pure correlated case, $D(P(|\Theta\rangle))$; the second, weak complementarity again holds when D is that of the mixed correlated case [239].

Finally, consider the *maximum* distinguishability, $\max D(\bar{\mathcal{P}})$, for all preparations $\bar{\mathcal{P}}$ determining a given statistical operator ρ and any vector $|\Theta\rangle$ in $\mathcal{H}_I \otimes \mathcal{H}_{II}$ that yields this ρ for the system-I ensemble, that is, a *purification* of this statistical operator.[42] One finds that

$$\max D(\bar{\mathcal{P}}) = D(|P(\Theta)\rangle) . \tag{1.42}$$

Complementarity relations for distinguishability and interference visibility continue to hold for the extended class of preparations, namely,

$$[\max D(\bar{\mathcal{P}})]^2 + v^2(\bar{\mathcal{P}}) = 1, \tag{1.43}$$

and

$$[D(\bar{\mathcal{P}})]^2 + v^2(\bar{\mathcal{P}}) \leq 1 ; \tag{1.44}$$

see [239]. These single-system relations prove useful in practical applications of qubit interferometry and signal detection in quantum cryptography; see Section 9.4. Recall that the phenomena discussed in this section involve only the self-interference of a single-qubit system, despite the consideration that it might be entangled with a second, ancillary system.

1.6 The Mach–Zehnder interferometer

A double-slit-like arrangement where only two directions are available to the self-interfering system is realized in the Mach–Zehnder interferometer, shown in Fig. 1.6 below, wherein the exit ports of a beam-splitter act as "slits." In this interferometer, a quantum system, most commonly taken to be a photon, enters from the left and/or from below into a beam-splitter with two exit paths. It provides a *spatial qubit*, consisting of occupation of one and/or the other interior beam path. Each path then encounters a mirror, a phase shifter, a second beam-splitter, and finally a particle detector.[43] One can also use

---

[42] Any mixed quantum state ρ can be purified in a larger Hilbert space where the system in question is considered to be a subsystem of a bipartite composite system in a pure state $|\Theta\rangle$; also see Sect. 9.7.

[43] Phaseshifters may also be placed in the paths leaving the exit ports to be used along with the transmittance of the final beam-splitter—together comprising a *transducer* T, as in each of the wings of the complex interferometer shown in Fig. 3.2—to prepare a more general pure state of this qubit and to implement specific single-qubit logic operations or phase encoding for quantum key distribution when connected to a quantum communication line [101, 239, 473].

this interferometer to prepare a *phase qubit* by selecting only those systems entering a single initial input port and exiting a single final output port.

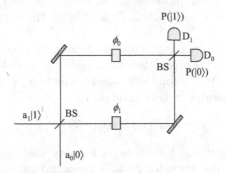

Fig. 1.6. The Mach–Zehnder interferometer providing a range of qubit states as the input qubit amplitudes a_i and phase shifts ϕ_i are varied, with detectors providing count rates proportional to the probability of lying in the output computational-basis states described by state-projectors $P(|0\rangle)$ and $P(|1\rangle)$. The corresponding probabilities of detection for input amplitudes $a_0 = 0$, $a_1 = 1$ are $p(0) = \sin^2[(\phi_0 - \phi_1)/2]$ and $p(1) = \cos^2[(\phi_0 - \phi_1)/2]$; see also Fig. 13.1.

For specificity, let us now take the system in question to be a photon. The beam-path state of a photon exiting the initial beam splitter can be proportional to

$$\text{either} \quad |0\rangle + i|1\rangle$$
$$\text{or} \quad i|0\rangle + |1\rangle \ ,$$

being the former if the photon were input from the left, or the latter if it were input from below; see Fig 1.6. The beam-splitters can each be said to implement the $\sqrt{\text{NOT}}$ quantum gate; see Section 1.4, above. This is so in the sense that, by taking a null shift to occur in the phaseshifter, the *second* beam-splitter has a similar effect to the first, with the net effect on the photon that it exits the interferometer with the *opposite* qubit value from that input; as a result of the destructive interference in one final exit path and constructive interference in the other, depending on the beam-splitter port initially entered, the two beam-splitters together acting as a NOT gate operation (up to a phase factor) in the quantum computational basis, as the particle will exit in the opposite path from which it entered.[44] However, when $-\pi/2$ phaseshifters are placed in the paths $|i\rangle$ $(i = 0, 1)$ *before* and *after* each beam-splitter,

---

[44] A phase difference may also be introduced by a difference of path length, for example, in an unbalanced configuration where the location of the second beam-splitter plus detector pair and the location of a mirror relative to those as shown in

the resulting subapparatus about each beam-splitter can be seen to perform a *Hadamard* transformation on the qubit; for example, see [101]. The result of *two* such net transformations—in the absence of the introduction of any phase difference between paths—is to leave the qubit state unchanged, because $H^2 = I$. Introducing changes of phases and/or input amplitudes gives rise to interference patterns at the detectors, which allows one to study quantum effects in these spatial qubits.

1.7 Quantum coherence and information processing

Although important single-qubit quantum information-processing tasks exist, such as quantum key distribution (QKD) using highly attenuated laser light, where apparatus such as the Mach–Zehnder interferometer adequately induce the relevant quantum phenomena, more interesting quantum information-processing tasks require more than one qubit to be present on which logic operations can be carried out so that quantum entanglement may be involved. These multiple-qubit gates induce additional, yet more subtle phenomena that are manifested only in more complex apparatus to be discussed in subsequent chapters. For example, although QKD can be performed with individually en-coded single qubits, QKD under the Ekert protocol requires a system of two qubits. In particular, genuine quantum computing requires the maintenance of highly coherent superpositions of computational basis states of multiple qubit systems in order to be effective.[45] A brief discussion of the relationship between individual-qubit and multiple-qubit descriptions is therefore called for here before we consider the important subject of quantum measurement, which is also required for quantum computation.

Whenever one is dealing with a quantum system composed of two or more subsystems, the Hilbert space of the system is the tensor product of the Hilbert spaces of the subsystems.[46] N classical bits give rise to 2^N possible classical computational states parameterized by N-bit strings $x_i \in GF(2)^N$. By contrast, the pure-state space for a system of N qubits is the Hilbert space

$$\mathcal{H}^{(N)} = \mathbb{C}^2 \otimes \mathbb{C}^2 \otimes \cdots \otimes \mathbb{C}^2 . \tag{1.45}$$

Fig. 1.6 are interchanged, as in each of the two wings of the Franson interferometer illustrated in Fig. 3.3. This path length can also be varied, as is often done in phase-encoded quantum-key-distribution apparatus.

[45] Nonetheless, see [120]. Quantum computation can be viewed as in essence multi-particle interferometry and vice-versa [151].

[46] The tensor product is described in Appendix A. The prescription for its use in the quantum mechanics is given by Postulate IV of standard quantum mechanics, discussed in Appendix B.1. When notating quantum states associated with tensor product spaces in Dirac notation, the tensor product symbol "$\otimes$" is often omitted but implied, as when $|\psi\rangle \otimes |\phi\rangle$ is written $|\psi\rangle|\phi\rangle$ which is sometimes also written simply $|\psi\phi\rangle$, as in Eq. 1.46 below.

As a result, an N-qubit pure state is parameterized by $2^N - 1$ *complex num-bers*. For example, there are 2^N complex components of a vector $|\Psi\rangle \in \mathcal{H}^{(N)}$ written as a superposition state in the computational basis, which are then reduced by one by fixing the value of its unphysical global phase and normal-izing its length. It is important to note that the information representable in N qubits in general cannot be represented in a polynomial number of classical bits, preventing quantum systems from being efficiently simulated by classical computation.[47] The *computational basis* for this space, namely, $\{|\mathbf{x}_i\rangle\}$ can, nonetheless, be labeled by the 2^N possible N-bit strings $\mathbf{x}_i$, which can be viewed as eigenvalues corresponding to the computational basis of eigenvec-tors.[48] A generic N-qubit (register) state-vector $|\Psi\rangle \in \mathcal{H}^{(N)}$, according to the superposition principle, can be written in the computational basis as

$$|\Psi\rangle = \sum_{i=0}^{2^N-1} a_i |\mathbf{x}_i\rangle , \qquad (1.46)$$

where the sum is taken over all 2^N strings of N bits, $a_i \in \mathbb{C}$, the global phase angle is set to zero, and the total probability is unity,

$$\sum_{i=0}^{2^N-1} |a_i|^2 = 1 . \qquad (1.47)$$

Standard quantum computation is described by the evolution of a multiple-qubit state, typically beginning from the initial fiducial state $|x_0\rangle \equiv |00\ldots0\rangle$, according to a transformation that is decomposable as a series of unitary multiple-qubit gate operations, followed by a measurement readout project-ing the unitarily transformed state onto the computational basis.[49] Thus, quantum algorithms are ultimately *irreversible* and *probabilistic* in nature, though the unitary (logical) portions of the evolution are themselves *deter-ministic* and *reversible*; see Chapter 14.[50]

Quantum algorithms benefit from a unique form of computational par-allelism arising from the presence of quantum superpositions within these very large Hilbert spaces. This parallelism allows, in a particular sense, the

---

[47] Indeed, the inability of classical computers to *efficiently* simulate quantum sys-tems was one of the original motivations for the exploration of quantum comput-ing. However, the difference between cases is not as great as sometimes presented; see the discussion of the Knill–Gottesman theorem in the box in Sect. 13.5.

[48] It is this correspondence that allows for classical readout of the result of quantum computation, though not classical treatment of the quantum computation *itself*.

[49] See Chapter 13 for a description of how general unitary logic operations may be performed using a finite number of quantum gates.

[50] Note, however, that practical models of quantum computation involve ancillae that are measured for the purpose of error-correction and that have nondeter-ministic and irreversible evolutions. Alternative, "one-way" quantum computa-tion has also been defined, which operates somewhat differently [82].

evaluation of a given function *for many values at once*. Unlike classical computational parallelism, in which *several* circuits are required to operate at one time, quantum parallelism operates with a *single* circuit on a number of computational-basis vectors in a quantum superposition; this occurs because the state of a quantum register can be a superposition of computational basis states on which an algorithm can be realized as a unitary transformation up until the readout stage. *Multiple streams of data are, in essence, represented together in a single quantum data set acted on by a single quantum circuit.* Due to the size of the Hilbert space available to a number of qubits, a single quantum circuit operates, in effect, on an exponentially large data set.

Uniquely quantum states of multiple-qubit systems exist within these large Hilbert spaces that exhibit a number of phenomena that have no local realistic mechanical explanation due to their entanglement, which can be exploited to help quantum computing surpass its classical counterpart in efficiency.[51] However, multiple-qubit states are also very fragile, being susceptible to decoherence effects [317, 470]. After a short period of time, the pure quantum states described by Eq. 1.46 are inevitably altered by interactions with their environments and must then be described instead by a *mixed* quantum states of the form

$$\rho(t) = \sum_{i=0}^{2^N-1} \sum_{j=0}^{2^N-1} \rho_{ij} |\mathbf{x}_i\rangle \langle \mathbf{x}_j| \ . \tag{1.48}$$

Interestingly, a similar process is a natural element of the quantum measurements that are essential to quantum computation, as they are necessary for computational readout; although decoherence must be avoided in the *middle* of quantum computation, it plays a role at the *end* of quantum computation when *classical* information must be extracted by measurement of the computing system [475].[52]

Finally, as an example of quantum state decoherence, consider a state of a multiple-qubit system presented in the standard form of Eq. 1.46 in contact with a thermal bath at temperature T. The density matrix of the state will evolve toward the diagonal form

$$\rho(t) = \sum_{i=0}^{2^N-1} \rho_{ii} P(|\mathbf{x}_i\rangle) \ , \tag{1.49}$$

where $\rho_{ii} = \exp(-E_i/kT)/\sum_{j=0}^{2^N-1} \exp(-E_j/kT)$, k being the Boltzmann factor and E_i energy eigenstates, because the off-diagonal matrix elements describing the coherence tend toward zero, a process that takes place over a timescale dependent on the system–environment interaction [153, 319].

---

[51] The question of local realistic descriptions of quantum states is immediately addressed in the following chapter. Entanglement itself is the focus of Chapters 6 and 7.

[52] Quantum decoherence is addressed in detail in Chapter 10. Quantum measurement is discussed in detail in the following chapter.

2

Measurements and quantum operations

Quantum states can undergo two distinct sorts of transformation: in addition to unitary transformations such as the quantum gates discussed in the previous chapter, non-unitary transformations can take place, measurements being the most significant of these. Complete quantum information-processing tasks generally involve a measurement step because quantum measurements are required for classical information to be read out from quantum states. Here, before describing the general class of state transformations via the operations formalism and before placing standard and generalized measurements within it, a number of important early contributions to quantum measurement theory are surveyed. These provide important distinctions and clarify the meaning of the terms *measurement, preparation, and selection* in the quantum context. However, because the issues with which these contributions are concerned are often rather subtle, on a first reading one may wish to proceed directly to Sections 2.3–5, in which quantum expectation values, projection postulates, and reduced states are characterized, after which generalized operations and measurements are discussed.

Quantum measurements must be performed in order to determine the properties of a quantum system, which may have been prepared in an incompletely known state.[1] They involve the physical coupling of an apparatus to

---

[1] Henry Margenau distinguished quantum measurement from quantum state preparation as follows. State *preparation* determines the state of a physical system *"but leaves us in ignorance as to the incumbency of that state after preparation,"* whereas *measurement* "certifies that *some system responded to a process, even though we are left in ignorance as to the state*; after measurement, for example, the measured system may have been destroyed." [Margenau's emphases.] The latter situation is the case in photon counting, for example. See also Pauli's distinction between first-kind and second-kind measurements in Sect. 2.2. Margenau pointed out that there are numerous processes in which both of these characteristics are combined, that is, that are both preparations *and* measurements [295]. In practice, preparation is often simply the selection of part of the output of a nondestructive measurement device, for example, passage of a photon through

the system to be measured in order to provide results given as the registration of an appropriate property, sometimes referred to as a *pointer observable*. In the quantum information context, typically the object system of interest is a number of qubits serving as a *quantum register* and the pointer property of the apparatus takes bit-string values, although pointer-property values can differ from the computational-basis eigenvalues for the register, in which case there must be a well-defined *pointer function* serving to bring the elements of the two sets of values into one-to-one correspondence.

Regardless of the measurement model assumed, one also implicitly or explicitly assumes a formal relationship between system properties and eigenstates, namely that a system property is attributed a definite value when and only when the system is in an eigenstate of the operator corresponding to that property; when in other states no such value is attributed. This is sometimes called the *eigenvalue–eigenstate link* [348].[2] Similarly, a *calibration postulate* is sometimes introduced, requiring that if a system is in an eigenstate of an operator corresponding to a property then a measurement of this property leads *with certainty* to an outcome indicating that the system is in this eigenstate [303]. Typically, a quantum state is taken to be defined by its preparation (or measurement), which is used either to *select* (or *predict*) or to *post-select* (or *retrodict*) the state at times afterward or beforehand, respectively. Quantum probabilities specified by the state therefore have an implicitly *conditional* character.[3]

A fundamental difficulty arises when one attempts to provide a fully internal deterministic description of measurement in quantum mechanics, due to the unitary character of the standard time-evolution (the *Schrödinger evolution* described by the operator U) of closed systems, that is, systems not interacting with anything outside of themselves, and the superposition principle. This difficulty is often referred to as the *quantum measurement problem* and can be seen to arise as follows [90]. Consider a measuring apparatus (often assumed to be macroscopic), initially in an eigenstate $|p_0\rangle$ of the pointer property, and a system to be measured (often assumed to be microscopic) having a property of interest and corresponding operator O with discrete nondegenerate eigenvalues $\{o_j\}$ that is taken to be in a specific eigenstate $|o_i\rangle$ before the measurement process begins and that is assumed to remain *unchanged* during the measurement process. With these assumptions, the measurement process would result in the transformation

---

    a given port of a polarizing beam-splitter, the output of the other port being disregarded. By virtue of such selection, the apparatus acts as an analyzer; see Fig. 2.2.2.

[2] One can think of this assumption as a means of making more explicit the meaning of Postulate I of standard quantum mechanics given in Sect. B.1.

[3] Note, however, that the relative character of quantum states does not render them *essentially* epistemic in nature; see the discussion of the ignorance interpretation of quantum probability of Sect. 1.1 and hidden variables in Sect. 3.1. Note also that closed-system quantum evolution is time-symmetric.

$$|\Psi_j^{(i)}\rangle \equiv |p_0\rangle|o_j\rangle \tag{2.1}$$

$$\rightarrow |\Psi_j^{(f)}\rangle \equiv |p_j\rangle|o_j\rangle , \tag{2.2}$$

for each value of j that is a possible measurement outcome. However, suppose that the state of the system to be measured were instead initially not an eigenstate of O but rather a *superposition* state $\sum_j a_j|o_j\rangle$, which is also an allowed system state by the superposition principle. In that case, taken to be a unitary operation acting linearly on state-vectors, the measurement process would be described by the pure state transformation

$$|\Psi\rangle \equiv |p_0\rangle \sum_j a_j|o_j\rangle \tag{2.3}$$

$$\rightarrow |\Psi'\rangle \equiv \sum_j a_j|p_j\rangle|o_j\rangle . \tag{2.4}$$

In the statistical operator description, the transformation of this initial pure state to a final pure state would then be

$$\rho \rightarrow \rho' = U\rho\, U^\dagger , \tag{2.5}$$

where

$$\rho = P(|\Psi\rangle) , \quad \rho' = P\left(\sum_j a_j|p_j\rangle|o_j\rangle\right) , \tag{2.6}$$

because unitary transformations preserve state purity. What one *needs* from a successful measurement, however, is a transformation of the initial pure state, ρ, of the complete system to a final state of the form

$$\rho^{(f)} = \sum_j |a_j|^2 P(|\Psi_j^{(f)}\rangle) \tag{2.7}$$

that is a *mixed* state for an ensemble of situations occurring with probabilities corresponding to the probabilities of measurement outcomes, $|a_j|^2$, only *one* of which is obtained upon measurement.

The result of the system's evolving according to the unitary time-evolution prescribed by the postulates of quantum mechanics is a coherent superposition involving *several* distinct measuring system states. In particular, we see that it is a coherent superposition of measurement states of the prescribed form shown in Eq. 2.2 rather than a mixed state. Thus, unitary evolution alone yields a measurement of the quantity O that remains *indefinite* in outcome and so provides an inadequate description of the required measurement process. The failure of the unitary evolution to provide a definite measurement outcome is the quantum measurement problem, sometimes also called the *macro-objectification problem*.[4] Nonetheless, this conceptual difficulty does not prevent one from making practical use of the quantum formalism.

---

[4] D'Espagnat identifies five aspects of the measurement problem, including this one [128]. It is valuable to compare and contrast the composite system state in Eq.

2.1 The von Neumann classification of processes

The quantum measurement problem has resulted in a long, though somewhat uncomfortably held, fundamental distinction in quantum mechanics between measurements and other processes. In particular, John von Neumann distinguished two types of quantum state change, or *intervention*: the first type being the sort taking place during measurement involving a process of subjective perception of the measurement result by a percipient outside of the quantum description, the second type being those taking place otherwise [444]. With such a distinction, one can consistently treat measurement devices themselves as quantum systems as above. By contrast, the approach of Niels Bohr in the *Copenhagen interpretation* of quantum mechanics is to consider measurement apparatus as classical systems *not* described by quantum mechanics [310].[5] The discontinuous change of quantum state at the end of the measurement process—that is, upon its coming to be known by a percipient—was taken by von Neumann to be accompanied by a change in the state of the measured system such that an immediate repetition of the measurement would with certainty yield the *same* result as the initial measurement. This is sometimes referred to as *the repeatability hypothesis*. Von Neumann accordingly invoked what is now known as the (traditional) *projection postulate* used by Dirac, Heisenberg, Pauli and others beginning in the late 1920s.[6]

2.6 with the subsystem reduced state given by Eq. 1.49 which can result from the unitary evolution of a composite system. A more detailed treatment of this problem can be found in [380]. For a statistical perspective on the topics discussed in this chapter one may wish to refer to [219].

[5] Another alternative also not further discussed here due to its highly unpleasant metaphysical implications but popular with some physicists and mathematicians investigating quantum computing, most notably David Deutsch, is the so-called *many-worlds* interpretation of quantum mechanics. In this interpretation, measurement is to be fully described by a unitary evolution in the joint Hilbert space of the measured system, the measuring system, and the entire environment of the two, producing a *von Neumann chain* of superposition states of all these systems of the sort given by Eq. 2.6, ultimately resulting in a "wavefunction of the *universe*" in an elaborate superposition state that never collapses; those portions of the resulting ramifying set of situations in which *different* sets of measurement outcomes are obtained by measurers are assumed in some way to be *inaccessible* to one another; see [303] for a discussion of measurement under this interpretation and [381] for a discussion of metaphysical implications of the interpretation. In essence, this interpretation attempts to circumvent the quantum measurement problem by metaphysical fiat at the level of the entire universe, rather than invoking the subjective perception that is naturally present in any measurement process the outcome of which comes to be known by a subject, as von Neumann did. This idea was first carefully investigated by Hugh Everett III and John A. Wheeler [134].

[6] The name "projection postulate" itself was first given by Margenau in 1958 [295]. For his part, Dirac dictated that "a measurement always causes the system to jump into an eigenstate of the dynamical variable being measured" [136].

In von Neumann's treatment, the discontinuous change of state during measurement is expressed by the rule that, when subject to measurement, a quantum system initially in a pure state evolves *nonunitarily* into a mixed state. In particular, in accord with Eq. 2.7,

$$P(|\psi\rangle) \longrightarrow \rho' = \sum_i \left(\langle \psi_i | P(|\psi\rangle) | \psi_i \rangle \right) P(|\psi_i\rangle) , \qquad (2.8)$$

where $P(|\psi\rangle)$ is a projector onto the initial state $|\psi\rangle$, the projectors $P(|\psi_i\rangle)$ onto the eigenvectors $|\psi_i\rangle$ sum to the identity, and the weights $\langle \psi_i | P(|\psi\rangle) | \psi_i \rangle$ sum to unity and correspond to probabilities that the values of the property being measured are found to be those of the subensembles corresponding to the projectors $P(|\psi_i\rangle)$, when the system is known to initially have been in the state $|\psi\rangle$; see Fig. 2.1. This stage of state evolution is sometimes also referred to as *pre-measurement*, because no particular measurement outcome is selected by it. For projectors, $P(|\psi_i\rangle)P(|\psi_j\rangle) = \delta_{ij}P(|\psi_i\rangle)$, guaranteeing that if the same subspace is projected on immediately and repeatedly then such measurements always return the same value.

Because the set of vectors $\{|\psi_i\rangle\}$ form a basis for $\mathcal{H}$, the matrix elements of the final statistical operator are, in an arbitrary basis $\{|\alpha_i\rangle\}$,

$$[\rho']_{ij} = \sum_k \langle \alpha_i | \psi_k \rangle \langle \psi_k | \psi \rangle \langle \psi | \psi_k \rangle \langle \psi_k | \alpha_j \rangle \qquad (2.9)$$

$$= \sum_k w_k [\rho'_k]_{ij} , \qquad (2.10)$$

that is, the process takes pure states to *mixtures* described by the weights w_k, as required. This discontinuous process is von Neumann's *type-I process*. When this process gives rise to any change in state it is irreversible [444]. When a subensemble, corresponding to a given value of k, constituting a proportion w_k of the total normalized ideal ensemble, is then also *selected*—in the case of an individual system by its being actualized, which happens with probability w_k—one has

$$P(|\psi\rangle) \rightarrow P(|\psi_k\rangle)|\psi\rangle , \qquad (2.11)$$

the r.h.s. generally having nonunit norm. The resulting pure ensemble state can then be renormalized, so that the statistical operator of the particular selected pure subensemble has trace one and can then be described simply by the statistical operator $P(|\psi_k\rangle)$. The pair of processes described by Eqs. 2.9–11 constitute a *selective measurement*.

Von Neumann's *type-II process* is the usual continuous (*automatic*) process described by the Schrödinger evolution

$$\rho \rightarrow \rho' = U \rho \, U^\dagger , \qquad (2.12)$$

where U is the unitary operator describing temporal evolution discussed in Section 1.1. In this process, the purity and trace of the statistical operator remain unchanged, so that no renormalization of state is required.

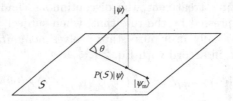

Fig. 2.1. Projection of a quantum state-vector $|\psi\rangle$ into a vector subspace $\mathcal{S}$ by a projector $P(\mathcal{S})$. Specifically, the projection of $|\psi\rangle$ onto a ray corresponding to $|\psi_m\rangle$, with which it makes an angle θ, is shown here; the probability for this transition to occur is $\cos^2\theta$. A von Neumann-Lüders measurement corresponds to the set of possible projections onto a complete orthogonal set of subspaces, not necessarily rays, spanning the Hilbert space of the system being measured; see Sects. 2.3–4.

2.2 The Pauli classification of measurements

At a finer level of detail, Wolfgang Pauli classified measurements themselves into two categories, those of the *first kind* and those of the *second kind*, as follows [324].

(1) Measurements that, when they are known to have been performed on a quantum system with outcomes that remain *unknown*, result in probabilities for the quantity measured having definite values, which have become determinate as a result of the measurement, that are *equal immediately before and after* it, though the state of the system will have been changed and may subsequently provide different measurement results, are *measurements of the first kind*. In this case, the state is changed only to the extent necessary for a measurement to be performed, leaving unchanged the particular *property* measured. Pauli further distinguished *repeatable measurements*, which are those first-kind measurements that, when repeated, *cannot* lead to a new result, which are the sort of measurements prescribed by the von Neumann projection postulate. The Stern–Gerlach spin measurement is the archetypical example of such a measurement.[7] An example from linear optics is the measurement of linear polarization by a birefringent crystal; see Fig. 2.2-1.

(2) Measurements in which the state of the measured system is controllably changed in such a way that repeating them will lead to statistical results *different from* those of the initial measurement. When definite conclusions can be made regarding the quantity being measured in this way, the measurements are *measurements of the second kind*. The measurement of light polarization by a polarizer passing only one linear polarization state is an example of such a measurement, because the state is required to come to match the known state of the polarizer in this case; see Fig. 2.2-2. Photon counting wherein

---

[7] For a description of the Stern–Gerlach measurement, see Sect. 1.1 of [359]; for an analysis of behavior of coherence in the Stern–Gerlach apparatus, see [162].

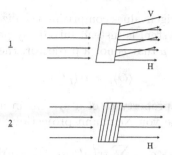

Fig. 2.2. Examples of apparatus performing first kind and second kind measurements, respectively, according to the Pauli classification—light polarization: (1) refracted by a Rochon prism, and (2) selectively absorbed using a dichroic polarization analyzer.

photons are destroyed is another example of a second-kind measurement (*cf.* Footnote 1).

Pauli and von Neumann both emphasized that, when the interaction between a measurement apparatus and a system being measured is analyzed, the linearity of the Schrödinger equation describing the state evolution of composite system formed by these two (sub)systems provides consistency between alternative descriptions of system behavior in which the division (or "cut") between measuring system and measured system is made differently. Pauli viewed the need for a nondeterministic projection as arising naturally from the fact that the interaction between the measuring system and the measured system is "in many respects intrinsically uncontrollable."[8]

2.3 Expectation values and the von Neumann projection

The *expectation value* of a property represented by an Hermitian operator O of a quantum system in a pure state given by a state-vector $|\psi\rangle$ is

$$\langle O \rangle_{|\psi\rangle} = \langle \psi | O | \psi \rangle \tag{2.13}$$

$$= \sum_i o_i |\langle o_i | \psi \rangle|^2, \tag{2.14}$$

---

[8] For an example treatment of irreversible transformations in an experimental situation involving Stern–Gerlach apparatus, see [372]. A contemporary formulation of this measurement taxonomy that addresses subtle additional issues that arise when various different interpretations of the quantum formalism are introduced—such as the Copenhagen, minimal statistical, "realistic," and many-worlds interpretations—can be found in [303]. The standard model of quantum measurement theory is surveyed in [93].

where $\{o_i\}$ is the set of eigenvalues comprising the *eigenvalue spectrum* of O;[9] when the state of the system is instead mixed, by necessity being described by a statistical operator ρ that is not a projector, the expectation value is

$$\langle O \rangle_\rho = \mathrm{tr}(\rho O) \ . \tag{2.16}$$

The measurement of a property O of a system in a general state (pure or mixed), according to the von Neumann projection rule, is described by the transition

$$\rho \to \rho' = \sum_i \mathrm{tr}\big(\rho P(|o_i\rangle)\big) P(|o_i\rangle) \ . \tag{2.17}$$

A measurement of a property is said to be *maximal* (or *complete*) when it provides fully distinct values for the quantity measured, so that no more information can be obtained by further measurements of the property. For such measurements, the above projection rule is entirely adequate. If, instead, the measurement performed is capable of discriminating only sets of values, the measurement is said to be *nonmaximal*; in that case, it provides *incomplete* information about the property. Consider, for example, the measurement of a *qutrit*, a quantum system possessing a trivalent property O with values $o_i = -1, 0, 1$. A maximal measurement will have three possible outcomes, one for each of the possible values. By contrast, a measurement with only two outcomes, say "$-$" for system property values -1 or 0, and "$+$" for system property value $+1$, is a nonmaximal measurement. An example situation wherein the latter would be realized involves an imperfect Stern–Gerlach type apparatus acting on a spin-1 system such that a particle with z-spin $+1\hbar$ enters a distinct spatial beam downstream from the magnet but particles with spins $0\hbar$ or $-1\hbar$ are not allowed to separate, entering only a common, second beam.

For measurements of properties whose operators have degenerate, that is, nonunique eigenvalues, as in the above example, this projection rule can be improved upon, as we show in the following section.

---

[9] Expectation values thus take the form of *average values* for measurements on ensembles of quantum systems prepared in the same state under statistically ideal circumstances. A related mathematical theorem central to quantum mechanics is the *spectral theorem*: each Hermitian operator, O, can be written

$$O = \sum_i o_i P(|o_i\rangle) \ , \tag{2.15}$$

where $P(|o_i\rangle)$ is the projector onto the finite Hilbert subspace spanned by $|o_i\rangle$. Equation 2.15 provides the *spectral decomposition* (or *eigenvalue expansion*) of the operator O. This theorem does *not* hold for operators in *infinite-dimensional* Hilbert spaces, even when there exists a countably infinite set of basis vectors. Such a decomposition does not exist in general in that case because there may not exist a countably infinite set of *eigenvectors* that form a basis. There do exist topologies on infinite-dimensional spaces for which the theorem in a generalized form (the *nuclear spectral theorem*) *does* hold, however; for example, see [64].

2.4 The Lüders rule

If the projection operator corresponding to an outcome for a measurement of a property O projects onto a subspace of finite dimension *greater than one*, then the (original form of the) von Neumann projection postulate, including subensemble selection and renormalization, prescribes that the measurement process be described by the process

$$\rho \longrightarrow \rho' = P_k \, , \tag{2.18}$$

where here the projector is written P_k. This rule yields a system state after measurement that is *independent* of the details of the state before the measurement, beyond those pertinent to the measurement outcome itself, as can be seen by noting that ρ *does not appear in the description of* ρ'. Accordingly, the von Neumann prescription fails to maintain the distinction between initially pure states and initially mixed states and fails to preserve coherence of pure states in nonmaximal measurements.

A more general prescription of projective state-change as a result of such a selective quantum measurement is the *Lüders projection* (*Lüders rule*),

$$\rho \longrightarrow \rho' = \frac{P_k \rho P_k}{\mathrm{tr}(\rho P_k)}, \tag{2.19}$$

under which the state after measurement clearly is dependent on the state of the system beforehand. The values of successive measurements under this rule will coincide when another measurement is made between successive measurements of O of a property *compatible with* O in the sense that the corresponding operators commute, unlike in the case of the measurements as characterized by the von Neumann projection described above [250, 290]. Thus, under the Lüders prescription for state projection, if one prepares two ensembles of systems in the state ρ, the first measured for some property compatible with O and the second having O itself measured *first*, the relative frequencies of the values of the compatible observable are the same for those two cases, yielding pure subensembles from pure ensembles. Furthermore, when the initial state of the system is pure, the Lüders rule is the *only* projection rule for which this is true [400]. The Lüders prescription is itself a consequence of the Feynman rules for computing quantum probability amplitudes, which are based on the concept of the indistinguishability of processes [399]. The Lüders rule is now commonly considered to be the appropriate general description of a "von Neumann" (read *precise*) measurement, as distinct from generalized (POVM) measurement; see Section 2.7, below.[10] Note that both of the above prescriptions are descriptions of *selective* measurements.

---

[10] Note that von Neumann measurements of a discrete ordinary observable are repeatable, but a repeatable measurement of such an observable need not be a von Neumann measurement; for example, see [94]. For continuous variables, see [92].

Nonselective measurements, by contrast, of systems in initial states ρ that are not necessarily pure are described by the transition

$$\rho \to \rho' = \sum_i P_i \rho P_i \,, \tag{2.20}$$

where P_i is a projector onto the (not necessarily one-dimensional) eigensubspace corresponding to the outcome i, under the Lüders prescription.

The Lüders rule describes measurements that are *minimally disturbing* (or *coherent*) in the sense that they project an initially pure state onto eigensubspaces with a weight proportional to the square of the projection onto each subspace. By contrast, the original von Neumann measurement rule describes measurements that are *maximally disturbing* (or *incoherent*) relative to other first-kind measurements.[11] A distinction between these rules has also been made as follows. The original von Neumann measurement is intended to describe measurements of ensembles, whereas the Lüders rule is intended to describe individual measurements [185].

2.5 Reduced statistical operators

Describing measurement quantum mechanically involves the examination of the interaction between a measuring apparatus and the system it measures.[12] The *joint state* of the measurement apparatus and the system, initially a pure state, can be considered to remain pure and described by the standard unitary evolution throughout measurement. However, each of the *subsystems*, considered alone, enters a *mixture* described by the reduced statistical operators obtained by partial tracing out parameters describing the other subsystem, as the joint state becomes entangled. The quantum information as measured by the quantum entropy of the state of each subsystem accordingly decreases, as shown in Chapter 5. Such a description also applies to system–environment interactions, which are described in Chapter 10.

Let $\{|u_i\rangle\}$ and $\{|v_j\rangle\}$ be bases for Hilbert spaces $\mathcal{H}_1$ and $\mathcal{H}_2$ of countable dimension, describing two subsystems 1 and 2, respectively, forming a composite system in state ρ. The set of vectors $\{|u_i\rangle \otimes |v_j\rangle\}$ $(i = 1, 2, \ldots; j = 1, 2, \ldots)$ is then a basis for the Hilbert space of the total system, $\mathcal{H} = \mathcal{H}_1 \otimes \mathcal{H}_2$. Any operator O on $\mathcal{H}$, such as ρ, can be written in the form

$$O = \sum_{ij,kl} \big\{ |u_i\rangle |v_j\rangle O_{ij,kl} \langle u_k| \langle v_l| \big\}, \tag{2.21}$$

where $O_{ij,kl}$ are scalars (the matrix elements corresponding to O). Finding the partial trace is somewhat like finding the marginal distribution of a component

---

[11] For this reason the original, von Neumann projection rule is sometimes referred to as the "clumsy experimenter's rule." See also [126].

[12] For descriptions of the measurement process including details of measurement interaction see, for example [6, 330]. For more tensor products, see Sect. A.5.

of a two-dimensional random variable from the probability distribution of the latter in classical probability theory: the *partial trace* of O, with respect to the first subsystem, for example, is

$$\text{tr}_1 O \equiv \sum_i \langle u_i | O | u_i \rangle . \tag{2.22}$$

In particular, the result of partial tracing the statistical operator ρ of the combined system over each of the subsystems individually is the pair of *reduced statistical operators*

$$\rho_1 = \text{tr}_2 \rho , \tag{2.23}$$
$$\rho_2 = \text{tr}_1 \rho , \tag{2.24}$$

each describing the state of one subsystem, for example, in the case of the *dismissal* of the other subsystem. The reduced statistical operator is the only statistical operator providing correct measurement statistics for subsystems [268]. When the overall state ρ is an entangled pure state, the reduced states ρ_1 and ρ_2 describing the component systems are *mixed* rather than pure, a situation not arising for marginal distributions in classical mechanics.

2.6 General quantum operations

As we saw in the description of the measurement process above, it is often valuable to describe transformations of quantum states besides those described by the standard unitary evolution of closed systems yet still allowed by quantum mechanics. Important examples of these include the evolution of open quantum systems, which may also lie outside the class of transformations described by the projections considered above. It is therefore useful to consider the class of completely positive trace-preserving (CPTP) linear transformations,

$$\rho \to \mathcal{E}(\rho) , \tag{2.25}$$

often called *operations*, taking statistical operators to statistical operators, each described by a *superoperator*, $\mathcal{E}(\rho)$, satisfying the following conditions.

(i) $\text{tr}[\mathcal{E}(\rho)]$ is the *probability* that the transformation $\rho \to \mathcal{E}(\rho)$ takes place;

(ii) $\mathcal{E}(\rho)$ is a linear convex map on statistical operators, that is,

$$\mathcal{E}\left(\sum_i p_i \rho_i\right) = \sum_i p_i \mathcal{E}(\rho_i), \tag{2.26}$$

p_i being probabilities. ($\mathcal{E}(\rho)$ then extends uniquely to a linear map.)

(iii) $\mathcal{E}(\rho)$ is a completely positive (CP) map.

A linear map $L : B(\mathcal{H}) \to B(\mathcal{H})$, where $B(\mathcal{H})$ is the space of bounded linear operators on $\mathcal{H}$, is said to be *positive* if $L(\mathcal{O}) \geq \mathbb{O}$ for all $\mathcal{O} \geq \mathbb{O}$, that is, all $\mathcal{O} \in B(\mathcal{H})$ for which $\langle\psi|\mathcal{O}|\psi\rangle \geq 0$ for all $|\psi\rangle \in \mathcal{H}$;[13] such a positive L is *completely positive* (CP) if, in addition, any $\mathbb{I}_N \otimes L \in B(\mathbb{C}^N \otimes \mathcal{H})$ is positive, for all $N \in \mathbb{N}$. Note that matrix transposition in any basis,

$$\mathcal{T} : |i\rangle\langle j| \rightarrow |j\rangle\langle i| , \tag{2.27}$$

for example, the computational basis, is a positive map that is *not* completely positive. A CPTP map is just a CP map that is also TP.

Operations $\mathcal{E}(\rho)$ satisfy the above three conditions if and only if they are such that

$$\mathcal{E}(\rho) = \sum_i K_i \rho K_i^\dagger , \tag{2.28}$$

for some set $\{K_i\}$ of Hilbert-space operators for which $\mathbb{I} - \sum_i K_i^\dagger K_i \geq \mathbb{O}$ [261]. The elements K_i are sometimes called *decomposition operators* (or *operation elements* or *Kraus operators*) and the set $\{K_i\}$ called the *operator decomposition*.[14] Equation (2.28) provides the *operator-sum* representation for the operation $\mathcal{E}(\rho)$. The *trace preserving* (TP) property for $\mathcal{E}(\rho)$, $\mathrm{tr}(\mathcal{E}(\rho)) = \mathrm{tr}(\rho)$, translated in terms of the decomposition operators $\{K_i\}$, is the property

$$\sum_i K_i^\dagger K_i = \mathbb{I} , \tag{2.29}$$

which is a *completeness relation* that, because K_i and $K_i^\dagger$ do not necessarily commute, may differ from the condition

$$\sum_i K_i K_i^\dagger = \mathbb{I} , \tag{2.30}$$

which is required for a CP map to be a *unital* map, that is, a map for which $\mathcal{E}(\mathbb{I}) = \mathbb{I}$. One example of such a map is the qubit-depolarizing channel; a negative example is provided by the amplitude-damping channel; see Section 9.6. If the operator decomposition of a CP map satisfies both these conditions the map is *doubly stochastic*.

The operator decomposition of an operation is *not unique*. In particular, any two sets of operators $\{K_i\}$ related to each other by unitary transformations equally well represent the same operation $\mathcal{E}(\rho)$. A decomposition is said to be *minimal* if there exists no decomposition into a smaller set of operators than it contains, which is the case if and only if its operators are linearly independent. An operation is said to be *pure* if there is a decomposition of it that contains only one operator; the standard quantum-mechanical closed-system

---

[13] A linear map L taking $\rho \mapsto L(\rho)$ is one such that for $\rho = p_1\rho_1 + p_2\rho_2$, $L(\rho) = p_1 L(\rho_1) + p_2 L(\rho_2)$.

[14] These decomposition operators K_i are not necessarily Hermitian; for example, see the operator elements of the amplitude-damping channel discussed in Sect. 9.6.

time-evolution operation (the *Schrödinger evolution*) is an example of a pure operation.

Any CPTP map can be understood in terms of a unitary transformation acting in a Hilbert space larger than the Hilbert space $\mathcal{H}_1$ associated with the relevant statistical operator. In particular, for a state $\rho \in \mathcal{H}_1$ in the larger space $\mathcal{H}_1 \otimes \mathcal{H}_2$, there is a state $\rho' \in \mathcal{H}_2$ such that

$$\rho = \mathrm{tr}_2(\rho \otimes \rho') \ . \tag{2.31}$$

Any CPTP map can be viewed as the transformation

$$\mathcal{T}(\rho) = \mathrm{tr}_2\big(U(\rho \otimes \rho')U^\dagger\big) \ , \tag{2.32}$$

for some unitary operator, U [261]. Some other examples of CPTP maps are the *nonselective measurement* $\rho \to \sum_i P(|\psi_i\rangle)\rho P(|\psi_i\rangle)$ and the *composition* of quantum systems $\rho \to \rho \otimes \rho'$, such as when an ancilla is *added* to a system.

2.7 Positive-operator-valued measures

Consider, for a moment, measurement in the broadest setting, wherein the eigenvalue spectrum of the Hermitian operator O representing a physical property may be continuous, so that a measurement might place its value within a Borel set $\Delta \in \mathbb{R}$ and leave the state of the system with support (O, Δ) with respect to O. A projector $P_O(\Delta)$ from the spectral decomposition of O might describe the (quantum mechanically) maximally specified state of such a system. There are significant difficulties arising from such a description; for example, see [92, 398]. The generalized measurements we now consider provide one well-defined way of describing of such situations, although those will not be explicitly dealt with here since we consider only discrete properties.

Generalized measurements are the class of quantum operations that are described by positive-operator-valued measures (POVMs) [122]. Given a nonempty set S and a σ-algebra Σ of its subsets X_m, a *positive-operator-valued measure* E is a collection of operators $\{E(X_m)\}$ satisfying the following conditions.[15]

(i) *Positivity*: $E(X_m) \geq E(\emptyset)$, for all $X_m \in \Sigma$.

(ii) *Additivity*: for all countable collections of disjoint sets X_m in Σ,

---

[15] See Sect. A.7 for the definition of σ-algebra. A *Borel σ-algebra* is the σ-algebra generated by the open intervals (or the closed intervals) on a topological space— for example, in $\mathbb{R}$—which are the *Borel sets*. The set S is often a standard measurable space, that is, a Borel subset of a complete separable metric space. Because such spaces of each cardinality are isomorphic, they are all measure-theoretically equivalent to Borel subsets of the real line, $\mathbb{R}$; see [217].

$$E(\cup_m X_m) = \sum_m E(X_m) \ . \tag{2.33}$$

(iii) *Completeness*: $E(S) = \mathbb{I}$.

In the operator space of quantum mechanics, POVMs are the natural correspondents of probability measures. If the *value space*, (S, Σ), of a POVM E is a subspace of the real Borel space $(\mathbb{R}, \mathcal{B}(\mathbb{R}))$, then E provides a unique Hermitian operator on $\mathcal{H}$, namely

$$\int_{\mathbb{R}} \mathrm{Id} \ dE \ , \tag{2.34}$$

where Id is the identity map. The probability of outcome m upon a generalized measurement of a pure state $|\psi\rangle$ is given by

$$p(m) = \langle\psi|E(X_m)|\psi\rangle \ ; \tag{2.35}$$

when the state is mixed, this probability is instead given by

$$p(m) = \mathrm{tr}(\rho E(X_m)) \ ; \tag{2.36}$$

see also the discussion of Gleason's theorem and its extension in Section 3.4.

The effect of a POVM measurement of the initial state ρ is exhibited by post-measurement states ρ'_m and corresponding outcome probabilities $p(m)$. The positive operators $E(X_m)$ in the range of a POVM are referred to as *effects* and represent the events associated with outcomes of generalized measurements. A collection of effects is said to be *coexistent* if the union of their ranges is contained within the range of a POVM. The post-measurement state of a system initially described by a statistical operator ρ under a POVM $\{E(X_m)\}$ is often taken to be

$$\rho'_m = \frac{M_m \rho M_m^\dagger}{\mathrm{tr}(M_m \rho M_m^\dagger)} \ , \tag{2.37}$$

where each of the $E(X_m)$ can be written $M_m^\dagger M_m$, M_m being called a *measurement decomposition operator* (*cf.* [315]); in the special case that the M_m are projectors, this expression coincides with the Lüders–von Neumann measurement rule given by Eq. 2.19—this can be seen by recalling that projectors are Hermitian and idempotent.[16] When, and only when, the measurement operators M_m are projectors—so that the POVM is a *projection-operator-valued measure* (PVM)—are they *identical* to decomposition operators $E(X_m)$, in which case they are also multiplicative, that is, $E(X_m \cap X_n) = E(X_m)E(X_n)$ for all countable subsets of the corresponding set—equivalently, $E(X_m)^2 =$

---

[16] For a more general extension of state projections for POVMS, see Sect. 3.3 of [127].

$E(X_m)$. As noted above, von Neumann measurements of a discrete PV measurement are always repeatable, although the converse is not true.[17] Any POVM can be given as an isometric embedding into a larger Hilbert space together with a PV measurement, so that one can consider von Neumann measurements in a total state space of a composite system consisting of the measured system *together with* an ancilla.

POVM measurements play an important role in quantum key distribution, where they have been used for a variety of quantum signal detection and estimation tasks. POVM elements, when providing positive outcomes, allow one to eliminate quantum states from consideration as describing the measured system. An example of a POVM used for such a purpose is the following [37]. Given the two projectors

$$P(\neg|\phi\rangle) \equiv \mathbb{I} - P(|\phi\rangle) , \qquad (2.38)$$
$$P(\neg|\phi'\rangle) \equiv \mathbb{I} - P(|\phi'\rangle) , \qquad (2.39)$$

where $\langle\phi|\phi'\rangle = \sin 2\theta$, one can construct a POVM $\{E_m\}$ with the following elements.

$$E_1 = P(\neg|\phi\rangle)/(1 + |\langle\phi|\phi'\rangle|) . \qquad (2.40)$$
$$E_2 = P(\neg|\phi'\rangle)/(1 + |\langle\phi|\phi'\rangle|) . \qquad (2.41)$$
$$E_3 = \mathbb{I} - (E_1 + E_2) . \qquad (2.42)$$

POVM measurements using $\{E_1, E_2, E_3\}$ prove more efficient for QKD and eavesdropping thereon than traditional measurements described by the projectors $\{P(\neg|\phi\rangle), P(\neg|\phi'\rangle)\}$. POVMs similarly sometimes allow quantum state tomography to be performed with improved efficiency; see Chapter 8 and [349].

For example, in quantum key distribution under the B92 protocol, the sender of cryptographic key bits uses the nonorthogonal states

$$|\phi\rangle = \cos\theta|\bar{0}\rangle + \sin\theta|\bar{1}\rangle \qquad (2.43)$$
$$|\phi'\rangle = \sin\theta|\bar{0}\rangle + \cos\theta|\bar{1}\rangle \qquad (2.44)$$

to send a random binary sequence to the receiver, $|\phi\rangle$ encoding bit 0 and $|\phi'\rangle$ encoding 1. The receiver (*or an eavesdropper*) can perform measurements of bits that sometimes fail to find the desired bit, but when they succeed *always* provide the bit correctly. In particular, when used as arguments in the POVM described by Eqs. 2.40–42 the first two elements correspond to definite signal state identifications and the third element corresponds to an *indefinite* result; see Sections 1.5, 9.4, 12.5, and, for example, [152]. The probability of succeeding and obtaining any given bit correctly is $1 - \sin 2\theta$ in this case. An experimental realization of such a POVM in linear optics is shown in Fig. 2.3, below.

---

[17] For more detailed information-theoretical treatments of quantum measurement, see [208] and [267].

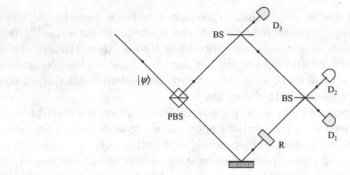

Fig. 2.3. The realization of a projection-operator-valued measurement in linear optics. PBS indicates a polarizing beam-splitter such as a Wollaston prism, BS indicates an ordinary 50–50 beam-splitter, and R indicates a polarization rotator. Such an apparatus can be used as a quantum key bit receiver, where D_3 provides indefinite-bit-value detections and D_1 and D_2 provide definite-bit-value detections. See [76], and Sect 9.4 where an ancilla is used for similar purposes.

3

Quantum nonlocality and interferometry

The most strikingly quantum-mechanical situations phenomena involve entangled states with components located in spacelike-separated regions, customarily referred to as *laboratories*. Each laboratory is taken to contain a physical *agent* capable of performing quantum operations on subsystems within it and potentially communicating with agents in other laboratories, via either one-way or two-way classical and quantum communication channels. When considering two-component systems described by bipartite statistical operators, ρ_{AB}, the corresponding two agents are customarily referred to as "Alice" and "Bob," with the labels A and B indicating the corresponding subsystems or laboratories. Entanglement-based quantum-key-distribution systems are practical examples in which this convention accurately reflects the experimental situation. Locality considerations are often explicitly brought into play when studying entanglement. However, it should also be kept in mind that it is by no means obvious that violations of locality conditions in the traditional sense are sufficient for the characterization of entanglement, despite their value in the investigation of entangled quantum states. The distinction between locality violation and quantum state entanglement should therefore be kept in mind. This chapter focuses on entanglement more in relation to local operations themselves than in relation to information, which relation is the focus of Chapter 6; the intervening chapters introduce various information-theoretic concepts and quantities, both classical and quantum, that will prove essential for the important and often subtle discussion taken up there.

Operations on composite quantum systems are classified as follows. The class of local operations ("LO") is that of operations that are carried out on individual subsystems located within the laboratories of their corresponding agents. The operations of classical communication ("CC") are information transfer acts between agents in separate laboratories carried out via non-quantum means, and may be in one or two directions—these are discussed in detail in the following chapter. Local operations together with those of classical communication ("LOCC") are operations on quantum systems by agents who are also capable of communicating classically. The distinction between

LOCC and mere LO is particularly important in that classical communication between agents allows the local operations of an agent to be *conditioned on outcomes of previous measurements* carried out by other agents; using LOCC, the actions of Alice and Bob can be correlated in a way explicable as global operations in both their laboratories that are not necessarily describable as direct products of local operations. In this chapter, it is also important to keep in mind that, although LOCC allows correlated mixed states to be created from previously uncorrelated states, LOCC is not sufficient for the creation of entangled states.

LOCC operations include local unitary operations, local measurement operations, and the addition or disposal of parts of the total system, all of which are independently addressed in preceding. Accordingly, a quantum operation O_{AB} carried out by Alice and Bob is implementable by a pair of parties via LOCC when it can be written as a convex sum of local operations,

$$O_{AB} = \sum_i p_i A_i \otimes B_i \ , \tag{3.1}$$

so that operations by Alice and Bob are carried out *independently* in the two laboratories with probabilities p_i. In the case of operations on a number of copies of a quantum system for any of the above classes, the adjective "collective" is added to the above-mentioned classes, and the above acronyms are given the prefix "C." In cases where transformations are not achievable deterministically, but rather only with some probability, they are considered *stochastic operations* and the adjective "stochastic" is added, so that the above acronyms are generally given the prefix "S," as in SLOCC. These various state transformations are discussed in greater detail in Chapters 6 and 7, where bipartite and multipartite entanglement, respectively, are discussed.

The investigation of entanglement has long been bound up with the investigation of locality and realism and their relation to the quantum-mechanical description of composite systems. In particular, one is interested in the question of whether there may exist adequate *local realistic descriptions* of entangled systems and the question of the whether any such description is compatible with the postulates of quantum theory. These questions have often been approached by considering the possibility of "hidden variables."

3.1 Hidden variables and state completeness

Hidden-variables approaches to explaining the behavior of microscopic systems are predicated on the possibility that the quantum-mechanical specification of physical states is in some sense *incomplete*. The first hidden-variables model in the quantum context was that proposed by Louis de Broglie in the mid-1920's [124, 125] and more completely developed by David Bohm in the early 1950's [65]. Hidden-variables treatments of quantum phenomena are based on the consideration of a putative *complete state*, λ, which is often not specified in detail, underlying the (therefore inherently statistical) quantum state, ρ. The "hidden" variables of these theories are those parameters not in the quantum state that ostensibly *complete* the specification of the full set of system properties. Statistical principles, namely, the preservation of functional subordination in the space of quantum properties and the preservation of the convex structure of the set of quantum states, impose some conditions on any hidden-variables model. Despite their being called "hidden," these variables are *not* assumed to be in principle empirically inaccessible. They are often taken to be fully contained in the complete state, which is sometimes also called a *dispersion-free state*. The simplest sort of hidden-variables model is that in which λ provides definite values to all quantum system properties that correspond to the projectors described in Chapter 1. Such models, sometimes referred to as *noncontextual* hidden-variables theories, determine the value of a quantity obtained by measurement, regardless of what other quantities are simultaneously measured along with it, and specify the complete state of the overall system composed of the measured system together with the measurement apparatus.

John S. Bell provided the following example of a noncontextual hidden-variables model for the qubit [31].[1] In this model, the qubit is described in the spinorial representation, ψ, together with a real parameter $l \in [-\frac{1}{2}, \frac{1}{2}]$ that serves to complete the specification of the dispersion-free state λ. The qubit properties are represented by matrices in $H(2)$ of the form $\alpha\sigma_0 + \sum_{i=1}^{3} \beta_i\sigma_i$, having eigenvalues

$$\alpha \pm |\boldsymbol{\beta}| \tag{3.2}$$

and expectation values

$$\left\langle \alpha + \sum_{i=1}^{3} \beta_i\sigma_i \right\rangle = \left(\psi, \left(\alpha\sigma_0 + \sum_{i=1}^{3} \beta_i\sigma_i \right)\psi \right) , \tag{3.3}$$

where $\boldsymbol{\beta}$ is a three-component real vector and the σ_μ ($\mu = 0, 1, 2, 3$) are the Pauli matrices. $\boldsymbol{\beta}$ is taken to have the component values $\beta_1, \beta_2, \beta_3$ when the qubit is in the zero computational-basis state. Measurement of the property $\alpha\sigma_0 + \sum_{i=1}^{3} \beta_i\sigma_i$ provides eigenvalues

$$\alpha + |\boldsymbol{\beta}|\text{sign}\left(\lambda|\boldsymbol{\beta}| + \frac{1}{2}|\beta_3| \right)\text{sign}\mathcal{X} , \tag{3.4}$$

where $\mathcal{X} = \beta_3$ if $\beta_3 \neq 0$, $\mathcal{X} = \beta_1$ if $\beta_3 = 0$ and $\beta_1 \neq 0$, and $\mathcal{X} = \beta_2$ if $\beta_3 = 0$ and $\beta_1 = 0$; the sign function is defined by the conditions that $\text{sign}F = +1$ if $F \geq 0$, and $\text{sign}F = -1$ if $F < 0$. In this model, one finds that the quantum-mechanical expectation values are indeed recovered by taking a uniform average over the range of values of the hidden variable l.

---

[1] Kochen and Specker also produced an explicitly noncontextual hidden-variables model [260].

The more subtle *contextual* hidden-variables models, also introduced by Bell, require not only λ but also *other* relevant parameters associated with the conditions (or *context*) of their measurement to assign each projector a definite value. This idea was further developed by Stanley Gudder, who considered the context to be a maximal Boolean subalgebra of the lattice of quantum Hilbert subspaces [202].[2] *Algebraic contextuality* involves the specification of any other quantities that are measured jointly with the quantity of interest. *Environmental contextuality* involves there being some nonquantum-mechanical interaction between the system subject to measurement and its environment that occurs before measurement and influences the value of the measured properties. An even weaker class of hidden-variables models, the *stochastic hidden-variables theories*, requires the hidden variables and experimental parameters to specify merely the *probabilities* of measurement outcomes corresponding to projectors. Finally, a distinction is made between local and nonlocal hidden-variables theories that becomes more clear as one progresses through the series of significant results below; see, for example Section 3.5. In the case of *nonlocal* hidden-variables theories, the action on a subsystem of a composite system may have an immediate effect on another, spacelike-separated system.

The results we examine now pertain to hidden-variables models, either directly or indirectly, and their relation to quantum statistics. These results and associated empirical tests weigh mainly *against* the existence of hidden variables, but are ultimately insufficient to entirely eliminate the nonlocal type of hidden-variables theory.[3] However, because the appeal of quantum cryptography, for example, lies in the hope of *absolute security* in the sense of security "guaranteed by the laws of nature" when eavesdroppers are allowed *unlimited* technological capacities, such exotic hidden-variables theories remain important to quantum information science and have recently begun to be explicitly considered in regard to quantum cryptographic protocols; see, for example, [5].

3.2 Von Neumann's "no-go" theorem

The von Neumann "no-go" theorem explores dispersion-free states, thereby addressing hidden-variables theories that might enable them as well [444]. One can imagine a situation wherein the measurement of a given quantity attributed to an ensemble of systems gives different values even though all members of the ensemble have the same specification. Then, either there exist different subensembles distinguished by some hidden variable outside of the

---

[2] See Sect. A.9 for associated mathematics.

[3] In addition to the results described here, the Kochen–Specker theorem is discussed briefly in Sect. A.8. A particularly noteworthy unified treatment of no-hidden-variables theorems by N. David Mermin focusing on the Kochen–Specker theorem should also be consulted [297].

quantum description, or the measured dispersion of values arises from Nature. In the former case, there must be as many subensembles as there are different results. Von Neumann sought to show that no dispersion-free descriptions exist that enable a hidden-variables description of quantum phenomena. A state ψ is *dispersion-free* when the dispersion, $\mathrm{Disp}_{\psi}O$, of all properties, O, is zero when the system is in it.[4] The von Neumann "no-go" theorem makes the following assumptions about operators in relation to physical properties of the system to be explained by any putative hidden-variables model.

(i) Any real linear combination of three or more Hermitian self-adjoint operators represents a measurable quantum property;

(ii) The corresponding linear combination of expectation values is the expectation value of that combination of operators.

Von Neumann's no-go theorem is that no such hidden-variables model exists.

This result is less than definitive regarding the existence of a successful hidden-variables theory for the following reason. Though the second condition seems natural to impose on the dispersion-free states because it is satisfied by quantum-mechanical operators, there is no a priori reason to expect that this condition must be satisfied for *individual* dispersion-free states, because these must be averaged over to be compared with the statistical behavior described by traditional quantum mechanics and the statistics measured in experiments on quantum systems. For example, Bell pointed out for his model described in the previous section that von Neumann's assumptions require expectation values to be linear functions of *both* α and β and a dispersion-free state to have the expectation value of a quantum property equal to one of its eigenvalues; however, in that simple hidden-variables model the expectation value is *not* a linear function of β [31].

3.3 The Einstein–Podolsky–Rosen argument

An early thought-provoking analysis of quantum composite-system states explicitly pointing out the surprising nature of entangled quantum states that also introduced considerations of locality and realism in regard to microscopic physical systems was made by Albert Einstein, Boris Podolsky and Nathan Rosen (EPR), who provided a specific argument for the incompleteness—though importantly, not the *incorrectness*—of the quantum-mechanical description of the microscopic world [147]. The conditions imposed by EPR, here tailored to the case of two particles viewed as qubits the states of which can be found by measurements along particular directions—for example, measurement of photon polarization states by polarizers oriented along directions

---

[4] See Sect. B.2 for the quantum-mechanical expression for dispersion.

normal to the axis of photon counter-propagation (*cf.* Fig. 3.1)—are as follows.[5]

(i) *Perfect correlation*: When the states of qubits A and B are measured along the same direction, the corresponding outcomes will be opposite.

(ii) *Locality*: Since at the time of measurement the two systems no longer interact, no real change can take place in the second system in consequence of anything that may be done to the first system.

(iii) *Reality*: If, without in any way disturbing a system, we can predict with certainty (*i.e.*, with probability equal to unity) the value of a physical quantity, then there exists an element of physical reality corresponding to this physical quantity.

(iv) *Completeness*: Every element of the physical reality must have a counterpart in the physical theory.

The initial EPR argument was given in the context of *continuous* properties of quantum systems but is readily and perhaps better suited to physical systems involving discrete properties. In particular, without loss of essential characteristics, one can consider the simple two-qubit system first brought forward for this purpose by David Bohm in the state now recognized to be central to quantum information processing, namely, the singlet state

$$|\Psi^-\rangle = \frac{1}{\sqrt{2}}\left(|01\rangle - |10\rangle\right), \tag{3.5}$$

which has the important property of remaining of the same form when re-expressed in *any* orthonormal basis obtainable from the computational basis by rotating the basis of a subsystem Hilbert space by an arbitrary nonzero angle ξ [66]. With this simplification, the EPR argument with the above assumptions may be presented as follows (*cf.* [383]).

(i) If an agent can perform an operation that permits him to predict with certainty the outcomes of a measurement without disturbing the measured qubit, then the measurement has a definite outcome, whether this operation is *actually* performed or not.

(ii) For a pair of qubits in the state $|\Psi^-\rangle$, there is an operation that an agent can perform allowing the outcome of a measurement of one subsystem to be determined *without disturbing* the other qubit.

An agent can find, by measuring the quantity corresponding to $P(|0\rangle)$ for one qubit, the value of the quantity corresponding to $P(|1\rangle)$ as well. Thus, by (ii), she can obtain the values of the same two properties of the *second* qubit *without disturbing it*, by virtue of the perfect anti-correlation between qubits in the joint state $|\Psi^-\rangle$. By (i), these values of the second qubit are

---

[5] The first condition has been adapted to the case of qubits. Conditions (ii)–(iv) are stated exactly as in original text of the EPR paper. For a modern version of the EPR argument based on the logic of quantum conditionals, see [385].

definite. However, the agent could just as well have checked the values of the quantities corresponding to a *different* basis, say the diagonal basis represented by $P(|\nearrow\rangle)$ and $P(|\searrow\rangle)$. But then these other values must *also* be definite. Thus, the value of the states of *both systems* for *all* values of ξ must be definite. The description of the system of particles by the quantum state $|\Psi^-\rangle$ is in this way argued to be *incomplete.*

3.4 Gleason's theorem

A definitive theoretical result regarding the hidden-variables question and the completeness of the quantum-mechanical description of physical systems is Gleason's theorem. This result is often considered to be the most significant technical advance in the foundations of quantum mechanics to be obtained after von Neumann's initial investigation of the hidden-variables question. The theorem demonstrates a sense in which the quantum statistical operators *do* provide complete state descriptions [189]. It also clarifies the difficulties associated with putative hidden-variables descriptions touched on by von Neumann without requiring the second of his assumptions, which is often viewed as unwarranted.

A vital lemma underlying Gleason's theorem is the following ([189]). Let $|\phi\rangle$ and $|\psi\rangle$ be two state-vectors in a Hilbert space $\mathcal{H}$ of dimension at least 3, such that for a given system state $\langle P(|\psi\rangle)\rangle = 1$ and $\langle P(|\phi\rangle)\rangle = 0$. Then $|\phi\rangle$ and $|\psi\rangle$ cannot be arbitrarily close to one another. In particular, $\|\,|\phi\rangle - |\psi\rangle\,\| > \frac{1}{2}$.

The physical system state is here taken to provide a map from each projector, P_i, to a real number, $p(P_i)$, between 0 and 1, $p : P_i \mapsto p(P_i)$, such that $p(\mathbb{O}) = 0$ and $p(\mathbb{I}) = 1$ where $\mathbb{O}$ is the projector onto the zero vector $\mathbf{0}$ and $\mathbb{I}$ projects onto all of $\mathcal{H}$, and such that $P_1 P_2 = 0$ implies $p(P_1 + P_2) = p(P_1) + p(P_2)$; p is also taken to be a countably additive probability measure.

Gleason's theorem: All probability measures that can be defined on the lattice of quantum propositions P_i from the quantum statistical operators, that is, all quantum probabilities, are of the form

$$p(P_i) = \mathrm{tr}(\rho P_i) , \tag{3.6}$$

for some statistical operator ρ on $\mathcal{H}$, for all $\mathcal{H}$ of dimension greater than two.[6]

---

[6] That is, the values corresponding to mutually orthogonal projectors are derivable using a Born-type rule; see Postulate II of quantum mechanics in Appendix B [68]. For a discussion of the lattice of quantum propositions formed from the projectors P_i, which represent bivalent quantities, see Appendix A. Gleason's lemma as presented above conforms to Bell's re-derivation [32]. Gleason's theorem can be seen to provide a generalization of the Radon–Nikodym theorem. The trace measure assigns to each projector the dimension of its range, which can then be normalized by the dimension of the pertinent (finite-dimensional) Hilbert space;

Gleason's theorem shows that every probability measure over the set of projectors arises from a statistical operator on the Hilbert space of the system of interest, as expressed in Eq. 3.6. The relation of Gleason's lemma to the question of the possibility of hidden variables is the following. Consider putative *dispersion-free states* for which projectors would take expectation values of either 0 or 1 under the mapping. The condition $\sum_i \langle P(|\phi_i\rangle)\rangle = 1$ implies that *both* values must occur, because there are no other possible values for satisfying the condition and neither alone suffices. In this case, there must be arbitrarily close pairs $|\psi\rangle, |\phi\rangle$ having *different* expectation values, 0 and 1 respectively. However, such pairs *cannot* be arbitrarily close, by the lemma. Therefore, *there are no dispersion-free states*. Thus, no theory is capable of adequately reproducing quantum statistics via hidden variables parameterizing dispersion-free probability measures [32].

Gleason's results show the set of quantum states to be *complete* in the sense that they yield the probability measures definable on the lattice of quantum propositions corresponding to the projectors.[7] This result is still open to a reasonable objection, however. Namely, it can be considered unnatural to require dispersion-free states to provide nontrivial relationships between experiments that cannot be made *simultaneously*.[8]

3.5 Bell inequalities

John Bell famously further advanced the investigation of quantum behavior by deriving a theorem in the form of a general *inequality relation* providing a clear borderline between local classically explicable behavior and less intuitive forms of behavior, such as nonlocality and contextuality as described in Section 3.1.[9] Following the lead of EPR, Bell defined local models as follows. A *local hidden-variables theory* for experimental situations of the EPR type is one such that every complete state assigns a definite probability to a positive measurement outcome for a bivalent property of one subsystem when the hidden parameter describing it—taken to be capable of taking at least two values—takes a given value independently of measurements performed on the other subsystem.[10]

---

it is thus obtainable by considering ρ to be the maximally mixed state on the space; see Sect. 1.5 of [348].

[7] An extension of Gleason's theorem to the setting of POVMs having implications for the interpretation of quantum probabilities has recently been proven by Paul Busch [91].

[8] A natural, weaker requirement would be merely to require that quantum mechanical *averages* over them do so. For a more detailed discussion of this argument, see Ch. 1 of [30].

[9] It is interesting to note that Bell had himself officially listed as a "quantum engineer" in the CERN personnel directory. For a detailed survey of the work of Bell, see [231].

[10] John Jarrett showed Bell's locality condition to be the conjunction of two independent conditions [243], later named *parameter independence* and *outcome*

This inequality is the prototype of the collection of inequalities now typically referred to as *the Bell inequalities*.

Bell arrived at his crucial inequality for realistic hidden-variables theories of the type that EPR had suggested might exist, as follows. He began by considering a putative complete state, λ, describing a pair of particles, that at a given instant fully specifies *all* the elements of physical reality present in the pair.[11] Any such state capable of giving rise to perfect correlations will predetermine the outcomes of joint measurements of the component of spin of these particles along a given direction, $\mathbf{n}_i$, for each particle i. Bell considered a probability measure, $\mu(\Lambda)$, on the entire space Λ of parameters providing complete states. Expectation values, $E^{\mu(\Lambda)}$, of the relevant physical quantities were taken to be of the form

$$E^{\mu(\Lambda)}(\mathbf{n}_1, \mathbf{n}_2) = \int_\Lambda A_\lambda(\mathbf{n}_1) B_\lambda(\mathbf{n}_2) d\mu(\lambda) , \qquad (3.7)$$

where $\lambda \in \Lambda$, and $A_\lambda(\mathbf{n}_1)$ and $B_\lambda(\mathbf{n}_2)$ indicate *measurement results* along directions $\mathbf{n}_i$ on the two different systems of the arrangement. He then arrived at an inequality of the form

$$|E^{\mu(\Lambda)}(a, b) - E^{\mu(\Lambda)}(a, c)| \leq 1 + E^{\mu(\Lambda)}(b, c) , \qquad (3.8)$$

where $\{a, b, c\}$ is any set of three angles specifying directions of measurement in planes normal to the line of counter-propagation of the particles; see Fig. 3.1 [30, 32].

Issues regarding the assumptions used to derive this inequality, noted by Bell himself and others, subsequently led to a search for other related inequalities now also referred to as Bell inequalities (or Bell-type inequalities), based on weaker assumptions.[12] In particular, the *Clauser–Horne* (CH) *inequality* resulted from this investigation: *classical probabilities* must obey the relation

$$-1 \leq p_{13} + p_{14} + p_{23} - p_{24} - p_1 - p_3 \leq 0 , \qquad (3.9)$$

as well as all the inequalities resulting from permutations of indices, where p_1 and p_3 are the probabilities that the first particle is found along the *first* of the

---

independence by Abner Shimony [382]. The term "Bell's theorem" refers to a collection of results having in common the demonstration of the impossibility of a Local Realistic interpretation of quantum correlations.

[11] The complete state as originally specified by Bell "determines the results of measurements on the system, either by assigning a value to the measured quantity that is revealed by measurement regardless of the details of the measurement procedure, or by enabling the system to elicit a definite response whenever it is measured, but a response which may depend on the macroscopic features of the experimental arrangement or even on the complete state of the measured system together with that arrangement" [386].

[12] The full details of these other inequalities and assumptions can be found, for example, in [32, 387].

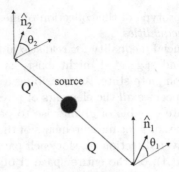

Fig. 3.1. Geometry of an apparatus for performing a test of Bell-type inequalities by two-qubit polarization interferometry. Two nonorthogonal states of qubits Q and Q' are measured, parameterized by angles θ_1 and θ_2, respectively, in the plane *normal* to the direction of qubit counter-propagation. (Compare this arrangement with that of the two-qubit spatial interferometer of Fig 3.2, below.)

four directions $\{a, b, c, d\}$ and the second particle is found along the *third* direction; p_{ij} stands for the joint probability of finding the first particle along the direction i *and* the second particle along direction j, $1 \leq i, j \leq 4$.[13] Because in actual experimental situations it is generally impossible to have control of the complete state of the total system, one assumes that the experimental arrangement prescribes a *probability distribution* over state specifications that provides the above probabilities through averages over Λ. No special restriction is placed on Λ or the probability distribution used in the derivation of the above result; indeed, the inequality follows from the elementary algebra of numbers lying between 0 and 1, as probabilities must by definition.

In order to lend greater practicality to explorations of issues of hidden-variables and locality, allowing them to be precisely probed by experiment, John Clauser, Michael Horne, Abner Shimony, and Richard Holt (CHSH) then also modified Bell's original treatment so as to be applicable in any *practical* experimental arrangement sufficiently similar to that of the two-spin atomic system that had been considered in experimental tests of locality-related inequalities up until that time, arriving at what is now known as the *CHSH inequality*:

$$|S| \leq 2 , \tag{3.10}$$

$$\text{for} \quad S \equiv E(\theta_1, \theta_2) + E(\theta_1', \theta_2) + E(\theta_1, \theta_2') - E(\theta_1', \theta_2') , \tag{3.11}$$

where the Es are expectation values of the products of measurement outcomes given parameter values θ_i and θ_i' (the angles shown in Fig. 3.1) of the two different directions $\hat{n}_i$ for the same laboratory i relative to a reference direction

---

[13] Recall that these particles correspond to qubit-pair systems.

[108].[14] The CHSH inequality is the Bell-type inequality now most commonly referred to in the literature. The correlation coefficients contributing to S can be expressed in terms of experimental detection rates as

$$E(\theta_i, \theta_j) = \frac{C(\theta_i, \theta_j) + C(\theta_i^\perp, \theta_j^\perp) - C(\theta_i, \theta_j^\perp) - C(\theta_i^\perp, \theta_j)}{C(\theta_i, \theta_j) + C(\theta_i^\perp, \theta_j^\perp) + C(\theta_i, \theta_j^\perp) + C(\theta_i^\perp, \theta_j)} , \qquad (3.12)$$

where the $C(\cdot, \cdot)$ are, in particular, *coincidence detection count rates*, i is the index for particle 1, j the index for particle 2, and the parameter $\theta^\perp$ indicates a parameter corresponding to the direction perpendicular to that specified by θ in the plane normal to the direction of particle propagation.[15]

According to the predictions of quantum mechanics for noiseless quantum channels, a maximum violation of this inequality by a factor of $\sqrt{2}$ can be achieve when, for example, one prepares the quantum state $|\Phi^+\rangle = \frac{1}{\sqrt{2}}(|00\rangle + |11\rangle)$, where $|0\rangle$ indicates photon polarization oriented along one of the orthogonal axes of the plane indicated in Fig. 3.1 and $|1\rangle$ indicates polarization oriented along the other, and performs measurements with $\theta_1 = \frac{\pi}{4}$, $\theta_1' = 0$, $\theta_2 = \frac{\pi}{8}$, and $\theta_2' = \frac{3\pi}{8}$, steps of $\frac{\pi}{8}$ radians, where the two angles in each lab (that is, side) differ by $\frac{\pi}{4}$ radians, corresponding to $\frac{\pi}{2}$ radians in the Poincaré–Bloch sphere; see [386] for an explicit calculation and Section 12.4 for an application. Since its introduction, the observed value of S has served experimentalists as a figure of merit for the quantum nature of sources of entangled quantum systems in such "Bell tests." It is useful in this context to introduce the so-called *Bell operator*

$$\mathcal{B} \equiv \hat{\mathbf{a}} \cdot \boldsymbol{\sigma} \otimes (\hat{\mathbf{b}} + \hat{\mathbf{b}}') \cdot \boldsymbol{\sigma} + \hat{\mathbf{a}}' \cdot \boldsymbol{\sigma} \otimes (\hat{\mathbf{b}} - \hat{\mathbf{b}}') \cdot \boldsymbol{\sigma} , \qquad (3.13)$$

where $\hat{\mathbf{a}}, \hat{\mathbf{a}}', \hat{\mathbf{b}}, \hat{\mathbf{b}}'$ are unit-vectors defining the directions of the pertinent qubit measurements, that is, the directions $\hat{\mathbf{n}}_i$ and $\boldsymbol{\sigma} = (\sigma_1, \sigma_2, \sigma_3)$. In particular, the Bell operator can be used to provide a compact operator form of the CHSH inequality via its expectation value, namely

$$\langle \mathcal{B} \rangle = \mathrm{tr}(\rho \mathcal{B}) \leq 2 . \qquad (3.14)$$

The Bell operator is also an "entanglement witness"; see Section 6.7. Quantum mechanics provides an experimentally well confirmed value near $\langle \mathcal{B} \rangle = 2\sqrt{2}$.

A further Bell-type inequality having a particularly simple proof assuming, unlike the proof of the CHSH inequality, perfect anticorrelations for measurements along parallel axes, was first given by Eugene Wigner that has since come to be known as the *Wigner inequality*, namely,

$$p_{++}(a, b) + p_{++}(b, c) - p_{++}(a, c) \geq 0 , \qquad (3.15)$$

---

[14] The first experiments to be studied to find nonclassical behavior related to investigations of quantum nonlocality were those published in 1950 by Wu and Shaknov [463], with spin qubits in a singlet state [66].

[15] Note that the denominator simply provides normalization.

where one agent, say that of a quantum-key distributor Alice, chooses between two polarization measurements along directions a and b of one of two particles and another agent, say a quantum-key receiver Bob, similarly chooses between measurements along b or c of the other [456]. The direction *common* to the two parties, b, can be taken to be one of the elements of the reference basis $\{|0\rangle, |1\rangle\}$ for the corresponding optical polarization-coincidence experiment described by Fig. 3.1, where a qubit state parallel to a given polarizer state is considered a "+" result and a state orthogonal to the polarizer state is a "−" result. If every particle were to have hidden variables determining the outcomes of the measurements on the particles, the probabilities of "+" measurement outcomes on both sides, $p_{++}(i,j)$, would obey this inequality.

The assumptions of the derivation of Wigner's inequality are that measurement outcomes on the two particles in identical directions are anti-correlated and that measurement outcomes on the two particles are independent of each another, with the background assumption that the probabilities involved refer to the statistics of an ensemble of identically prepared particle pairs. The quantum-mechanical probability for such a result along arbitrary directions, θ_1 and θ_2, for a pair of particles in the Bell singlet state $|\Psi^-\rangle$, namely,

$$p_{QM}(|\Psi^-\rangle) = \frac{1}{2}\sin^2(\theta_1 - \theta_2) , \tag{3.16}$$

produces a maximal deviation from the satisfaction of this inequality when, for example, $a = -\frac{\pi}{6}, b = 0$ and $c = \frac{\pi}{6}$, three directions $\frac{\pi}{6}$ radians apart, which provides a value of $-\frac{1}{8}$ for the left-hand-side of Eq. (3.15). The Wigner inequality has been recently used to perform entangled-state quantum key distribution in practice; see [449].

Such violations of Bell-type inequalities by quantum mechanics have by this time been studied in great generality. Because they rely on fundamental properties of probability, the expressions bounding the probabilities and expectation values in these inequalities can be derived by, for example, enumerating all conceivable classical possibilities. These can be viewed as extreme points spanning the classical correlation polytopes, the faces of which are expressed by Bell-type inequalities; see Section A.8. All Bell-type inequalities involve sums of (joint) probabilities and expectation values. To show the incompatibility of the predictions of quantum mechanics with these inequalities, the quantum counterparts and expectation values can be substituted for the probabilities and expectation values appearing in them. The results systematically show the violation of such local-realistic bounds by quantum-mechanical predictions.[16] Bell-type inequalities for pairs of systems of arbitrarily high dimension have also been found [112].

---

[16] All expressions entering the quantum expression corresponding to the pertinent part of any Bell-type inequality are self-adjoint. Because the norm of the self-adjoint transformation appearing in the inequality obeys the min-max principle, finding the maximal violation of Bell inequalities corresponds to the solution of a quantum eigenvalue problem, such as that for the Bell operator above. For a

3.6 Interferometric complementarity

Tests of the above Bell-type inequalities illustrate the importance of quantum interference for probing nonlocal properties of quantum systems, particularly the interference of qubit pairs. Nonlocal interferometric behavior can be examined in other illuminating ways as well. For example, there is a general quantum interferometric complementarity relation between single-qubit interference visibility, v_1, and two-qubit interference visibility, v_{12}, further illustrating the surprising nature of quantum correlations exhibited in two-particle interferometry [239]. In this regard, it was first explicitly noted in the late 1980's that when the two-particle interference visibility is unity the one-particle visibility is zero, and conversely [220]. In a doubled two-slit experiment with a source of generic two-particle states, the former case can be understood in terms of the washing out of photon self-interference due to uncertainty of the initial direction of individual particles; see Fig. 3.2. In the latter case, one notes that the arrival of one particle at one screen allows, in effect, two "virtual slits" to exist for the other due to the correlation between them, corresponding to reduced relative uncertainties, giving rise to single-particle interference [197]. A systematic investigation of intermediate cases was carried out to further explore this relationship, demonstrating that such a general complementarity relation holds for a large family of pure states $|\Theta\rangle$, defined below [234].

A schematic illustration of the class of experimental arrangements in which this complementarity can be exhibited is given in Fig. 3.2, namely, a doubled version of the discrete two-beam experiment described in Section 1.5, where the particle source produces *generic* pure two-particle states emerging in beam pairs and transducers (variable beam-splitters together with sets of phaseshifters) capable of exploring the full set of local unitary transformations of two-qubit states (described by the group $SU(2) \times SU(2)$) are introduced (rather than merely 50–50 beam-splitters and single phaseshifters), followed by pairs of particle detectors in two laboratories. Particle A is taken to be that in beams 0 and/or 1, and similarly for particle B. Each pair contributing to the output ensemble is produced by the source (say, by filtered spontaneous parametric down-conversion; see Section 6.16) in a two-qubit pure state

$$|\Theta\rangle = \gamma_1|0\rangle_A|0'\rangle_B + \gamma_2|0\rangle_A|1'\rangle_B + \gamma_3|1\rangle_A|0'\rangle_B + \gamma_4|1\rangle_A|1'\rangle_B , \qquad (3.17)$$

with $\gamma_i \in \mathbb{C}$ such that

$$|\gamma_1|^2 + |\gamma_2|^2 + |\gamma_3|^2 + |\gamma_4|^2 = 1 , \qquad (3.18)$$

and $|0\rangle_A$ and $|1\rangle_A$ being basis vectors in the Hilbert space $\mathcal{H}_A$ of the first particle corresponding to propagation in the beams 0 and 1, and $|0'\rangle_B$ and $|1'\rangle_B$ being similar vectors in the Hilbert space $\mathcal{H}_B$ of the second particle.

_____

more detailed exploration of this approach to Bell inequalities see, for example, [336].

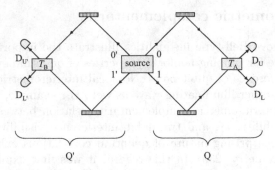

Fig. 3.2. An interferometer containing two spatial qubits, Q and Q'. T indicates a transducer capable of performing all local unitary transformations of a single qubit. D indicates a particle detector. The laboratory of Alice is to the right and that of Bob is to the left. (Compare with the arrangement of Fig. 3.1 describing interferometry with polarization qubits under a more restricted class of measurements.)

Beams 0 and 1 are brought together in lab A at a transducer, T_A, corresponding to a single-party unitary gate producing two output beams U and L, and similar beams in lab B are brought together into another transducer, T_B, implementing a similar gate that produces output beams, U' and L', as indicated. The output beams are assumed to be equipped with ideal particle detectors, D. As the transducers T_A and T_B are varied, the probability $P(UU')$ of coincidence detection in beams U and U', similar joint-detection probabilities $P(UL')$, $P(UL')$, $P(LL')$, and single-detection probabilities $P(U), P(L), P(U'), P(L')$, corresponding to particle coincidence-detection and single-detection rates, respectively, are modulated and can be used to provide interferometric visibilities, as described below.

Given that $|\Theta\rangle = \alpha|\bar{0}\rangle_A|\bar{0}\rangle_B + \beta|\bar{1}\rangle_A|\bar{1}\rangle_B$, where α and $\beta \in \mathbb{C}$ with $|\alpha|^2 + |\beta|^2 = 1$, the vectors $|\bar{0}\rangle$ and $|\bar{1}\rangle$ being orthonormal (*cf.* Sect. 6.2), the most general single-qubit local unitary transformation (LUT), T_A, can in this context be described as acting on particle A, providing an output spatial-qubit state that can be written

$$T_A |\bar{0}\rangle_A = ae^{i\phi_1}|U\rangle + be^{i\phi_1}|L\rangle , \tag{3.19}$$

$$T_A |\bar{1}\rangle_A = -be^{-i\bar{\phi}_1}|U\rangle + ae^{-i\phi_1}|L\rangle , \tag{3.20}$$

where a and b are real numbers the squares of which together sum to unity (*cf.* Eqs. 1.12 and 1.30) and ϕ_1 and $\bar{\phi}_1$ being phase angles; similarly, for particle B, providing a second output spatial-qubit state that can be written

$$T_B |\bar{0}\rangle_B = ce^{i\phi_2}|U'\rangle + de^{i\bar{\phi}_2}|L'\rangle , \tag{3.21}$$

$$T_B |\bar{1}\rangle_B = -de^{-i\bar{\phi}_2}|U'\rangle + ce^{-i\phi_2}|L'\rangle , \tag{3.22}$$

c and d also being real numbers, the squares of which together sum to unity, and ϕ_2 and $\bar{\phi}_2$ are phase angles. The joint local operation of this pair of transducers is described by the general pair of local unitary operations induced by them separately, namely,

$$T = T_A \otimes T_B \; . \tag{3.23}$$

Two-qubit interferometric behavior can then be studied via the modulation of single-detection and joint-detection probabilities as T is varied over the full range of parameters for the two LUTs, altering the above amplitudes and phases [29]. From the maximum and minimum probabilities of detection, one can calculate visibilities characterizing the interference. One is particularly interested in the one-qubit interferometric fringe visibility

$$V_i = \frac{[P(Y)]_{\max} - [P(Y)]_{\min}}{[P(Y)]_{\max} + [P(Y)]_{\min}} \; , \tag{3.24}$$

where $i = A, B$, $Y = U, L$, and in V_{12}, which is the two-qubit interferometric visibility, in the sense of variations of detection probability as gates are varied, calculable from the probabilities $P(YY')$ of occupation of the joint-paths YY',[17] generalizing the case of the single paths Y above giving rise to the single-qubit visibilities V_i; for example,

$$V_{12} = \frac{[\bar{P}(UU')]_{\max} - [\bar{P}(UU')]_{\min}}{[\bar{P}(UU')]_{\max} + [\bar{P}(UU')]_{\min}} \; , \tag{3.25}$$

where

$$\bar{P}(UU') = P(UU') - P(U)P(U') + \frac{1}{4} \tag{3.26}$$

represents *nonaccidental* coincidence probabilities and similarly for the three other possible pairs of paths [234].[18]

The remarkable phenomena that take place in two-qubit interferometry result from the fact that, when the joint state $|\Theta\rangle$ is entangled, it can be the case that

$$P(UU') \neq P(U)P(U') \; , \tag{3.27}$$

and likewise for the other joint probabilities $P(UL')$, $P(LU')$, and $P(LL')$. That is, highly correlated behavior of particles A and B arises due to quantum entanglement. A strong complementarity relation, taking the form of an *equality* [234], holds for all $|\Theta\rangle$, namely,

$$V_{12}^2 + V_A^2 = 1 \; , \tag{3.28}$$

$$V_{12}^2 + V_B^2 = 1 \; , \tag{3.29}$$

---

[17] Consider, as explicit examples, single and joint probabilities of the form of those in Eqs. 3.36–37 for the related (Franson) configuration of Sect. 3.7. (Also see Fig. 1.6 and Footnote 44 of Ch. 1.)

[18] The constant term $\frac{1}{4} = (\frac{1}{2})(\frac{1}{2})$ added here compensates for the over-subtraction of the constant "background" in the product of single-qubit probabilities, so that only accidental *modulation* is subtracted from the "raw" coincidence probability.

as has been explicitly experimentally confirmed [2, 360].[19]

The two complementarities we have by this point discussed, that between path distinguishability and single-qubit visibility of Eq. (1.40) and that directly above, are closely related. In particular, the more entangled $|\Theta\rangle$ is, the tighter the bound on the single-qubit visibility is. This result can be understood as follows. The information between basis vectors, $|0\rangle$ and $|1\rangle$, of $\mathcal{H}_A$ is related by the vectors in $\mathcal{H}_B$ to which they can be correlated; observations made on only *one* particle of each pair cannot fully extract this information. Similarly, a high degree of entanglement entails high two-qubit interferometric-fringe visibility, and permits good inferences about the path of particle A associated with spatial qubit Q from the results of measurements of particle B associated with spatial qubit Q'. Indeed, V_{12} is sometimes referred to as the *entanglement visibility* (*cf.* [203]). The one-qubit interference visibility thus enters into *at least two* complementarity relations, that between single-qubit interference visibility and single-qubit bit-distinguishability (*i.e.* path distinguishability, *cf.* Sect. 9.4) and that between single-qubit interference visibility and two-qubit interference visibility (*i.e.* entanglement).

A distinction can be made between the "classical" and "nonclassical" correlations of two qubits, which are manifested in the coincidence interference visibility as described above. Consider a bipartite quantum system with Hilbert space $\mathcal{H} = \mathcal{H}_1 \otimes \mathcal{H}_2$ and described by a statistical operator ρ. Recall that such a state is *uncorrelated* if one can write $\rho = \rho^{(1)} \otimes \rho^{(2)}$, where $\rho^{(i)}$ are the statistical operators on the $\mathcal{H}_i$ ($i = 1, 2$). The expectation values of products of bounded linear operators $A^{(i)}$ on subsystems, such as the probabilities of the form $P(YY')$ above, can then be factored, that is,

$$\mathrm{tr}\big(\rho(A^{(1)} \otimes A^{(2)})\big) = \mathrm{tr}\big(\rho(A^{(1)} \otimes \mathbb{I})\big)\mathrm{tr}\big(\rho(\mathbb{I} \otimes A^{(2)})\big) \qquad (3.30)$$

$$= \prod_{i=1}^{2} \mathrm{tr}\big(\rho^{(i)} A^{(i)}\big) . \qquad (3.31)$$

In this case, outcomes of measurements of the $A^{(i)}$ are such that the probabilities of joint measurement outcomes are simply products of the probabilities of outcomes of the measurements performed in the two laboratories. By contrast, a statistical operator describing an ensemble wherein the quantum states of the two portions of the total system are *correlated*, in that the subsystems $\rho^{(i)}$ are in the same state ρ_j ($j = 1, \ldots, n$) with probabilities p_j, can be written as a convex combination of separable states,

$$\sum_{j=1}^{n} p_j \rho_j \otimes \rho_j . \qquad (3.32)$$

In that case, the expectation values of measurements of the properties $A^{(i)}$ are of the form

---

[19] The first experiments explicitly confirming this complementarity relation were performed in the Quantum Imaging Lab at Boston University [2, 360].

$$\sum_{j=1}^{n} p_j \prod_{i=1}^{2} \mathrm{tr}\big(\rho_j^{(i)} A^{(i)}\big) \; . \tag{3.33}$$

Any system with a density matrix of the form shown in Eq. 3.32 (with $j \geq 2$) is said to be *classically correlated*, even if formed by mixing entangled states [450]. The greatest two-particle interference visibility that can be obtained with classically correlated states in the arrangement of Fig. 3.2 is 0.5 [353]. Nonclassically correlated states (*i.e.* entangled states) can give rise to higher values of visibility of two-particle interference; the CHSH Bell-type inequality of Eq. 3.10 can be violated once the visibility surpasses $1/\sqrt{2} \approx 0.71$.

3.7 The Franson interferometer

Similarly to the above two-spatial-qubit interferometer, the Franson interferometer is a distributed interferometer composed of two single-qubit Mach–Zehnder interferometers. In the Franson interferometer, the interference of particle pairs with themselves is possible because there are alternative paths of different length in both sub-interferometers, giving rise to a "temporal" (time-bin, or phase) qubit in each due to the corresponding pair of alternative times of arrival at each final beam-splitter; see Fig. 3.3. This interferometer corresponds to a (constrained) temporal-qubit version of the two-spatial-qubit interferometer considered above and shown in Fig. 3.2, with only the relative phase between two-qubit alternative processes free to be altered (rather than the general pair of local unitary transformations of Eq. 3.23) in order to produce an interferogram in coincidence counts; it is designed to realize the limiting case of *maximal* two-particle interference visibility. The path-length difference, $\Delta l = d_{\mathrm{long}} - d_{\mathrm{short}}$, in each of the two single-qubit interferometers, arranged to be the same for both, corresponds to the transit-time difference between paths $\Delta T = \Delta l / c$ in each and is arranged to be greater than the single-particle coherence length, δl, corresponding to a single-particle coherence time $\tau_1 = \delta l / c$, precluding single-particle self-interference in either wing. In practice, the particles used in this interferometer are now typically photons produced by spontaneous parametric down-conversion.[20]

Thus, conditions are imposed so that there is no fixed phase-relation between amplitudes for single-photon passage along short and long paths at the final beam-splitter of either interferometer as described above, ensuring that the entanglement of the two qubits is as strong as possible, in accordance with the interferometric complementarity relations described by Eqs. 3.28–29.[21] The transit-time difference ΔT is also kept shorter than the correlation time τ_2 of the two photons, still allowing *two-photon interference* to be

---

[20] This process is described in detail in Sect. 6.16. When James Franson introduced this interferometer, he envisioned a source based on an atomic cascade [174].

[21] Explicit entanglement measures beyond the visibility of entanglement are discussed in Chapter 6, below.

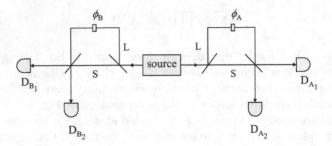

Fig. 3.3. The Franson interferometer for two phase-time qubits. The unique portions of the short and long paths of each wing are indicated by "S" and "L," respectively. The ϕ_A, ϕ_B indicate phase shifts in labs A and B, respectively [174].

observed. The effective quantum state in the interferometer is

$$|\Psi\rangle = \frac{1}{2}\big(|\text{short}\rangle|\text{short}\rangle - e^{i(\phi_A + \phi_B)}|\text{long}\rangle|\text{long}\rangle\big)\,, \qquad (3.34)$$

where "short" and "long" are shorthand notation for the alternative (multiple segment) photon paths available and ϕ_A and ϕ_B are the variable phase shifts of the two interferometers, because this temporal constraint ensures that there are no other contributions to the state, such as $e^{i\phi_A}|\text{short}\rangle|\text{long}\rangle$ and $e^{i\phi_B}|\text{long}\rangle|\text{short}\rangle$, that would otherwise be present. ΔT is kept significantly longer than the photon detector resolution dt (which is generally on the order of 1 ns duration) allowing for the observation of the possibilities of passage of pairs of photons in pairs of paths $|\text{short}\rangle|\text{short}\rangle$ and $|\text{long}\rangle|\text{long}\rangle$ above, and their interference. Overall, then, one requires that

$$\tau_2 > \Delta T > \tau_1\,, \qquad (3.35)$$

with ΔT arranged to be, for example, of an order greater than that of 1 ns. This results in the selection of the large, central interferometric feature of the three features that can appear in the temporal coincidence interferogram in an experimental configuration such as that shown in Fig. 3.3 [174].

Joint detection in distant wings A and B of the interferometer of Fig. 3.3, at pairs of detectors D_{X_i} (where $X = A, B$), thus occurs with the probabilities

$$P(D_{A_i} D_{B_j}) = \frac{1}{4}\big(1 + \cos(\phi_A + \phi_B)\big)\,, \qquad (3.36)$$

$$P(D_{A_k} D_{B_k}) = \frac{1}{4}\big(1 - \cos(\phi_A + \phi_B)\big)\,, \qquad (3.37)$$

where $i, j, k = 1, 2$ and $i \neq j$, similarly to those appearing in the two-spatial qubit interferometer discussed in the previous section. The probabilities $P(D_{X_i})$ of single-photon counts are just $\frac{1}{2}$, being marginal probabilities obtained by adding the joint probabilities (3.36) and (3.37) in which the corresponding detector appears; the positive and negative cosine modulation terms simply cancel each other out. That is, there is *no interference of single-photon possibilities*, the counts of which occur at random, regardless of how the individual phases ϕ_A and ϕ_B—or the joint phase $\phi_A + \phi_B$, for that matter—are varied.

The Franson interferometer has been used for entangled-photon QKD, where component interferometers lie one each in the separated laboratories of Alice or Bob; for example, see [188, 422].

3.8 Two-qubit quantum gates

The transformations carried out by the general single-qubit transducers (local unitary gates) T_i of Section 3.6 have been used in the various above-described situations to realize *independent* pairs of local, single-qubit operations described by Eq. 1.31, each acting on one of two qubits. Such local unitary transformations allow the exhibition of the effect of existing entanglement in quantum interferometers. More general two-qubit operations also exist, which play an essential role in quantum information processing under the quantum circuit model and can in some cases *create* entangled states rather than merely aid in exhibiting *existing* entanglement.

Let us now consider the creation of entanglement through the use of the very important two-qubit example of the C-NOT (or XOR) quantum gate.

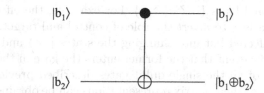

Fig. 3.4. Action of the C-NOT gate. The computational-basis states are labeled by binary values b_i ($i = 1, 2$); in the ket at lower-right, $\oplus$ signifies addition mod 2.

The quantum C-NOT gate corresponds to a unitary operation of the form $|0\rangle\langle 0| \otimes \mathbb{I} + |1\rangle\langle 1| \otimes (|0\rangle\langle 1| + |1\rangle\langle 0|)$. The C-NOT gate can be used to create entanglement between previously unentangled qubits and is represented by the matrix

$$C - NOT \doteq \begin{pmatrix} 1 & 0 & 0 & 0 \\ 0 & 1 & 0 & 0 \\ 0 & 0 & 0 & 1 \\ 0 & 0 & 1 & 0 \end{pmatrix}.$$

The first qubit of a C-NOT gate is called the "control qubit" and the second the "target qubit." From the point of view of the computational basis, the control bit is fixed and the target bit is changed *if and only if* the control bit takes the value 1, as can be seen by noting that the lower-right submatrix, corresponding to that condition, is that of the quantum bit-flip operator σ_1, whereas the upper-left submatrix block, corresponding to the opposite condition, is that of the identity operator, $\mathbb{I}$. The gate is usually drawn as shown in Fig. 3.4, with the control bit on top and the target bit on the bottom.

This gate can be used together with the single-qubit Hadamard gate to create Bell states from two independent qubits initially described collectively by an unentangled product state, as described by the quantum circuit shown in Fig. 3.5.

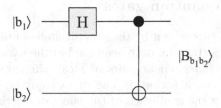

Fig. 3.5. A quantum circuit for the creation of the (entangled) Bell states $|B_{b_1 b_2}\rangle$ defined by Eqs. 6.13–14 from a product state of two qubits, $|b_1\rangle|b_2\rangle$.

It is important to note that the distinction between the control and target qubits is relative to the choice of basis. Consider the effect of a quantum gate that acts as a controlled gate in the computational basis in a different basis, say the diagonal basis $\{|\nearrow\rangle, |\searrow\rangle\}$. For example, the effect of the C-NOT gate in this basis is to *invert* the role of control and target, leaving the "target qubit" unaffected but interchanging the states $|\nearrow\rangle$ and $|\searrow\rangle$ of the "control qubit" in the event that the former enters the gate in the state $|\searrow\rangle$. Controlled versions of all the single-qubit gates described previously can be similarly implemented; their matrix representations can be obtained from that of the C-NOT gate above by simply replacing the lower-right 2×2 submatrix with that describing the single-qubit gate.[22] That is, one constructs operators of the form $|0\rangle\langle 0| \otimes \mathbb{I} + |1\rangle\langle 1| \otimes U$, where U is the operation to be conditionally performed.

One similarly obtains *multiple-qubit controlled gates* by generalizing this construction. For example, in order to extend the controlled-NOT gate so as to perform the NOT operation on a target qubit conditional on the state of *two* control qubits, one performs a unitary operation represented by the matrix

[22] Explicit matrix representations for such gates in the case of path and polarization degrees of freedom are readily worked out, and have been provided in full detail in the literature; for example, see [161, 315].

$$C - C - NOT \doteq \begin{pmatrix} 1 & 0 & 0 & 0 & 0 & 0 & 0 & 0 \\ 0 & 1 & 0 & 0 & 0 & 0 & 0 & 0 \\ 0 & 0 & 1 & 0 & 0 & 0 & 0 & 0 \\ 0 & 0 & 0 & 1 & 0 & 0 & 0 & 0 \\ 0 & 0 & 0 & 0 & 1 & 0 & 0 & 0 \\ 0 & 0 & 0 & 0 & 0 & 1 & 0 & 0 \\ 0 & 0 & 0 & 0 & 0 & 0 & 0 & 1 \\ 0 & 0 & 0 & 0 & 0 & 0 & 1 & 0 \end{pmatrix}.$$

This is the control-control-NOT (or Toffoli) gate, whereby the third target bit is flipped if and only if the *both* control bits take the value 1: the upper-left block is the two-qubit-space identity whereas the lower-right block is the matrix representation of a control-NOT. The quantum circuit for this gate is shown in Fig. 3.6.

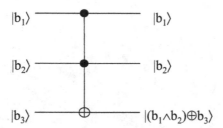

Fig. 3.6. Action of the C-C-NOT (or Toffoli) gate on three qubits. In the ket at lower-right, $\wedge$ signifies the AND operation and $\oplus$ indicates addition mod 2.

4

Classical information and communication

Information theory has, until relatively recently, almost exclusively focused on what is now considered classical information, namely, information as stored in or transferred by classical mechanical systems. Quantum information theory has also largely consisted of the extension of methods developed for classical information to analogous situations involving quantum systems. Furthermore, because measurements of quantum systems produce classical information, traditional information-theoretical methods play an essential role in quantum communication and quantum information processing because measurements play an essential role in them. Accordingly, a concise overview of the elements of classical information theory is provided in this chapter. Several specific classical information measures are discussed and related here, as are classical error-correction and data-compression techniques. The quantum analogues of these concepts and methods are discussed in subsequent chapters.

Classical information theory is based largely on the conception of information developed by Claude Shannon in the late 1940s, which uses the *bit* as the unit of information [378].[1] Entropies hold a central place in this approach to information, following naturally from Shannon's conception of information as the *improbability* of the occurrence of symbols occurring in memories or signals. The choice of a binary unit of information naturally leads to the choice of 2 as the logarithmic base for measuring information. Any device having two states stable over the pertinent time scale is capable of storing one bit of information. A number, n, of identical such devices can store $\log_2 2^n = n$ bits of information, because there are 2^n states available to them as a collective. An example of such a set of devices is a memory register. A variable taking two values 0 and 1 is also referred to as a bit and can be represented by $x \in GF(2)$; similarly, *strings* of bits can be represented by $\mathbf{x} \in GF(2)^n$.

Consider first the properties of a string of characters that are produced and sequentially transmitted in a classical communication channel. The Shannon information content of such a string of text can be understood in terms of how

---

[1] John W. Tukey first introduced the term *bit* for the binary information unit [12].

improbable the string is to occur in such a channel. The information content (or *self-information*, or *surprisal*) associated with a given signal *event* $\mathbf{x}$ is

$$I(\mathbf{x}) = \log_2 \frac{1}{p(\mathbf{x})} = -\log_2 p(\mathbf{x}) \,, \tag{4.1}$$

where $p(\mathbf{x})$ is the *probability of occurrence* of the event; the logarithm appears as a matter of mathematical convenience because the number of available states can be very large. The simplest events are the occurrence of individual symbols, in which case one is concerned with their probabilities, as in the case of bit transmission when one considers $p(0)$ and $p(1)$. The information associated with an event is considered *obtained* when the event actually occurs. This is consistent with the idea that no information is gained by learning an event that occurs with certainty whereas learning, for example, the outcome of a fair coin-toss provides *all* the information essential to the toss and entirely eliminates previous uncertainty as to its outcome. One generally considers a number of such events associated with a given signal.

A finite number of mutually exclusive events together with their probabilities constitutes a *finite scheme*. To every such scheme there is an associated uncertainty, because only the probabilities of occurrence of these events is known.[2] This uncertainty is captured by the Shannon entropy, which is introduced below. The information associated with the joint occurrence of two independent events, which happens with a probability given by the product of those of the individual events, can therefore be written as a *sum* of the associated information values.[3]

4.1 Communication channels

The fundamental task of communication is to obtain, at a remote destination and with the greatest possible accuracy, information sent from a given initial source through a *channel* joining their locations. One way of defining a classical information source is as a sequence of probability distributions over sets of strings produced in a number of emissions by a transmitter into such a communication channel to a receiver. The output at the end of the channel endows an agent at the destination with a given amount of information about its source, provided the information is not altered during transmission, that is, provided the channel is *noiseless*. The essential requirement on a means of communication is that any message sent via it belong to a set of possible messages that *could have been* sent by a source in this way, because the actual message being transmitted is not known *a priori*. Assuming a finite number of such possible messages, *any* monotonic function of the number of messages is a good measure of the information produced when one message is selected

---

[2] Various of measures of uncertainty are discussed in [427].

[3] This follows from Bayes' theorem; see Sect. A.8.

from this set. The logarithmic function above is a natural choice because, in addition to being mathematically convenient, it can also be retained when the number of possible messages is infinite.

Communication channels can be attributed *capacities* for transmitting information. By comparing the entropy of a source with the capacity of a channel, one can determine whether the information produced by the source can be fully transmitted through the channel. A very simple description of a re-

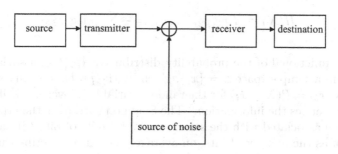

Fig. 4.1. The additive noise channel.

alistic classical communication channel is that of the *additive noise channel* wherein the transmitted *signal*, $s(t)$, is influenced by additive random *noise*, $n(t)$; see Fig. 4.1. Due to the the presence of this noise, the resulting signal, $r(t)$, is given by

$$r(t) = s(t) + n(t) . \qquad (4.2)$$

The primary model of a discrete, memoryless noisy channel is the *binary*

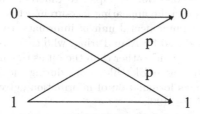

Fig. 4.2. Schematic of the binary symmetric channel (BSC). A transition between values of any given bit occurs with probability p, known as the *symbol-error probability*. Initial bit values appear at left, final bit values at right. Bit values therefore remain unchanged with probability 1−p.

symmetric channel (BSC), which is schematized in Fig. 4.2. In this channel there is a probability p that noise can introduce a *bit error*, an unintended change of value of any given bit during transmission.[4] Error correction, which is discussed in Sections 4.5, is the process of eliminating such errors.

4.2 Shannon entropy

The traditional measure of classical information is the *Shannon entropy*,

$$H(X) \equiv H[p_1, p_2, \ldots, p_n] = -\sum_i^n p_i \log_2 p_i \ , \tag{4.3}$$

which is a functional of the probability distribution $\{p_i\}_{i=1}^n$, associated with events x_i in a sample space $S = \{x_i\}$, given by the *probability mass function* as $p_i \equiv p_X(x_i) = P(X = x_i)$ for the random variable X over n possible values that characterizes the information.[5] The Shannon entropy is thus the *average information* associated with the set of events, in units of bits.[6] It can also be thought of as the number of bits, on average, needed to describe the random variable X.

If one considers a sequence of n independent and identically distributed (i.i.d.) random variables, X^n, drawn from a probability mass function $p(X = x_i)$, the probability of a typical sequence is of order $2^{-nH(X)}$; there are, accordingly, roughly $2^{nH(X)}$ possible such sequences.[7] The latter can be seen by noting that typical sequences will contain $p(x_i)n$ instances of x_i, so that the number of typical sequences is $n! / \prod_i^n (np(x_i))!$ which, under the Stirling

---

[4] More detailed models of classical communication channels can be found, for example, in [344].

[5] See Sect. A.2 for pertinent definitions. The events associated with the random variable X here correspond, for example, to different numbers or letters of an alphabet on the sides of a die appearing face-up in a toss.

[6] Such a measure and a fundamental unit of information were also independently (and previously) introduced by Alan Turing, who chose 10 as a base rather than Shannon's 2, and so the "ban" rather than the bit as the fundamental information unit, while he was working at Bletchley Park during the Second World War. For a description of Turing's formulation of information entropy, which involved "weight of evidence," see [190].

[7] This is known as the *asymptotic equipartition property* (AEP). One can distinguish two sets of sequences, the typical and atypical sequences, being complements of each other, where the *typical* sequences are those with probability close to $2^{-nH(X)}$; in particular, the typical set is that of sequences $\{x_1, x_2, ..., x_n\}$ for the random variable X such that $2^{-nH(X)-\epsilon} \leq p(x_1, x_2, ..., x_n) \leq 2^{-nH(X)+\epsilon}$ for all $\epsilon > 0$. The AEP is a consequence of the weak law of large numbers, namely that, for i.i.d. random variables, the average is close to the expected value of X for large n. The *theorem of typical sequences* can also be proven, which supports the notion that in the limit of large sequence length *almost all* sequences produced by a source belong to this set.

approximation, is $2^{nH(X)}$. In the limit of large n, the probability of atypical sequences is negligible.

The Shannon measure satisfies two requirements, *invariance under permutations of probabilities* p_i, and *additivity*,

$$H[p_1, p_2, \ldots, p_n] = H[p_1 + p_2, p_3, \ldots, p_n] + (p_1 + p_2)H\left[\frac{p_1}{p_1 + p_2}, \frac{p_2}{p_1 + p_2}\right].$$
(4.4)

Referring back to the expression of the information content associated with an event given in Eq. 4.1, we see that the above expression is just the expected value of the information content, which one can view as the expected information gain in coming to know the associated events. From Eq. 4.3, we see that, for a bit, the proper form of Shannon entropy is the *binary entropy*

$$H(p)_{\text{binary}} = -p \log_2 p - (1 - p) \log_2(1 - p) , \qquad (4.5)$$

where p is (without loss of generality) the probability of the bit value 0 and $1 - p$ is the probability of the alternative bit value, 1. When $p = 1/2$, one finds that $H(p) = 1$. Shannon's notion of entropy is similar to the familiar physical notion of entropy in statistical mechanics, which serves as a measure of uncertainty or disorganization in a physical system; the second law of thermodynamics requires that the entropy of a closed dynamical system be *nondecreasing*. Shannon entropy has the *concavity* property

$$H\big(px + (1 - p)x'\big) \geq pH(x) + (1 - p)H(x') , \qquad (4.6)$$

where p, x, x' lie in the interval $[0, 1]$.

For a *pair* of random variables, A and B, one can also define the *joint entropy* of the pair as

$$H(A, B) = - \sum_{a,b} p(a, b) \log_2 p(a, b) , \qquad (4.7)$$

where $p(a, b) \equiv P_{AB}(A = a, B = b)$ are the joint probabilities that $A = a$ and $B = b$, and sums are taken over the two sample spaces associated with both A and B. One finds that $H(A) \leq H(A, B)$, meaning that one cannot be *more* uncertain of the state of single physical system characterized by A than one is about the *joint state* of two systems described by A and B.[8]

A useful method for comparing two *different* discrete probability distributions is provided by introducing a *relative* entropy function: given two probability distributions, $p(a) = \{p(a_1), \ldots, p(a_n)\}$ and $p(b) = \{p(b_1), \ldots, p(b_n)\}$, the Shannon *relative entropy* (or *discrimination*)[9] between them is

---

[8] This property turns out *not* to hold for the *quantum* (von Neumann) *entropy*—as we show later in the next chapter—marking a significant difference between the classical and quantum cases, and so the unique character of quantum information.

[9] The relative entropy was first introduced by Kullback and Leibler, and is therefore often referred to simply as the *Kullback–Leibler distance* [263].

$$H[p(a)\|p(b)] \equiv \sum_i p(a_i) \log_2 \frac{p(a_i)}{p(b_i)} , \tag{4.8}$$

known as the *Kullback–Leibler distance* between $p(a)$ and $p(b)$, with the conventions that $0 \log_2 \frac{0}{p(b_i)} = 0$ and $0 \log_2 \frac{p(a_i)}{0} = \infty$. However, $H[p(a)\|p(b)]$ is not a *metric* because it is *not* symmetric in a and b. This distance is useful for distinguishing statistical behaviors and states. The relative entropy satisfies the *Gibbs inequality*,

$$H[p(a)\|p(b)] \geq 0 , \tag{4.9}$$

which is an equality only when $p(a) = p(b)$. The sum of the Shannon entropy of a random variable and the relative entropy of that variable under two distributions,

$$H(A) + H[p(a)\|p(b)] , \tag{4.10}$$

is sometimes referred to as the *inaccuracy*, because it characterizes the ignorance as to the correct distribution of A as produced by some communication source.

The *conditional entropy* of a random variable A is its entropy conditional upon knowledge of another random variable B:

$$H(A|B) \equiv H(A, B) - H(B) . \tag{4.11}$$

Imagine that one wishes to infer the value of random variable A from knowledge of a random variable B. One can then use the *Fano inequality*

$$H_{\text{binary}}(p_{\text{error}}) + p_{\text{error}} \log_2(|A| - 1) \geq H(A|B) , \tag{4.12}$$

where p_{error} is the probability of making this inference *incorrectly* and $|A|$ is the size of the sample space associated with A, for example, the number of words in a code. This bound captures the intuition that a large conditional entropy $H(A|B)$ corresponds to a large probability of an *erroneous* inference of A, given B. It is often relevant in channel coding; see Section 4.6 below.

The Shannon *mutual information* between two random variables, A and B, described by the joint probability distribution $p(a, b) = \{p(a_i, b_j)\}$ and marginal distributions $p(a) = \{p(a_i)\} = \sum_j p(a_i, b_j)$ and $p(b) = \{p(b_j)\} = \sum_i p(a_i, b_j)$, respectively, is

$$I(A : B) \equiv H[p(a)] + H[p(b)] - H[p(a, b)] . \tag{4.13}$$

This quantity can be understood as describing the *degree of correlation* between the two variables: the amount of information about A that is acquired by determining the value of B, as well as the degree of distinguishability of a given correlated situation from a fully uncorrelated situation, so that one may also write

$$I(A : B) = H[p(a, b)\|p(a)p(b)] . \tag{4.14}$$

Classical *information processing* can be studied from the point of view of Markov chains, that is, sequences of random variables. A *Markov process* is such a sequence with the property that each random variable in the sequence is independent of all preceding members of the sequence. Markov chains obey the *data-processing inequality*

$$H(A) \geq I(A:B) \geq I(A:C) \,, \tag{4.15}$$

for a Markov chain $A \to B \to C$, where the first inequality is an equality if and only if given B, A can be reconstructed. Thus, if a given random variable B is obtained from random variable A, due to noise, data-processing cannot *increase* the amount of mutual information between the input and output variables. This captures the fact that though information can be lost, it cannot arise out of *nowhere*. The *reverse data-processing inequality*,

$$I(A:C) \leq I(B:C) \leq H(C) \,, \tag{4.16}$$

describes the phenomenon that information processed in a second processing step exceeds that processed overall. A third inequality, namely, the *data-pipelining inequality* follows from recognizing that given a Markov chain $A \to B \to C$, $C \to B \to A$ is also a Markov chain, and is written

$$I(C:B) \geq I(C:A) \,, \tag{4.17}$$

which captures the intuition that any information shared by A and C is *also* shared by C and B.

As we have seen, communication channels are such that their outputs depend probabilistically on their inputs; a channel can be studied via the distribution of its output given the possible input. The *information channel capacity* is defined as the maximum mutual entropy over all possible inputs described by probabilities $p_A(a_i)$. The *operational channel capacity* is defined as the greatest bit-rate at which input information can be transmitted with arbitrarily low error. The *noisy channel coding theorem* shows these two quantities to be equal: the capacity of a discrete, memoryless communication channel is

$$\mathcal{C} = \max_{\{p_A(a_i)\}} I(A:B) \,, \tag{4.18}$$

where A characterizes the input to the channel and B characterizes its output; the units of channel capacity are bits-output-per-symbol-input. For a binary channel, the capacity lies in the range $[0, 1]$. For the binary symmetric channel, the capacity is simply $1 - H(p)$. In the case of a *noiseless* such channel, any transmitted bit is received at the destination without error; each transmission carries a bit to the receiver with certainty. The channel capacity is accordingly

1 bit-output-per-symbol for a noiseless channel. If the transmission rate is *less than* the channel capacity, then for any $\epsilon > 0$ there is a code having a block length large enough that the error probability is less than ϵ. Codes exist, therefore, that allow *error-free* communication at rates below this (Shannon) channel capacity. At rates above this capacity, some errors are guaranteed to exist. This result is known as the *Shannon–Hartley theorem*.

4.3 Rényi entropy

The Rényi entropy is a useful generalization of the Shannon entropy measure. The *Rényi entropy of order* r, in the case of a discrete probability distribution, is defined as

$$H_r(A) = \frac{1}{1-r} \log_2 \sum_{i=1}^{n} p^r(a_i) , \qquad (4.19)$$

for $0 < r < \infty$ and $r \neq 1$ [352]. $H_r(A)$ is a continuous positive decreasing function of r. One obtains the Shannon entropy from the Rényi in the limit $r \to 1$, so that

$$H_1(A) = H(A) . \qquad (4.20)$$

In the limit $r \to \infty$, one obtains the *min-entropy*

$$H_\infty(A) = -\log_2 \max_{a_i} p(a_i) . \qquad (4.21)$$

The Rényi entropy of order two is known as the *extension entropy*; the *inverse participation ratio*, $R(A)$, is its exponentiation

$$R(A) = \exp\big(H_2(A)\big) , \qquad (4.22)$$

the inverse of which is *index of coincidence*, which in turn is the complement of the *linear entropy*,

$$L(A) = 1 - \frac{1}{R(A)} . \qquad (4.23)$$

The Rényi entropy has proven useful in security analyses of quantum cryptosystems, for example.[10]

4.4 Coding

A particularly useful method of encoding a number of strings of k symbols is to map each string into an n-element string of symbols (each taken from a set of q symbols, possibly different from those used in the original string), taken as a block, that corresponds to an n-dimensional vector in a linear space $V(n, q)$.[11] The result of such an encoding of a number of such strings into a

---

[10] A specific of such an application is discussed in Chapter 12. Elsewhere in physics, the Rényi entropy has been applied to the study of multi-fractal structures.

[11] Any code composed entirely of codewords that are n-element strings is a *block code of length* n.

single such linear space V is a code that spans a k-dimensional subspace of the space. If the codewords for a set of such strings are taken from the Galois field $GF(2)^n$, then the code is referred to as a *binary linear code*.[12] Such *linear codes* (or *parity-check codes*) are characterized by the dimension, n, of the space V and an $n \times k$ binary matrix, the *generator* G, that describes the encoding of messages according to a rule

$$\mathbf{x} \mapsto G\mathbf{x} \,. \tag{4.24}$$

A linear code that uses n bits to encode k-bit blocks is called an $[n, k]$ code. The "extra" n-k bits can serve as "parity-check" bits assisting in the correction of errors. The sum of arbitrary number of code words of a linear code is also a word in the code. Linear codes are thus easily specified: only the kn bits describing the generator (or, alternatively, the corresponding parity-check matrix, M, see below) need be given to provide the code, out of a possible $n2^k$ bits of the exhaustive list of codewords that in some cases of nonlinear codes must be provided. For example, if one were to use three-bit coding to encode a single bit one, would be using a $[3,1]$ code with the generator

$$G = \begin{pmatrix} 1 \\ 1 \\ 1 \end{pmatrix} \tag{4.25}$$

to encode a bit $\mathbf{x}$ into the two codewords $w_i = Gx_i$—the corresponding parity-check matrix is given in the following section. This is an important example of a *repetition code*.

The distance between two words in a linear code, that is, the number of bits in which the words differ, can be used as a measure of their distinguishability. The *Hamming weight* of a string is given by its distance from the n-bit string consisting entirely of zeroes. The *Hamming distance*, d, of a code is the minimum of the distances over the set of pairings of codewords within it [205]. The above example has code words $w_0 = (000)^{\mathrm{T}}$ and $w_1 = (111)^{\mathrm{T}}$, and so a Hamming distance $d = 3$: w_0 has Hamming weight 0 and w_1 has Hamming weight 3. Linear codes can accordingly be specified as $[n, k, d]$ codes, that is, $[n, k]$ codes with a Hamming distance d. Noise affects code words with the result that a given word, $\mathbf{w}$, is transformed in a way describable as

$$\mathbf{w} \to \mathbf{w}' = \mathbf{w} + \mathbf{e} \,, \tag{4.26}$$

where $\mathbf{w}'$ is the resulting word and $\mathbf{e}$ characterizes the bit error induced by the noise. Such a code allows for the correction of m bit errors if and only if its Hamming distance is *larger than* $2m$.

---

[12] In the general case, a Galois field $GF(q)$, where q is prime, is used, the code being referred to as q-ary.

For a code of length n, there is a *parity-check matrix*, M, satisfying (mod 2) the *error-check* property

$$M\mathbf{w} = \mathbf{0} , \qquad (4.27)$$

for every symbol $\mathbf{w}$. The parity-check matrix M and generator matrix G are related by G's being orthogonal to the columns of the *transpose* of M, that is, the inner product between G and the columns of M^{T} is 0 mod 2. The error-check property allows the receiver of an encoded message to discover the bit errors induced by noise during transmission because it implies that

$$M\mathbf{w}' = M\mathbf{e} , \qquad (4.28)$$

provides the *error syndrome* for every correctable error $\mathbf{e}$. The error syndrome supplies information as to the particular error that must be corrected.[13] If there exists an ordering of code bits such that a linear code has a parity-check matrix that is cyclic, the code is known as a *cyclic code*. The codewords created by such a code are cyclic as well, in that cyclic permutations of code words are code words.[14]

> It is also often advantageous to use codes of variable lengths, as in the case of Morse code. In particular, *Huffman coding* is such a method that approaches the minimum number of bits allowable without resulting in a loss of information [227]. The method is based on the use of a frequency-sorted binary tree. It is effective because, although information is generally presented as a sequence of symbols representable as a string of n bits, all possible 2^n combinations of n bits will not generally be used with the same probability. Huffman coding replaces the presented symbols by a binary code based on the *decreasing probability* of their appearance. Because the benefits of this method are sometimes offset by its tendency to produce long code strings, truncated versions in which only those more likely symbols are encoded in this way and the remainder are coded by fixed-length bit strings can be used to advantage. An important application of Huffman coding is lossless data compression.

---

[13] Corresponding quantum-coding and error-correction methods are discussed in Chapter 10. The quantum analogues of linear codes are the quantum stabilizer codes. For example, the seven-qubit quantum Steane code for the correction of an arbitrary error on a qubit is closely related to the [7,4,3] classical Hamming code for correcting classical bit errors; see [402, 403] and Chapter 10.

[14] Cyclic codes are central to algebraic error-correction coding methods. In general, linear codes may also be conventionally viewed from the graph-theoretical perspective; see, for example, [292].

4.5 Error correction

Let us now consider an explicit example of a linear error-correction code. Recall that such error-correction codewords are representable as vectors of a k-dimensional subspace, the *codespace*, in the n-dimensional vector space $GF(2)^n$. The quantity of greatest concern in the context of coding is the *loss*, L, from a channel. If an encoded message having length n is subject to loss, then for every output sequence through the channel the number of input sequences will then typically be 2^{nL}, which can render impossible the decoding of messages. Error correction uses codewords chosen so as to take the conditional entropy of this relevant ensemble to *zero*, so that there are effectively *no* losses. The Fano inequality, introduced in Section 4.2, has the consequence that the loss will go to zero if the error probability goes to zero, though the noise of the channel itself need *not* be zero.

Consider a particular bit which is susceptible to error, recalling that for classical binary information there is only *one* sort of error that can occur on a bit, the *bit flip*, such as occurs in the binary symmetric channel described in Section 4.1.[15] If, in a given situation, such errors are relatively rare, meaning they occur with a probability $p \ll 1$, they are easily corrected through the use of encoding based on redundancy, as in repetition codes. Let us consider in greater detail the repetition-based [3,1,3] code introduced above, specifically,

$$0 \mapsto 0_{\mathrm{L}} \doteq \begin{pmatrix} 0 \\ 0 \\ 0 \end{pmatrix} \tag{4.29}$$

$$1 \mapsto 1_{\mathrm{L}} \doteq \begin{pmatrix} 1 \\ 1 \\ 1 \end{pmatrix}, \tag{4.30}$$

where the subscript is used to indicate *logical bits*. A parity-check matrix for this repetition code is

$$M = \begin{pmatrix} 1 & 1 & 0 \\ 1 & 0 & 1 \end{pmatrix}. \tag{4.31}$$

Errors on individual bits, for example, single-bit errors on the first, second, and third bits, respectively, will change the sequence of components of 0_{L} and 1_{L} as

$$000 \mapsto 100 \qquad 111 \mapsto 011 \tag{4.32}$$
$$000 \mapsto 010 \qquad 111 \mapsto 101 \tag{4.33}$$
$$000 \mapsto 001 \qquad 111 \mapsto 110, \tag{4.34}$$

respectively. Such errors can be found and corrected by the *majority vote* method, in which one checks the three bits periodically; if there is an error

---

[15] This is not the case for qubit errors, as we show in Sect. 10.4.

one flips the bit that disagrees with the others, returning their state to one of the logical states. As long as $p \leq \frac{1}{3}$, the probability of a net error occurring using this method is reduced from its original value p to an improved value of $3p^2$. The price paid is the reduction of transmission rate by a factor of three, because three physical bits are used to transmit *each* logical bit. A quantum version of this method is considered later, in Section 10.4.

4.6 Data compression

Data compression is a method of encoding that reduces the length of the strings required to capture a quantity of information given some knowledge of the states provided, for example, by a transmitting source. The Shannon entropy, $H(A)$, of a random variable A provides a lower bound on the average length of its shortest description. A basic result of the theory of data compression is the *noiseless coding theorem*, which provides a lower bound on data compression by stating that a message cannot be compressed to *less* than its Shannon entropy per bit, as follows.

For any $\delta, \epsilon > 0$:

(i) With $H(A) + \delta$ available bits per signal, there exists a coding-decoding scheme with fidelity $F_M > 1 - \epsilon$, for all M sufficiently large;

(ii) With $H(A) - \delta$ available bits per signal, for any coding-decoding scheme, the fidelity $F_M < \epsilon$, for all M sufficiently large, where the fidelity is given by

$$F_M = \sum_{A_M} p(A_M) p_{\text{exact}}(A_M) \,, \qquad (4.35)$$

$A_M = a_{i_1} a_{i_2} \ldots a_{i_M}$ being a bitstring (block) with prior probability as distributed by the sender, Alice, $p(A_M) = p_{i_1} p_{i_2} \ldots p_{i_M}$, p_{i_J} being the probability of a given a_{i_J}.

This theorem provides a statistical justification for the Shannon entropy being considered a measure of uncertainty; see [379]. It also allows one to interpret the Shannon entropy as the mean number of bits needed to code the output of a source using an ideal code. The Shannon entropy can thus be viewed as a measure of the resources required to represent the information provided by a source. A quantum analogue of this result is discussed in Section 10.8.

Different methods of data compression operate with different efficiencies, depending on the statistical properties of the message. Generally, use of typical sequences is not the most efficient method of compressing information. The sender Alice can, for example, use block coding to compress information by jointly taking strings of M signals and coding them as shorter data sequences without the redundancies naturally contained in an arbitrary signal, as mentioned above. The receiver, Bob, can then decode (or *decompress*) these sequences, reconstructing them with any desired level of accuracy.

A *source code*, C, for a random variable A is a map from the domain of A to the set of finite-length strings in a given n-ary alphabet, $\mathcal{S} = \{0, 1, \ldots, n-1\}$. If the length of codeword $C(a)$ is $l(a)$ and the probability mass function is $p(a)$, then the *expected length* of the code is

$$L(C) = \sum_{a \in A} p(a) l(a) . \tag{4.36}$$

The *extension*, C^*, of the code C is provided by the concatenation of codewords

$$C^*(a_1, a_2, \ldots, a_n) = C(a_1) C(a_2) \ldots C(a_n) , \tag{4.37}$$

which is a mapping from finite strings in the range of A to $\mathcal{S}^*$, the set of finite strings of $\mathcal{S}$. A code, C, is said to be *nonsingular* if $a_i \neq a_j$ implies $C(a_i) \neq C(a_j)$, and *uniquely decodable* if its extension is nonsingular. C is a *prefix* (or *instantaneous* or *comma-free*) *code* if no codeword is a prefix of any other codeword. Transmitted codewords can be properly framed provided the signal is synchronized so that the beginning of the initial codeword can be identified. The expected length of any prefix code of any n-ary random variable A is greater than or equal to the base-n Shannon entropy of the code. The set of achievable codeword lengths is identical for prefix and decodable codes. *Shannon coding* is such a coding that uses codeword lengths of $\lceil \log \frac{1}{p_i} \rceil$.

Many optimal codes can be constructed. The Shannon entropy provides a limit on data compression and the number of bits required for the generation of random numbers. Huffman coding can be used to systematically find one such code by finding minimum expected description length assignments. It is a "greedy" coding method, in the sense that it replaces the two least likely symbols with *one* symbol at a given step. Huffman codes are competitively optimal: a number of fair coin flips given by the function H are required to generate a sample of a random variable having comparable entropy. A quantum analogue of Huffman coding can readily be carried out; see [80].

4.7 Communication complexity

Communication complexity can be used to investigate *distributed* tasks based on the following simple scheme. Consider two separated parties, Alice and Bob, *each* possessing an n-bit string and allowed to perform local computations and to communicate, so that one of them is able to announce the value of a given function, $f : X \times Y \to Z$, of these two strings to the other. This situation can also be generalized to any number of parties. Let Alice's string be x, and Bob's be y, with $x \in X = \{0,1\}^{\times n}$ and $y \in Y = \{0,1\}^{\times n}$, and $Z = \{0,1\}$. It is possible for Bob to determine $f(x, y)$ if Alice simply communicates the values of x to Bob. One desires, however, to *minimize* the amount of communication required between Alice and Bob to accomplish this task, rather than, say, the number of computational steps required.[16] For such a function, the *communication complexity*, $K(f)$, is the minimum number of *bits* necessarily communicated between Alice and Bob in order to determine $f(x,y)$.

The required computations can be chosen to be either deterministic *or* probabilistic in nature. A *promise* is also sometimes also added, which is a Boolean function $F(x,y)$ such that Alice and Bob are required to find $f(x,y)$ only when $F(x,y) = 1$. In a *deterministic protocol*, a communicated bit is a function only of previously sent bit-values of the input from the sender. One is interested in the number of bits sent in the worst case in the best possible correct deterministic protocol for computing the function f. By contrast, in a *nondeterministic protocol*, the bits to be communicated may depend on nondeterministic choices as well. A *nondeterministic protocol* for z is *correct* if it always returns $1 - z$ for $f(x,y) = 1 - z$ and for any x, y with $f(x,y) = z$, it returns z for *at least one* sequence of nondeterministic choices made. The *worst-case number of bits sent*, in the best possible correct nondeterministic protocol for z, is written $N^z(f)$.

The two complexity measures, K and N^z, are accordingly related as

$$K(f) \geq N^z(f) \ . \tag{4.38}$$

In the following chapter, quantum-information measures are introduced, many of which can be seen as extensions or analogues of the classical measures introduced above. Particularly interesting for such purposes is the case where Alice and Bob are allowed also to share *random variables*, because this situation is similar to the quantum situation where entangled quantum states are shared and corresponding qubit values are measured in the computational basis.

---

[16] See Sect. 13.1 for a discussion of computational complexity, which is more commonly considered.

5

Quantum information

The theory of quantum information, that is, of the transmission, storage, and processing of information using quantum-mechanical systems, is now well developed. It has been constructed largely by the generalization to this context of elements of traditional information theory, which have been designed to explain the transmission, storage, and processing information using classical means. The mathematical description of quantum systems differs fundamentally from that of classical states, as we saw in Chapter 1.[1] This difference lends quantum information a significantly different character from that of classical information.

Unlike in the classical case, most of the information stored in a generic quantum-mechanical system is stored in the form of *correlations between subsystems*. Not only is most quantum information stored in the form of such correlations, but these can be *extraordinarily strong* correlations, as we have seen in Chapter 3; fully entangled quantum states are the extreme cases. For example, for the Bell states the reduced states of single qubits are entirely indefinite, whereas the state of the qubit pair is *fully* correlated, that is, knowledge of the state of one qubit actualized through a quantum measurement is tantamount to knowledge of the other, as in the situation considered by Einstein, Podolsky and Rosen as framed by Bohm. The greatest difference in complexity between classical and quantum states arises when entanglement is present among components of composite quantum systems. In bipartite systems, the extraordinary correlations associated with entanglement are manifested, for example, in the violation of the Bell-type inequalities discussed in Section 3.5

---

[1] For example, the number of parameters needed to specify the state of a quantum information-bearing system grows exponentially with the number of its subsystems. This point is addressed in detail in Sect. 7.6. Note that distinctions between classical information and quantum information have been made in various ways—on this point see, for example [100, 145, 367]. In this book, a distinction is made on the basis of differences in the ability to transmit, store and process information in quantum systems versus in classical systems.

and the complementarity of self-interference visibility of subsystems and the total system self-interference visibility discussed in Section 3.6.

As we show in the following chapter, entanglement between even *two* subsystems provides a novel type of information-processing resource complementing that which can be provided by classical bits or unentangled quantum bits. This difference is even more pronounced in the case of larger multiple-qubit states, which can be distributed among a number of separate parties potentially participating in *distributed* information-processing. In this chapter, quantities characterizing both static and dynamical properties of quantum information are considered. These will prove useful for the understanding of quantum information processing tasks discussed in following chapters.

5.1 Quantum entropy

The standard measure of the information contained within a quantum system described by the statistical operator ρ is the *von Neumann entropy*

$$S(\rho) = -\mathrm{tr}(\rho \, \log_2\rho) \tag{5.1}$$
$$= -\sum_i \lambda_i \log_2\lambda_i \, , \tag{5.2}$$

where λ_i are the members of the set of eigenvalues of ρ and $0\log 0 \equiv 0$.[2] $S(\rho)$ is nonnegative, achieves its maximum value for the maximally mixed state, and is zero if and only if ρ is *pure*. For systems described by states in d-dimensional Hilbert spaces, $0 \leq S(\rho) \leq \log_2 d$, so that for qubits $0 \leq S(\rho) \leq 1$. $S(\rho)$ provides an information measure in units of *qubits* [367].

The von Neumann entropy plays a role in quantum information theory analogous to that played by the Shannon entropy in traditional information theory: the von Neumann entropy $S(\rho)$ measures the uncertainty of a quantum state associated with a quantum probability distribution.[3] However, the von Neumann entropy differs in important ways from the Shannon entropy. In the case of classical systems, the entropy can be viewed as the information gained by identifying the system state, whereas in general ρ cannot be fully identified by the observation of an event (*cf.* Section 2.2) so that $S(\rho)$ provides only a loose bound on this; see Section 9.3.

Writing the *quantum joint entropies*, $S(A, B) \equiv S(\rho_{AB})$, $S(A, B, C) \equiv S(\rho_{ABC})$, and so on—and for uniformity of notation, also taking $S(A) \equiv S(\rho_A)$—one finds that the von Neumann entropy has the following properties.

---

[2] Here, we have assumed the set of eigenvalues of ρ to be countable; see also [446].
[3] This quantity should, however, be distinguished from the uncertainty in values of incompatible quantum properties in the Heisenberg–Robertson uncertainty relation that exists even in the simplest pure state; see Section B.2.

(i) *Additivity for product states* (also a property of classical states):

$$S(\rho_A \otimes \rho_B) = S(A) + S(B). \tag{5.3}$$

(ii) *Strong subadditivity* for all quantum states:

$$S(A, B, C) + S(B) \leq S(A, B) + S(B, C) . \tag{5.4}$$

iii) *Concavity*:

$$S\left(\sum_i p_i \rho_i\right) \geq \sum_i p_i S(\rho_i) . \tag{5.5}$$

iv) *Invariance under unitary transformations of the quantum state*:

$$S(U\rho\, U^\dagger) = S(\rho) . \tag{5.6}$$

In the above, the p_i are the probabilities, summing to one, of a system being in the corresponding states ρ_i. The last property is related to the conservation of state purity discussed in Section 1.1. The Shannon entropy and the von Neumann entropy coincide only for an ensemble formed from *mutually orthogonal* pure quantum states. Thus, if one were to send a message encoded in a set of orthogonal qubit states, each pure, that can be described by an overall tensor-product state, the transmission would be equivalent to sending the same information as a set of *classical* bits, because each qubit is *perfectly distinguishable* once the encoding basis has been determined.[4]

By contrast to the effect of unitary transformations, which leave the quantum entropy unchanged, *measurements* can change it. For example, for a system initially described by the statistical operator ρ and by the operator ρ' after a measurement described by the Lüders rule the result of which is known, the final quantum entropy $S(\rho')$ is less than or equal to the initial quantum entropy, $S(\rho)$. As an extreme but important example, consider a qubit initially in the fully mixed state, say as the reduced statistical operator of one qubit of a pair in a Bell-state, $\rho = \frac{1}{2}\mathbb{I}$; for it, $S(\frac{1}{2}\mathbb{I}) = 1$. A precise measurement of the state of such a qubit will place it in a pure state $P(|\psi\rangle)$, at which point it will have quantum entropy $S\big(P(|\psi\rangle)\big) = 0$. Generalized measurements with unknown outcomes can similarly decrease the quantum entropy of a system. On the other hand, if a Lüders-type measurement is performed but the outcome remains unknown then the quantum entropy may *increase*.

---

[4] Indeed, a set of states from a known basis can be cloned with *perfect fidelity*, contrary to the general case which is imperfect and constrained as described by the quantum "no-cloning theorem"; were *unknown* quantum states perfectly distinguishable, they could be perfectly cloned; see Sect. 9.5.

The following triangle inequality, known as the *Araki–Lieb inequality*, relates the joint entropy and the subsystem entropies:

$$S(A, B) \geq |S(A) - S(B)| \ . \tag{5.7}$$

For a composite system in a pure state $|\Psi\rangle$, one has $\rho_{AB} = P(|\Psi\rangle)$, yielding $S(A, B) = 0$; so, it must also be that $S(A) = S(B)$ for any such system. Accordingly, for two individual quantum particles in a singlet state, $|\Psi^-\rangle$, $S(A, B) = 0$ whereas $S(A) = S(B) = 1$; that is, the correlation between the states of subsystems A and B is fully certain, whereas the individual states of subsystems A and of B are completely *uncertain*, as mentioned above. Such examples demonstrate the sensibility of the (subsystem) von Neumann entropy for describing entanglement, as we show in the next chapter.

The *joint entropy theorem*,

$$S\left(\sum_{i=1}^{n} p_i P(|i\rangle) \otimes \rho_i \right) = H(\{p_i\}) + \sum_{i=1}^{n} p_i S(\rho_i) \ , \tag{5.8}$$

also holds for a set of orthogonal states $\{|i\rangle\}$ of a system A and n statistical operators ρ_i of a second system B, both occurring with probabilities, p_i.

5.2 Quantum relative and conditional entropies

The *quantum conditional entropy* is given, analogously to the corresponding classical quantity, by

$$S(A|B) \equiv S(A, B) - S(B) \tag{5.9}$$

$$= S(\rho_{AB}) - S(\rho_B) \tag{5.10}$$

(*cf.* Eq. 4.11). However, unlike the classical conditional entropy, the quantum conditional entropy can become *negative*, indicating that it is possible for quantum systems to be more certain in the *joint state* of two component systems than in the states of its individual components, as again can be seen in the case of the singlet state $|\Psi^-\rangle$, the entropy values of which were given in the previous section.

The *quantum relative entropy* between the two states, ρ and σ, of a quantum system is defined as

$$S(\rho||\sigma) \equiv \mathrm{tr}\big(\rho(\log_2 \rho - \log_2 \sigma)\big) \ . \tag{5.11}$$

This quantity obeys *Klein's inequality*,

$$S(\rho||\sigma) \geq 0 \ , \tag{5.12}$$

which is an *equality* if and only if $\rho = \sigma$ [253].[5] Klein's inequality is analogous to Gibbs inequality for the classical relative entropy (*cf.* Eq. 4.9). Like the

---

[5] The quantum relative entropy was first introduced by Umegaki [428].

corresponding classical measure, this quantity is also not a metric due to a lack of symmetry with respect to its arguments. The quantum relative entropy characterizes the *distinguishability* of states defined in the same Hilbert space. In practice, one can use a POV measure to distinguish states, to the extent physically possible, based on the corresponding detection distributions.[6]

5.3 Quantum mutual information

The *quantum mutual information* between two subsystems described by states ρ_A and ρ_B of a composite system described by a joint state ρ_{AB} is

$$I(A:B) \equiv S(A) + S(B) - S(A,B) \tag{5.13}$$
$$= S(\rho_A) + S(\rho_B) - S(\rho_{AB}) , \tag{5.14}$$

also by analogy with the corresponding classical quantity (*cf.* Eq. 4.14). Note, however, that this quantum-mechanical quantity *exceeds* the bound for the classical mutual information. In particular, note that the quantum mutual information can reach *twice* the maximum value obtained in the corresponding classical mechanical situation:

$$I(A:B) \leq 2 \min\{S(A), S(B)\} , \tag{5.15}$$

which is a corollary of the Araki–Lieb inequality (Eq. 5.7) and implies that quantum systems can be *supercorrelated*, as mentioned previously. Note specifically that when a bipartite quantum system is in a pure state $I(A : B) = 2S(A) = 2S(B)$, as can be readily shown using the Schmidt decomposition; see Section 6.2, below.

The quantum mutual information has two different but related operational meanings [198, 370]. In particular, the total amount of correlation, as measured by the minimal rate of randomness that is required to completely erase all the correlations in a state ρ_{AB} (in a many-copy scenario), is equal to the quantum mutual information, which leads to the strong subadditivity of the von Neumann entropy [198]. The quantum mutual information can also be viewed as a type of *relative entropy*, in as much as

$$I(A:B) = S(\rho_{AB}||\rho_A \otimes \rho_B) \tag{5.16}$$

(*cf.* Eq. 4.14). The quantum mutual information also has the important property that it is nonincreasing under completely positive maps, which were introduced in Section 2.6 (*cf.* [432]).[7] As is the case for classical entropies, quantum

---

[6] Uses of POVMs for distinguishability are discussed in Sect. 1.6 in relation to the (limited) distinguishability of nonorthogonal states of a single qubit, and in Sect. 3.6 in relation to two-particle interference. POV measurements themselves are discussed in Sect. 2.7

[7] Note that the same symbol, I, has been used here for both the classical and quantum mutual information functions (*cf.* Section 4.2). Care should be taken in this regard.

entropies can be given for multipartite systems. For example, the quantum *conditional mutual information* within a tripartite system can be written

$$I(A : B|C) = S(A|C) - S(A|B, C) \qquad (5.17)$$
$$= S(A|C) + S(B|C) - S(A, B|C) . \qquad (5.18)$$

There also exist quantum chain rules analogous to classical chain rules for entropies.

An important theorem, known as Lieb's theorem, is the basis of many results related to quantum entropy measures. For example, the strong subadditivity inequality, the second property of von Neumann entropy listed above, is a very useful result that can be proven making use of it [369]. The strong subadditivity property of the von Neumann entropy allows one to demonstrate several useful properties of the entropies introduced in this section that also take the form of inequalities. For example, note that conditioning reduces entropy in the context of the tripartite division of a compound system:

$$S(A|B, C) \leq S(A|B) . \qquad (5.19)$$

Also note that discarding components of a compound system can decrease but *never* increase quantum *mutual* information: $I(A : B) \leq I(A : B, C)$, perhaps the most meaningful manifestation of the strong subadditivity of quantum entropy; see Eq. 5.4 and [198]. In the context of a four-component system, A, B, C, D, the quantum conditional entropy is subadditive,

$$S(A, B|C, D) \leq S(A|C) + S(B|D) , \qquad (5.20)$$

whereas the quantum mutual information is *not*.[8] Furthermore, the strong-subadditivity property of the von Neumann entropy allows one to show that the quantum *relative* entropy is nonincreasing under CPTP maps [283].

5.4 Fidelity and coherent information

A measure of the *fidelity of transmission* of a pure-state input $|\psi\rangle$ that produces final states σ_i with probabilities p_i in the statistical state $\rho = \sum_i p_i \sigma_i$ is

$$F(P(|\psi\rangle), \rho) = \langle \psi | \rho | \psi \rangle , \qquad (5.21)$$

or, in the case of a mixed-state input, ω,

$$F(\rho, \omega) = \left[\mathrm{tr}\left(\sqrt{\sqrt{\omega} \rho \sqrt{\omega}} \right) \right]^2 . \qquad (5.22)$$

---

[8] A proof of this subadditivity property can be found in Sect. 11.4.2 of [315].

The latter quantity is the maximum value attained by the pure-state expression over the set of pure states in a larger Hilbert space yielding ρ_i by partial tracing [244].[9] It is fundamental to the study of quantum communication.

The quantum analogue of the classical Fano inequality, the *quantum Fano inequality*, is

$$S(\rho, \mathcal{E}) \leq H_{\text{binary}}\big(\tilde{F}(\rho, \mathcal{E})\big) + \big(1 - \tilde{F}(\rho, \mathcal{E})\big)\log_2(d^2 - 1) , \tag{5.23}$$

for a quantum operation, $\mathcal{E}$, on a system with a Hilbert space of dimension $d \geq 2$, where the *entropy exchange* $S_e(\rho, \mathcal{E}) = -\text{tr}(W \log W)$ with the matrix W having elements

$$[W]_{ij} \equiv \text{tr}(E_i \rho E_j^\dagger)/\text{tr}\mathcal{E}(\rho) , \tag{5.24}$$

E_i being the elements of the operator-sum representation of the operation $\mathcal{E}$, which was first proved by Schumacher [368]. $S_e(\rho, \mathcal{E})$ quantifies the quantum entropy introduced into a system as a result of the operation $\mathcal{E}$ it has undergone.[10] $\tilde{F}(\rho, \mathcal{E})$ above is the "entanglement fidelity," which quantifies the degree to which entanglement between the system and another needed to form a pure total system is *preserved* under $\mathcal{E}$, namely, the particular state fidelity

$$\tilde{F}(\rho, \mathcal{E}) = \langle \Psi_{PQ} | \rho_{PQ'} | \Psi_{PQ} \rangle \tag{5.25}$$

between the initial and final pure states of such a total system, where $|\Psi_{PQ}\rangle$ is a pure state of the combined system of the input system Q and another "reference" system P yielding ρ as the reduced state for Q, and $\rho_{PQ'}$ is the state resulting from the effect of $\mathcal{E}$ on Q [369].

In the case of such a pair of systems P and Q forming a possibly entangled joint system PQ, the subsystem Q may be imperfectly isolated from its environment, the effect of which is described by $\mathcal{E}$. For example, Q may be sent through a quantum channel resulting in such a joint "output" state $\rho_{PQ'}$; if the input is pure the output generally will *not* be. One may then consider the *coherent information* between the subsystems,

$$I_{\text{coherent}}(\rho, \mathcal{E}) = S(Q') - S(P, Q') , \tag{5.26}$$

where the quantities on the right-hand side are von Neumann entropies of the transmitted subsystem and total system, respectively [369]. The coherent information, which can have *any* sign, in contrast to its classical analogue which is always negative, has useful properties. In particular, I_{coherent} is positive if and only if the output state is an *entangled* state, and can be viewed as a measure of nonclassical character by virtue of the preservation of quantum coherence by $\mathcal{E}$. I_{coherent} has the property that it cannot be increased by local operations on Q, so that

$$I_{\text{coherent}}(\rho, \mathcal{E}) \leq S(Q) \tag{5.27}$$

---

[9] Note that the fidelity is also not a metric, though it *is* symmetric in its arguments.
[10] Note that $\tilde{F}(\rho, \mathcal{E})$ is distinct from the *state* fidelity, F, defined in Eq. 5.21 above.

after the influence of the environment on Q. Because the reduction of the coherent information due to the effect of the environment is irreversible, a necessary and sufficient condition for the ability to perform perfect correction of errors introduced by the interaction of Q with the environment is that $I_{\text{coherent}}(\rho, \mathcal{E}) = S(Q)$.[11]

Consider a two-stage process formed by two subprocesses, $\mathcal{E}_1$ and $\mathcal{E}_2$, in series. As a result of the first process, $\rho \to \rho_{Q'} = \mathcal{E}_1(\rho_Q)$. The second process then results in

$$\rho_{Q'} \to \rho_{Q''} = \mathcal{E}_2(\rho_{Q'}) \tag{5.28}$$

$$= \mathcal{E}_{12}(\rho_Q), \tag{5.29}$$

where

$$\mathcal{E}_{12} = (\mathcal{E}_2 \circ \mathcal{E}_1)(\rho_Q) . \tag{5.30}$$

The coherent information then obeys the *quantum information-processing inequality*

$$S(Q) \geq I_{e1}(\rho_Q, \mathcal{E}_1) \geq I_{e12}(\rho_Q, \mathcal{E}_{12}) , \tag{5.31}$$

where $I_{e1}(\rho_Q, \mathcal{E}_1) = S(Q') - S_{e1}(Q)$ and $I_{e12}(\rho_Q, \mathcal{E}_{12}) = S(Q'') - S_{e12}(Q)$, where $S_{e_1}(Q)$ is the entropy exchange of the first stage and $S_{e12}(Q)$ is the entropy exchange for the composition of the two processes comprising the overall process $\mathcal{E}_{12}$. This is the quantum analogue of the classical data-processing inequality, Eq. 4.15.

5.5 Quantum Rényi and Tsallis entropies

Finally, note that a *quantum Rényi entropy*, analogous to the classical Rényi entropy introduced in Section 4.4, can be defined as well:

$$S_r(\rho) = \frac{1}{1-r}\text{tr}\rho^r , \tag{5.32}$$

where $r \geq 0$. Note, in particular, that

$$\lim_{r \to 1} S_r(\rho) = S(\rho) . \tag{5.33}$$

In addition, when $r = 0$,

$$S_0(\rho) = \log_2 R(\rho) , \tag{5.34}$$

where $R(\rho)$ is the rank of the corresponding density matrix. Furthermore,

$$\lim_{r \to \infty} S_r(\rho) = -\log_2 ||\rho|| . \tag{5.35}$$

---

[11] Quantum error detection and correction are discussed in detail in Chapter 10.

Like the von Neumann entropy, the quantum Rényi entropy has proven valuable for characterizing entanglement.[12] A related quantity is the *Tsallis entropy*, defined as

$$T_r(\rho) = \frac{1 - \text{tr}(\rho^r)}{(r-1)} \ , \tag{5.36}$$

which, like $S_r(\rho)$, also coincides with the von Neumann entropy in the limit $r \to 1$. Both entropies have conditional versions, namely,

$$S_r(B|A) = S_r(\rho) - S_r(\rho_A) \tag{5.37}$$

and

$$T_r(B|A) = \frac{\text{tr}(\rho_A^r) - \text{tr}(\rho^r)}{(r-1)\text{tr}(\rho_A^r)} \ , \tag{5.38}$$

where ρ_A is the statistical operator of a subsystem A within the compound system AB described by the state ρ. These conditional entropies are related to each other with respect to positivity; in particular,

$$T_r(B|A) \geq 0 \Leftrightarrow S_r(B|A) \geq 0 \ , \tag{5.39}$$

which is also equivalent to the positivity of the conditional von Neumann entropy $S(B|A)$ in the limit $r \to 1$ [442].

---

[12] For example, see Sect. 6.9.

6

Quantum entanglement

Quantum interference arises from the indistinguishability in principle, by precise measurement at a specified final time, of alternative sequences of states of a quantum system that begin with a given initial state and end with the corresponding final state. It is manifested, for example, in the two-slit interferometer and the double Mach–Zehnder interferometer discussed in Chapters 1 and 3, respectively. Most important, when the indistinguishability of alternatives for producing *joint events* arises, as in the latter apparatus, entanglement may be involved. Erwin Schrödinger, who first used the term "entanglement," called entanglement "the characteristic trait of quantum mechanics" [364, 365, 366]. The extraordinary correlation between quantum subsystem states associated with entanglement can be exploited by quantum computing algorithms using interference to solve computational tasks, such as factoring, far more efficiently than is possible using classical methods, as we show in later chapters. Entangled states are similarly exploitable by uniquely quantum communication protocols, such as quantum teleportation, superdense coding, and advanced forms of quantum key distribution, using local operations and classical communication (LOCC).

Entanglement is of perennial intrinsic interest because of the radically counter-intuitive behavior associated with the strong correlations it entails, that was discussed in Chapter 3. Albert Einstein, Boris Podolsky, and Nathan Rosen argued early on that quantum mechanics is incomplete if understood as a local realistic theory, based on the consideration of an (entangled) quantum state of the form

$$|\Psi(x_1, x_2)\rangle = \sum_{i=1}^{\infty} a_i |\psi(x_1)\rangle_i |\phi(x_2)\rangle_i \qquad (6.1)$$

[147]. David Bohm later explored entanglement in a far simpler context, that of a pair of spins in the singlet state

$$|\Psi^-\rangle = \frac{1}{\sqrt{2}}(|\uparrow\downarrow\rangle - |\downarrow\uparrow\rangle) , \qquad (6.2)$$

which has since been central to the investigation of the foundations of quantum mechanics and quantum information, wherein $\{|\uparrow\rangle, |\downarrow\rangle\}$ is typically taken as the computational basis and written $\{|0\rangle, |1\rangle\}$ [66].

Following these developments, John Bell greatly advanced the investigation of quantum entanglement by clearly delimiting the border between local classically explicable behavior and less intuitive sorts of behavior that are non-local, by deriving an inequality that must be obeyed by local realistic theories that might explain strong correlations between two distant subsystems forming a compound system, such as those arising in systems described in quantum mechanics by the singlet state [30]. Since these early investigations, the study of extraordinarily correlated behavior between subsystems within larger systems has been ongoing, as have efforts to put this unusual behavior to use. In this chapter, we consider the current understanding of quantum entanglement in *bipartite* quantum systems, which often uses the various quantum information measures introduced in the previous chapter.

6.1 Basic definitions

Under Schrödinger's definition, *entangled pure states* are simply those pure quantum states of multipartite systems that *cannot* be represented in the form of a simple tensor product of subsystem states

$$|\Psi\rangle \neq |\psi_1\rangle \otimes |\psi_2\rangle \otimes \cdots \otimes |\psi_n\rangle , \qquad (6.3)$$

where $|\psi_i\rangle$ are states of local subsystems, for example, spin states of fundamental particles [365, 366]. The remaining pure states of multipartite systems, which *can* be represented as simple tensor products of independent subsystem states, are called simply *product states*. The definition of entanglement can be extended to include *mixed states*, as follows. The mixed quantum states in which entanglement is most easily understood are states ρ_{AB} of bipartite systems, usually labeled AB with components labeled A and B in correspondence with the laboratories where they are located. Mixed states are called *separable* (or *factorable*) when they can be written as *convex combinations of products*,

$$\rho_{AB} = \sum_i p_i \rho_{Ai} \otimes \rho_{Bi}, \qquad (6.4)$$

where $p_i \in [0, 1]$ and $\sum_i p_i = 1$, ρ_A and ρ_B being statistical operators on subsystem Hilbert spaces, $\mathcal{H}_A$ and $\mathcal{H}_B$, respectively.[1] Entangled quantum states are simply those that are *inseparable*.

Separable mixed states contain no entanglement, as they are by definition the *mixtures* of product states and so can be created by local operations

---

[1] This definition extends beyond the statistical operators to other operators, generalizing the concept of entanglement beyond states.

and classical communication from pure product states: in order to create a separable state, an agent in one lab needs merely to sample the probability distribution $\{p_i\}$ and share the corresponding measurement results with an agent the another; the two agents can then create their own sets of suitable local states ρ_i in their separate labs.[2] However, by contrast, not all *entangled* states can be converted into each other in this way in the *multi-party* context, something that leads to distinct classes of entangled states and thus to different *sorts* of entanglement, as we show in the next chapter. In general, it is also not always possible to tell whether a given statistical operator *is* entangled. Given a set of subsystems, the problem of determining whether their joint state is entangled is known as the *separability problem*.

The simplest states within the class of separable states are the product states of the form $\rho_{AB} = \rho_A \otimes \rho_B$; ρ_A and ρ_B are then also the reduced statistical operators for the two subsystems and are *uncorrelated*. When there are correlations between properties of subsystems described by separable states, these can be fully accounted for *locally* because the separate quantum states ρ_A and ρ_B within spacelike-separated laboratories provide descriptions sufficient for common cause explanations of the joint properties of A and B such as that outlined above; also see [430]. In particular, the outcomes of local measurements on any separable statistical operator can be simulated by a local hidden-variables theory. The quantum states in which correlations between A and B *can* be seen to violate a Bell-type inequality, referred to as *Bell correlated* (or *EPR correlated*) states, cannot be accounted for by common cause explanations. If a *pure* state is entangled then it is Bell correlated.[3] Thus, pure entangled states do not admit a common cause explanation. However, this is *not* true for the mixed entangled states. For example, the Werner state,

$$\rho_W = \left(1 - \frac{1}{\sqrt{2}}\right)\frac{1}{4}\mathbb{I}\otimes\mathbb{I} + \frac{1}{\sqrt{2}}P(|\Psi^-\rangle)\,, \tag{6.5}$$

is *not* Bell correlated yet *is* entangled, because there is no way to write ρ_W as a convex combination of product states; in particular, it cannot be written in the form of Eq. 6.4 with only *one* nonzero p_i.[4]

The shortcoming of Bell-inequality violation as a necessary condition for entanglement is that it is unknown whether there exist Bell inequality violations for many nonseparable mixed states. In the presence of manipulations of such a state (or a collection of copies) by means of LOCC, some states *can* be made to violate a Bell-type inequality; those states that can be made to

---

[2] See Chapter 3 for a characterization of local operations.

[3] This was first pointed out by Sandu Popescu and Daniel Rohrlich [338] and Nicolas Gisin [186]. Note, however, that not all such states are *Bell states*, that is, elements of the Bell basis as, say, $|\Psi^-\rangle$ is; see Sect. 6.3, below [339].

[4] Note also that the Werner state is diagonal in the Bell-basis representation. An excellent review discussing the relationship between Bell inequalities and entanglement is [451].

violate a Bell inequality in this way are referred to as *distillable* states. The remaining, nondistillable states are known as *bound* states. What *is* clear is that state entanglement should not change under local operations and should not be increased by local operations together with classical communication, assumptions that play a central role in quantifying entanglement, as we show in Section 6.6 below. Let us first consider some fundamental tools in the study of entanglement.

6.2 The Schmidt decomposition

There exist special state decompositions that clearly manifest the correlations associated with entanglement. For pure bipartite states, the Schmidt decomposition serves this purpose well. Any bipartite pure state $|\Psi\rangle \in \mathcal{H} = \mathcal{H}_A \otimes \mathcal{H}_B$ can be written as a sum of bi-orthogonal terms: there exists at least one orthonormal basis for $\mathcal{H}$, $\{|u_i\rangle \otimes |v_i\rangle\}$ where $\{|u_i\rangle\} \in \mathcal{H}_A$ and $\{|v_i\rangle\} \in \mathcal{H}_B$ such that

$$|\Psi\rangle = \sum_i a_i |u_i\rangle \otimes |v_i\rangle \, , \tag{6.6}$$

$a_i \in \mathbb{C}$, referred to as a *Schmidt basis*. This representation is a *Schmidt (or polar) decomposition* of $|\Psi\rangle$, where the summation index runs up only to the *smaller* of the corresponding two Hilbert space dimensions, dim $\mathcal{H}_A$ and dim $\mathcal{H}_B$ [363]. It is often convenient to take the amplitudes a_i to be real numbers by absorbing any phases into the definitions of the $\{|u_i\rangle\}$ and $\{|v_i\rangle\}$. Unfortunately, the availability of this decomposition in multipartite systems is limited, being available with certainty only in the case of *bipartite* states.

For any entangled bipartite *pure* state, it is possible to find pairs of measurable quantities violating the Bell inequality. In particular, the *Schmidt observables*

$$U = \sum_i u_i |u_i\rangle\langle u_i| \, , \tag{6.7}$$

$$V = \sum_i v_i |v_i\rangle\langle v_i| \, , \tag{6.8}$$

are fully correlated when the system is in state $|\Psi\rangle$, providing such violations [186].

The number of nonzero amplitudes a_i in the Schmidt decomposition of a quantum state is known as the *Schmidt number* (or *Schmidt rank*), $\mathrm{Sch}(|\Psi\rangle)$. The Schmidt number proves useful for distinguishing entangled states. In particular, the Schmidt number of a state is greater than 1 *if and only if* it is entangled. It is useful as a (coarse) quantifier of the *amount* of entanglement in a system, in addition to serving as a criterion for entanglement.

The Schmidt number of a bipartite system is equivalently defined as

$$\text{Sch}(|\Psi\rangle) \equiv \dim \text{ supp } \rho_A = \dim \text{ supp } \rho_B \ , \tag{6.9}$$

where ρ_A and ρ_B are the reduced statistical operators for the two subsystems,

$$\rho_A = \sum_i |a_i|^2 |u_i\rangle\langle u_i| \ , \tag{6.10}$$

$$\rho_B = \sum_i |a_i|^2 |v_i\rangle\langle v_i| \ , \tag{6.11}$$

which are diagonal, possess *identical* eigenvalue spectra, and hence have identical von Neumann entropies. Furthermore, Schmidt number is preserved under local unitary state transformations.

Using the Schmidt decomposition, the *Schmidt measure (Hartley strength)* of the entanglement of pure states is defined as

$$E_S(|\Psi\rangle) \equiv \log_2\left(\text{Sch}(|\Psi\rangle)\right) \ , \tag{6.12}$$

providing entanglement in units of "e-bits," a term, like "qubit," introduced by Schumacher, where the Bell states correspond to one e-bit of entanglement. The probabilities that are the *squares* of the Schmidt coefficients a_i are precisely those quantities unchanged by unitary operations performed locally on the individual subsystems (LUT's). For this reason, it is reasonable to expect *any* more precise numerical measure of pure state entanglement to be calculable from the quantities $|a_i|^2$.[5] Because the statistical operator ρ of a bipartite system may have degenerate eigenvalues there is, however, not a truly unique Schmidt *basis*. For example, in the case of the Bell state $|\Psi^-\rangle$, the state takes the same form when represented in any other basis obtained from the computational basis representation (Eq. 3.5), which is of Schmidt form, by rotating the computational basis and performing a unitary transformation in the subspace of the first qubit and the conjugate transformation in that of the second qubit [154].

Again, the Schmidt decomposition is not always available beyond the case of bipartite systems. Consider the case of a system with *three* subsystems. If there existed such a decomposition, the measurement of one subsystem would provide the states of the remaining two; but, if these two are entangled, then the individual states *must* be indefinite.[6]

6.3 Special bases and decompositions

Basic examples of states in Schmidt form are the four elements of the *Bell basis*, which are the entangled states written

---

[5] One example of this is the concurrence, defined in Sect. 6.10, below.

[6] The generalization of this decomposition to special states of larger systems where such a decomposition does exist, such as the GHZ state $|GHZ\rangle = (1/\sqrt{2})(|000\rangle + |111\rangle)$, is discussed briefly later in Sect. 7.3.

$$|\Psi^{\pm}\rangle = \frac{1}{\sqrt{2}}(|01\rangle \pm |10\rangle) \qquad (6.13)$$

$$|\Phi^{\pm}\rangle = \frac{1}{\sqrt{2}}(|00\rangle \pm |11\rangle) \qquad (6.14)$$

in the computational basis and which are symmetrical or antisymmetrical under qubit exchange. These *Bell states* have played a central role in the investigation of quantum entanglement and tests of local realism, as shown in Chapter 3. The creation of the states of the Bell basis from a pair of unentangled qubits can be carried out by a process described by a quantum circuit involving only one Hadamard and one C-NOT gate; see Fig. 6.1. Bell states are also readily produced *ab initio* using spontaneous parametric down-conversion, which is discussed in Section 6.16. Bell states have the useful property that transforming the state of only one subsystem locally suffices for interconversion between them, which is not true, for example, of the two-qubit computational-basis states, which are of product form. Of particular interest is the singlet state, $|\Psi^{-}\rangle$, due to its great symmetry.

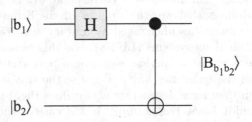

Fig. 6.1. A quantum circuit for the synthesis of Bell states, $|B_{b_1 b_2}\rangle$ from a product state. The input states are indicated by the bit values $b_i \in \{0, 1\}$, $i = 1, 2$: $b_1 b_2 = 00, 10, 01, 11$ yield $|\Phi^+\rangle, |\Phi^-\rangle, |\Psi^+\rangle, |\Psi^-\rangle$, respectively.

Another basis of entangled states for two-qubits, the so-called *"magic basis,"* is similar to the Bell basis but has different overall phases and norm,

$$|m_1\rangle = \frac{1}{2}(|00\rangle + |11\rangle), \qquad (6.15)$$

$$|m_2\rangle = \frac{i}{2}(|00\rangle - |11\rangle), \qquad (6.16)$$

$$|m_3\rangle = \frac{i}{2}(|01\rangle + |10\rangle), \qquad (6.17)$$

$$|m_4\rangle = \frac{1}{2}(|01\rangle - |10\rangle), \qquad (6.18)$$

and is a natural one for concurrence-based entanglement studies, discussed in Section 6.10, below [214].

Another useful basis is the *q-basis*,

$$|q_1\rangle = \sqrt{q}|00\rangle + \sqrt{1-q}|11\rangle , \qquad (6.19)$$

$$|q_2\rangle = \sqrt{1-q}|00\rangle - \sqrt{q}|11\rangle , \qquad (6.20)$$

$$|q_3\rangle = \sqrt{q}|01\rangle + \sqrt{1-q}|10\rangle , \qquad (6.21)$$

$$|q_4\rangle = \sqrt{1-q}|01\rangle - \sqrt{q}|10\rangle , \qquad (6.22)$$

which, for values $q \in [0, 1]$, interpolates between the (product) computational basis (for which $q = 0, 1$) and the (entangled) Bell basis (for which $q = 1/2$). Varying the value of q, say by taking $q = \cos\theta$ and varying θ, allows one to study the role of entanglement over this important range of pure states; for example, see Section 9.11 and [234].

A *Lewenstein-Sanpera (LS) decomposition* of a statistical operator $\rho \in \mathbb{C}^2 \otimes \mathbb{C}^2$ is one of the form

$$\rho = \lambda\rho_{sep} + (1-\lambda)P(|\Psi_{ent}\rangle) , \qquad (6.23)$$

with $\lambda \in [0, 1]$, where ρ_{sep} is separable and $P(|\Psi_{ent}\rangle)$ is the projector for a fully entangled state [282]. Such a decomposition exists for any two-qubit state. Although this decomposition is not unique, the decomposition for which λ takes an optimal value, λ_{max}, *is*. λ_{max} is sometimes referred to as the *degree of separability* and can be viewed as the degree of *classicality* of the state.[7] One example following from the LS decomposition is the Werner state (*cf.* Equation 6.5, above). Varying λ allows one to explore the role of entanglement over an important range of *mixed states*; for example, see [309].

Yet another useful class of basis is that of the *unextendable product bases*, which are sets of orthogonal product state-vectors such that there exists no additional product state-vector orthogonal to them in order to span the entire space in which they lie [45, 203]. A two-*qutrit* example is

$$|v_1\rangle = \frac{1}{\sqrt{2}}|0\rangle(|0\rangle - |1\rangle) , \qquad (6.24)$$

$$|v_2\rangle = \frac{1}{\sqrt{2}}(|0\rangle - |1\rangle)|2\rangle , \qquad (6.25)$$

$$|v_3\rangle = \frac{1}{\sqrt{2}}(|1\rangle - |2\rangle)|0\rangle , \qquad (6.26)$$

$$|v_4\rangle = \frac{1}{\sqrt{2}}|2\rangle(|1\rangle - |2\rangle) , \qquad (6.27)$$

$$|v_5\rangle = \frac{1}{3}(|0\rangle + |1\rangle + |2\rangle)(|0\rangle + |1\rangle + |2\rangle) . \qquad (6.28)$$

---

[7] Such a decomposition, which was anticipated by Shimony (see Sect. 6.15 and [383]), is known as the *best separable approximation*.

6.4 Stokes parameters and entanglement

As we saw in Chapter 1, qubits have a variety of representations, among these the real-valued one provided by the single-qubit Stokes parameters. Although they suffice when specifying individual qubits or several qubits in a separable state, the single-qubit parameters must be supplemented by additional parameters in order to describe *entangled systems*. Consider the general state of a pair of qubits. The *two-particle Stokes parameters*, $S_{\mu\nu} \equiv \mathrm{tr}(\rho\sigma_\mu \otimes \sigma_\nu)$ $(\mu, \nu = 0, 1, 2, 3)$, which are a generalization of the traditional Stokes parameters, are needed to describe entangled states, such as the Bell states, in the real representation, due to the increasing complexity of quantum states as number of qubits grows.[8] The two-qubit Stokes parameters, introduced by Ugo Fano just before 1950, can also be used to find the two-qubit statistical operator:

$$\rho = \frac{1}{4} \sum_{\mu,\nu=0}^{3} S_{\mu\nu}\sigma_\mu \otimes \sigma_\nu \,, \tag{6.29}$$

where $\sigma_\mu \otimes \sigma_\nu$ $(\mu, \nu = 0, 1, 2, 3)$ are tensor products of the identity and Pauli matrices [166];[9] the single-qubit Stokes parameters are recovered when either μ or ν is zero, so that the corresponding factor is an identity matrix.

The Stokes four-vector $[S_\mu]$ described in Section 1.3 is similarly generalized, as one can view the two-qubit Stokes parameters as forming a 16-element *Stokes tensor*, $[S_{\mu\nu}]$.[10] This tensor captures *all* the quantum correlations potentially present in a two-qubit system and plays a central role in the quantum state tomography of such a system, corresponding to a compendium of coincidence-measurement data.[11] For example, the Bell state $|\Psi^+\rangle$ corresponds to a Stokes tensor with $S_{00} = 1, S_{11} = -1, S_{22} = -1, S_{33} = 1$, the remaining parameters being zero. The Lorentz group invariant for the two-qubit Stokes tensor,

$$S_{(2)}^2\big(P(|\psi\rangle)\big) = \frac{1}{4}\bigg((S_{00})^2 - \sum_{i=1}^{3}(S_{i0})^2 - \sum_{j=1}^{3}(S_{0j})^2 + \sum_{i=1}^{3}\sum_{j=1}^{3}(S_{ij})^2\bigg) \,, \tag{6.30}$$

can be related to the entanglement of the two-qubit state, as we show in Section 7.4 [237].

---

[8] The practical value of the generalized Stokes parameters is manifest in their application to polarization-entangled photon pairs; for example, see [3].

[9] Recall that the Hilbert space for two-qubit systems is $\mathbb{C}^2 \otimes \mathbb{C}^2$. The two-qubit density matrices ρ are positive, unit-trace elements of the 16-dimensional complex vector space of Hermitian 4×4 matrices, $H(4)$. The operators $\sigma_{\mu\nu} \equiv \sigma_\mu \otimes \sigma_\nu$ provide a basis for $H(4)$, which is isomorphic to the tensor product space $H(2) \otimes H(2)$ of the same dimension, because $\frac{1}{4}\mathrm{tr}(\sigma_{\mu\nu}\sigma_{\alpha\beta}) = \delta_{\mu\alpha}\delta_{\nu\beta}$ and $\sigma_{\mu\nu}^2 = \mathbb{I}_2$.

[10] The term "Stokes tensor" was first applied to this structure in [240].

[11] Quantum state tomography is discussed in Chapter 8.

The elements of the general two-particle matrix $\rho = \left[\rho_{\mu\nu}\right]$ are related to the two-qubit Stokes tensor elements $S_{\mu\nu}$ by the following relations.

$$S_{00} = \rho_{00} + \rho_{11} + \rho_{22} + \rho_{33} \tag{6.31}$$
$$S_{01} = 2\text{Re}(\rho_{01} + \rho_{23}) \tag{6.32}$$
$$S_{02} = -2\text{Im}(\rho_{01} + \rho_{23}) \tag{6.33}$$
$$S_{03} = \rho_{00} - \rho_{11} + \rho_{22} - \rho_{33} \tag{6.34}$$
$$S_{10} = 2\text{Re}(\rho_{02} + \rho_{13}) \tag{6.35}$$
$$S_{11} = 2\text{Re}(\rho_{03} + \rho_{12}) \tag{6.36}$$
$$S_{12} = -2\text{Im}(\rho_{03} - \rho_{12}) \tag{6.37}$$
$$S_{13} = 2\text{Re}(\rho_{02} - \rho_{13}) \tag{6.38}$$
$$S_{20} = -2\text{Im}(\rho_{02} + \rho_{13}) \tag{6.39}$$
$$S_{21} = -2\text{Im}(\rho_{03} + \rho_{12}) \tag{6.40}$$
$$S_{22} = -2\text{Re}(\rho_{03} - \rho_{12}) \tag{6.41}$$
$$S_{23} = -2\text{Im}(\rho_{02} - \rho_{13}) \tag{6.42}$$
$$S_{30} = \rho_{00} + \rho_{11} - \rho_{22} - \rho_{33} \tag{6.43}$$
$$S_{31} = 2\text{Re}(\rho_{01} - \rho_{23}) \tag{6.44}$$
$$S_{32} = -2\text{Im}(\rho_{01} - \rho_{23}) \tag{6.45}$$
$$S_{33} = \rho_{00} - \rho_{11} - \rho_{22} + \rho_{33} \; . \tag{6.46}$$

6.5 Partial transpose and reduction criteria

In addition to the Schmidt number, $\text{Sch}(|\Psi\rangle)$, and Schmidt measure, E_S, for pure states described in Section 6.2 above, another simple quantity measuring entanglement for some *mixed* states is the *negativity*, $\mathcal{N}(\rho)$. This quantity involves the sum of the negative eigenvalues of the partial transpose of the density matrix of a bipartite system. It was first used to provide a criterion for entanglement by Asher Peres, who noted that when the partial transposition operation is performed on a separable mixed state the result is always another mixed state [329]. Partial transposition is matrix transposition relative to the indices of a subsystem; the matrix elements of the partially transposed density matrix are thus

$$\langle i_A j_B | \rho^{T_A} | k_A l_B \rangle \equiv \langle k_A j_B | \rho | i_A l_B \rangle \; . \tag{6.47}$$

Specifically, the "Peres–Horodečki (PH) criterion" for entanglement is the following: a state ρ is entangled if the partial transpose of the corresponding density matrix is *negative*. One can take

$$\mathcal{N}(\rho) = \frac{1}{2}\left(\|\rho^{T_A}\|_1 - 1\right) , \tag{6.48}$$

where $||\rho^{T_A}||_1$ is the trace-norm of the partial transpose matrix. Because $||O||_1 \equiv \mathrm{tr}\sqrt{O^\dagger O}$ for any Hermitian operator O, one can write

$$\mathcal{N}(\rho) = \left| \sum_i \lambda_i \right| , \tag{6.49}$$

where i runs over the *negative* values among the set of eigenvalues $\{\lambda_i(\rho^{T_B})\}$ of this density matrix.[12]

The negativity is readily computed and has been used to develop entanglement bounds. Its logarithm, the *logarithmic negativity*, is also sometimes considered, because it has operational interpretations such as an upper bound to the distillable entanglement considered, a bound on teleportation capacity, and an asymptotic entanglement cost under PPT; see Section 6.8 below and [17, 42].

The positivity of ρ^{T_A} (or ρ^{T_B}) is a necessary and sufficient condition for the *separability* of the statistical operator ρ for 2×2, 2×3 dimensional systems and for two continuous-variable systems (modes) in a Gaussian state [141]. For a related result not making use of a map between matrices that is linear, as partial transposition is, but rather a *nonlinear map* to solve the separability problem for Gaussian states of an arbitrary number of modes per site, see [181].

When applied to a Bell state, the result of partial transposition is a matrix with *at least one* negative eigenvalue. Positivity of the partial transpose is, in general, a necessary but *insufficient* condition for separability when subsystems with Hilbert spaces of higher dimension than that of a qubit are involved; for larger Hilbert spaces, there exist entangled states whose density matrices are *positive* under partial transpose (PPT). See Section 6.11 below for further discussion of the PH criterion and examples of states having PPT.

The "PPT preserving" class of quantum operations includes all bipartite quantum operations for which input states that are positive under partial transposition have output states that also have this property; these operations can produce only the *bound* variety of entanglement; see Section 6.8, below.

For the bound entangled states with PPT, *all* CHSH-inequalities are obeyed. The PH criterion implies another useful criterion, namely, both

$$\rho_A \otimes \mathbb{I} - \rho \geq 0 , \tag{6.50}$$

$$\mathbb{I} \otimes \rho_B - \rho \geq 0 , \tag{6.51}$$

---

[12] The eigenvalues of the density matrix are usually indicated in ascending order.

known as the *reduction criterion* for entanglement, which implies the recoverability of entanglement by distillation; also see Section 6.8, below. The violation of the reduction criterion is also sufficient for separability of ρ in the case of two qubits and the case of one qubit and one qutrit. Moreover, the criterion implies that the ranks of the reduced density matrices are less than or equal to that of the density matrix of the compound system [442].

6.6 The "fundamental postulate"

In addition to the conventional requirements that a measure of entanglement be nonnegative and normalized in the sense that it be unity for the Bell states, a fundamental pair of monotonicity conditions has been put forth for any candidate, below indicated generically as $E_X(\rho)$, to be good a measure of entanglement. These conditions define the class of entanglement monotones, which are functionals that characterize the strength of genuinely quantum correlations through the requirement that no state can be converted by local operations and classical communication (LOCC) to a state having a *higher* value of the monotone. In particular, a quantity $E_X(\rho)$ is called an *entanglement monotone* if it satisfies

$$E_X(\rho) \geq \sum_i p_i E_X(\rho_i) , \qquad (6.52)$$

and

$$E_X\left(\sum_i p_i \rho_i\right) \leq \sum_i p_i E_X(\rho_i) , \qquad (6.53)$$

for all local operations giving rise to states ρ_i with probabilities p_i, where at the end of the LOCC operation i, classical information is available with probability p_i and the state is ρ_i [437].

The first of the two conditions above, sometimes referred to as the *fundamental postulate*, requires *monotonicity on the average* for each local operation. The second condition requires $E_X(\rho)$ to be a convex function that is *monotonic under mixing*, that is, the discarding of information, providing mathematical convenience, which is sometimes relaxed. The above useful but limited entanglement measures, the Schmidt measure E_S and negativity $\mathcal{N}$, are examples of entanglement monotones for bipartite quantum systems.

Consider two sets of entanglement monotones, $E_l^{\Psi} = \sum_{i=1}^n |a_i|^2$ and $E_l^{\Phi} = \sum_{i=1}^n |b_i|^2$, where $l = 1, \ldots, n$, obtained from the Schmidt decomposition of two bipartite states $|\Psi\rangle, |\Phi\rangle$ having n components with Schmidt coefficients a_i and b_i respectively. The pure state $|\Psi\rangle$ can be transformed with certainty by local transformations to the pure state $|\Phi\rangle$ if and only if $E_l^{\Psi} \geq E_l^{\Phi}$ for all $l = 1, \ldots, n$ [439].

6.7 Entanglement monotones

Let us now explore the behavior of entanglement monotones in greater detail, by considering not only the basic requirements on them, but also those that relate to asymptotic behavior. The following conditions are those now commonly required of acceptable measures of bipartite entanglement E_X on all states ρ_{AB} of a pair of systems.

(i) $E_X(\rho_{AB}) = 0$ if ρ_{AB} is separable.

(ii) $E_X(\rho_{AB})$ is invariant under *all* local unitary operations $U_A \otimes U_B$, that is, $E_X(\rho_{AB}) = E_X\big((U_A \otimes U_B)\rho_{AB}(U_A \otimes U_B)^\dagger\big)$.

(iii) $E_X(\rho_{AB})$ cannot be increased by *any* LOCC transformation, that is, $E_X(\rho_{AB}) \geq E_X\big(\Theta(\rho_{AB})\big)$, where $\Theta(\rho_{AB})$ is a CPTP map.

The necessity of the first condition is obvious: separable states, specified by Eq. 6.4, are by definition *not* entangled. Conditions (ii) and (iii) are necessary for entanglement to be considered a collective, global property of quantum systems; they render impossible the creation or distribution of entanglement via LOCC *alone*. These conditions accord with each other because local unitary operations are CPTP maps that can be inverted by local unitary operations. The following further condition is sometimes also imposed.

(iv) The entanglement of n copies of a state ρ_{AB} is n times the entanglement of one copy,

$$E_X(\rho_{AB}^{\otimes n}) = nE_X(\rho_{AB}) , \tag{6.54}$$

in particular for the standard case of n Bell singlets, which are conventionally taken to have entanglement equal to unity, that is, are n "e-bits."

A continuity condition may also be imposed, namely,

(v) If $\langle \psi^{\otimes n} | \rho_n | \psi^{\otimes n} \rangle \to 1$ for $n \to \infty$, then

$$\frac{1}{n} \big| E_X\big(P(|\psi\rangle)^{\otimes n}\big) - E_X\big(\rho^{(n)}\big) \big| \to 0 , \tag{6.55}$$

for some joint state $\rho^{(n)}$ of n pairs of qubits [221].

With the fourth condition, known as *partial additivity*, and the fifth condition both in force, the pure state entanglement of bipartite quantum systems is *uniquely* described by

$$E(|\Psi\rangle_{AB}) \equiv S(\rho) = -\text{tr}(\rho \log_2\rho) , \tag{6.56}$$

the von Neumann entropy functional, where ρ is the (reduced) statistical operator of either one of the two subsystems of the compound system in state $|\Psi\rangle_{AB}$ [339].[13] The last two conditions are sometimes directly replaced by the condition that, for pure states, the measure reduces to this entropy.

---

[13] Full additivity would require that $E(\rho \otimes \sigma) = E(\rho) + E(\sigma)$. However, because bound entanglement may be activated, this condition is often viewed as unwarranted.

The von Neumann measure of entanglement has the property of being additive on pure states of the composite system: using the von Neumann entropy of the subsystem reduced states as E_X, and labeling the individual particles of Alice and Bob in the n copies as A_i, B_i, the *additivity property* is

$$E(|\Psi\rangle_{A_1 B_1} \otimes |\Psi\rangle_{A_2 B_2} \otimes \cdots \otimes |\Psi\rangle_{A_n B_n}) = \sum_{i=1}^{n} E(|\Psi\rangle_{A_i B_i})$$

for all pure $|\Psi\rangle_{AB}$.[14] In the case of *mixed* states, additivity is desirable but must be explicitly imposed, so in that context one refers to the *additivity conjecture*.[15] In the case of larger systems involving multiple parties, some of the above conditions must be slightly modified as discussed in the next chapter.

> In the simplest case, the *pure* states of two qubits A and B, entanglement is well quantified information-theoretically by the von Neumann entropy of either one of the single-qubit reduced statistical operators, which are identical; these operators are obtained by "tracing out" one qubit from the total system state described by the projector $P(|\Psi\rangle_{AB})$; see Section 2.5. Thus,
>
> $$E(|\Psi\rangle_{AB}) = S\big(\mathrm{tr}_A P(|\Psi\rangle_{AB})\big) = S\big(\mathrm{tr}_B P(|\Psi\rangle_{AB})\big) \ .$$

For *mixed* two-qubit states ρ_{AB}, a good entanglement measure is the (one-shot) *entanglement of formation*, E_f, defined via the convex-roof construction as the minimum average marginal entropy of the one-qubit reduced states for all possible decompositions of ρ_{AB} as a mixture of pure subensembles each described by a state $P(|\Psi_i\rangle_{AB})$, that is,

$$E_f(\rho_{AB}) = \min_{\{p_i, |\Psi_i\rangle\}} \sum_i p_i E(|\Psi_i\rangle) \ , \tag{6.57}$$

where $\{p_i, P(|\Psi_i\rangle)\}$ represents a decomposition of ρ_{AB}.[16] One can similarly define the *entanglement of assistance* as the corresponding *maximum* average

---

[14] The product symbol $\odot$ is sometimes substituted for $\otimes$ to emphasize that this tensor product is formed from *copies* of a state possessed by the same pair of agents, as opposed to distinct parties.

[15] Note, however, that a uniqueness theorem not assuming additivity has also been produced [437].

[16] This quantity is analogous to the total energy of thermodynamics, something discussed in greater detail in Sects. 6.12–13, below. Note that, although it can be expressed directly in terms of the von Neumann entropy $S(\rho_A)$, the form provided here allows for explicit reference to the states of the pertinent two-qubit pure subensembles.

marginal entropy. Pre-existing entanglement is not necessary to *distribute* entanglement so long as a quantum channel—a means of transmitting quantum systems—exists between Alice and Bob.

A related quantity is the *entanglement cost*, E_C, defined as the *smallest number of systems in Bell singlets*, per copy, needed to form copies of the given state ρ_{AB} by CLOCC operations $P(|\Psi^-\rangle)^{\otimes k'} \to \rho_{AB}^{\otimes n}$, in the limit as the number of shared pairs goes to infinity, which is

$$E_C(\rho_{AB}) = \lim_{n \to \infty} \frac{E_f(\rho_{AB}^{\otimes n})}{n}, \tag{6.58}$$

where the decomposition pertinent to the entanglement of formation here is that of $\rho_{AB}^{\otimes n}$.[17]

6.8 Distillation and bound entanglement

It is possible to obtain pure entangled states that violate Bell inequalities beginning with mixed states that do *not* violate a Bell inequality, by using entanglement distillation. Entanglement distillation, also known as *entanglement purification*, is the local processing of a number of copies of a quantum state so as to develop highly entangled states between parties, that is, the inverse process to that considered when finding the entanglement cost discussed above. Any local process by which the degree of entanglement between various subsystems of a larger overall quantum system is *increased* can be considered entanglement distillation. Such a process is valuable, for example, when channels of transmission of quantum information are noisy and degrade the entanglement resources needed to successfully carry out quantum information processing tasks. In particular, this process can assist in reducing the impact of quantum decoherence. Specific protocols for purifying entanglement are described in detail later in Section 9.11.

The associated functional, the *entanglement of distillation*, $D(\rho_{AB})$, is defined as the *maximum fraction of singlets* that can be extracted, that is, distilled from n copies of ρ_{AB} by the CLOCC transformation $\rho_{AB}^{\otimes n} \to P(|\Psi^-\rangle)^{\otimes k}$ in the asymptotic limit as $n \to \infty$:

$$D(\rho_{AB}) = \limsup_{n \to \infty} \left(\frac{k}{n} \right), \tag{6.59}$$

where k depends on n. This quantity can be viewed as analogous to thermodynamical free energy, and so is sometimes called the *free entanglement*. It expresses, for example, the utility of a given entangled mixed state for quantum teleportation.[18] However, $D(\rho_{AB})$ has none of the desirable convexity or additivity properties to be an information-theoretic measure [392].

---

[17] This result was first proven by Patrick Hayden, Michał Horodecki, and Barbara Terhal [209].

[18] See [47], for example, and Sect. 9.9, below for more on teleportation.

According to condition (iii) of the previous section, it must be the case that

$$D(\rho_{AB}) \leq E_f(\rho_{AB}) , \tag{6.60}$$

which reflects the irreversible character of state mixing. The distillable entanglement has been either evaluated or bounded for a variety of classes of mixed states. For mixed states ρ_{AB}, it is also natural to consider the difference $B(\rho_{AB}) = E_f(\rho_{AB}) - D(\rho_{AB})$, between the entanglement of formation and the entanglement of distillation, known as the *bound entanglement*. The bound entanglement is clearly nonnegative:

$$B(\rho_{AB}) \geq 0 . \tag{6.61}$$

Bound states are those statistical states that are *not* distillable yet the formation of a single copy of which requires entanglement, which can be viewed as a consequence of extreme state mixing.[19]

Entanglement distillation is irreversible, in the sense that more pure entanglement cannot be distilled from PPT states than may have been used to assist in their creation [438]. A discussion of general considerations surrounding the manipulation of entanglement considered as a quantum information resource is given below in Section 6.12.

6.9 Entanglement and majorization

For bipartite *separable* states, sometimes designated $\rho_{AB}^{(s)}$, the following relationships hold between the quantum Rényi entropies of systems and subsystems:

$$S_\alpha(\rho_A^{(s)}) \leq S_\alpha(\rho_{AB}^{(s)}) \tag{6.62}$$

$$S_\alpha(\rho_B^{(s)}) \leq S_\alpha(\rho_{AB}^{(s)}) , \tag{6.63}$$

where ρ_A and ρ_B are the reduced statistical operators of the components.

Similarly, for separable statistical operators

$$\lambda_{\rho_A^{(s)}} \prec \lambda_{\rho_{AB}^{(s)}} \tag{6.64}$$

$$\lambda_{\rho_B^{(s)}} \prec \lambda_{\rho_{AB}^{(s)}} , \tag{6.65}$$

where λ_ρ is the ordered vector of eigenvalues of the statistical operator ρ and $\succ$ denotes majorization, affirming that *separable states* are *at least* as mixed

---

[19] The first bound entangled states were discovered by Paweł Horodecki [223]. Note, however, that the existence of bound states does not preclude situations where forming a larger number of copies may require a *vanishingly small* amount of entanglement per copy; the states that do not violate the PH criterion form a known such class. Nonviolation of this criterion is preserved under LOCC.

globally as they are locally, as is clear in the case of the entropic characterization of entanglement described above; see Section A.3 and [316].[20] Indeed,
the majorization condition, known as *Uhlmann's relation*, implies the above
entropic inequalities for $r \geq 0$.[21]Nonviolation of the majorization condition is
not preserved under LOCC.

6.10 Concurrence

A practical measure of bipartite entanglement that has a geometrical meaning and can often be easily calculated is the *concurrence*. For pure states, this
quantity can be written $C(|\Psi_{AB}\rangle) = |\langle\Psi_{AB}|\tilde{\Psi}_{AB}\rangle|$, where $|\tilde{\Psi}_{AB}\rangle \equiv \sigma_2^{\otimes 2}|\Psi_{AB}^*\rangle$
which is referred to as the "spin-flipped" state-vector [460]. The concurrence
of a mixed two-qubit state, $C(\rho_{AB})$, can be expressed in terms of the mini-
mum average pure-state concurrence, $C(|\Psi_{AB}\rangle)$, where, as usual, the required
minimum is to be taken over *all possible* ensemble decompositions of ρ_{AB}.
The concurrence of a general state is then simply

$$C(\rho_{AB}) = \max\{0, \lambda_1 - \lambda_2 - \lambda_3 - \lambda_4\} \,, \qquad (6.66)$$

where the λ_i are the *square roots* of the matrix $\rho_{AB}\tilde{\rho}_{AB}$, indexed in order of
decreasing size, also known as the *singular values*, which are real and nonneg-
ative. For mixed states, the negativity is bounded by the concurrence:

$$\mathcal{N}(\rho_{AB}) \leq C(\rho_{AB}) \,, \qquad (6.67)$$

which inequality is an equality in the case of pure states [18]. The entanglement
of formation of a mixed state ρ_{AB} of two qubits can be expressed in terms of
the concurrence as

$$E_f(\rho_{AB}) = h(C(\rho_{AB})) \,, \qquad (6.68)$$

where $h(x) = -x\log_2 x - (1-x)\log_2(1-x)$ has the form of the (classical)
binary entropy function [460].[22]

---

[20] One vector of eigenvalues of the statistical operator (arranged in decreasing order)
majorizes another if its statistical operator is more mixed than that of the other.
To make the above comparison, one appends the required number of zero values
to the vector of subsystem eigenvalues.

[21] The quantum Rényi entropy is a concave function of these probabilities for $0 \leq r \leq 1$; see Sect. 5.5. The operators themselves are sometimes used in place of
the eigenvalue vectors in this notation. *Uhlmann's relation* states that, for two
Hermitian operators K and L, $K \succ L$ if and only if there exist unitary matrices U_i
and probabilities p_i such that $K = \sum_i p_i U_i L U_i^\dagger$, where the p_i are understood as
describing the *mixing* of operators obtained from L by the corresponding unitary
transformations.

[22] This result, obtained by William Wootters, relating geometric and entropic measures to entanglement, is conditional on a proof of the additivity conjecture for
entanglement of formation, which is supported by existing numerical evidence.

An analytic expression for the concurrence of bipartite systems of arbitrary finite dimensionalities d_A and d_B in pure states, the *I-concurrence*, is arrived at by generalizing the "spin-flip" operation and is written

$$C(|\Psi\rangle_{AB}) = \sqrt{2s_A s_B(1 - \mathrm{tr}(\rho_A^2))}\,, \qquad (6.69)$$

where s_A and s_B are scaling factors [358] and the labels A and B can be interchanged without effect. This quantity is readily seen as the square root of a scalar multiple of the mixedness of the subsystems. One obtains a measure applicable to *mixed* states of such systems by convex-roof extension:

$$C(\rho \doteq \{p_k, |\Psi\rangle_k\}) = \min_{\{p_k,|\Psi_k\rangle\}} \sum_k p_k C(|\Psi\rangle_k) \qquad (6.70)$$

$$= \min_{\{p_k,|\Psi_k\rangle\}} \sum_k p_k \sqrt{2s_A s_B\left(1 - \mathrm{tr}(\rho_A^{(k)2})\right)}\,,$$

where $\rho_A^{(k)}$ is the reduced state of the subsystem A within the pure subensemble k. The squares of these quantities are the *I-tangle* measures. The concurrence also has a geometrical interpretation, which is discussed later in Section 7.4, where a geometrical extension to any finite even number of subsystems is introduced.

One can also straightforwardly define the *concurrence of assistance*,

$$C_{\mathrm{assist}}(\rho_{AB}) = \max_{\{p_i, P(|\Psi_i\rangle)\}} \sum_i p_i C(|\Psi_i\rangle)\,, \qquad (6.71)$$

which is the *maximum* average concurrence of ensembles $\{p_i, P(|\Psi_i\rangle)\}$ capable of providing the state ρ_{AB} by mixing. The states from which the mixed state is formed can be taken, for example, to be those of a purification of ρ_{AB} in the presence of an ancillary system. Like the concurrence itself, this quantity has the advantage of being readily calculable via the trace operation.

6.11 Entanglement witnesses

An entanglement witness is defined as an Hermitian operator such that its expectation value is positive for every separable state but negative for some entangled states [415]. Given these properties, entanglement can be detected through the expectation value of an entanglement witness, because a *negative* value indicates the system is entangled. More precisely, a statistical operator ρ on the composite system space $\mathcal{H}_A \otimes \mathcal{H}_B$ is entangled if and only if there is an *entanglement witness*, an Hermitian matrix W such that $\text{tr}(\rho W) < 0$ but $\text{tr}(\rho^{(s)} W) \geq 0$ for *all* separable states $\rho^{(s)}$ [223]. Such an operator will have at least one negative eigenvalue.[23] The Bell operator, $\mathcal{B}$, is the most familiar such operator; see Section 3.5. By measuring the value of an entanglement witness for the state of a given system, one may be able to determine whether it is in an entangled state: if a negative expectation value is obtained it *cannot* be in a separable state.

Recall from our discussion of the Peres-Horodečki (PH) criterion in Section 6.5 above, that all positive maps on $\mathcal{H}_2 \otimes \mathcal{H}_2$ and $\mathcal{H}_2 \otimes \mathcal{H}_3$ can be written in the form $L_1 + L_2 \circ T$, where T is the general matrix transposition map; such maps are called *decomposable*. Then the relevant entanglement witness can be written in the form

$$W = R + (\mathbb{I} \otimes T)S \,, \tag{6.72}$$

where R and S are nonnegative Hermitian matrices. For larger Hilbert spaces, there exist nondecomposable positive maps, so that there exist *entangled* states for which the PH criterion *is* satisfied, namely, the bound entangled statistical states with positive partial transpose (PPT) [106].

6.12 Entanglement as a resource

Beyond simple collections of qubits, which serve as a resource for quantum communication tasks such as sending copies of pure states from transmitter to receiver for QKD, collections of shared entangled qubits allow one to perform a number of quantum information processing tasks and to implement uniquely quantum mechanical forms of communication, such as quantum dense coding and quantum teleportation.[24] Transmissible qubits constitute a *directed* resource, whereas entanglement is *undirected*, in that the direction of distribution leading to entanglement being shared is not relevant to its utility.

Entanglement (at least in its bipartite form) can be viewed as a physical resource similar to energy that can take several interchangeable forms and can be transferred between different sorts of quantum system. In order to find exactly how much of the resource of bipartite entanglement they share, two parties can concentrate Bell singlet states between them. In particular, they can distill, by collective LOCC (CLOCC) from a number, n, of copies of an initial bipartite pure (not necessarily *maximally*) entangled state $|\Phi\rangle_{AB}$, the greatest number $k < n$ of singlet states possible:

$$|\Phi\rangle_{AB}^{\otimes n} \rightarrow |\Psi^-\rangle_{AB}^{\otimes k} \,. \tag{6.73}$$

---

[23] Methods for constructing such operators have been given, for example, in [417].
[24] These tasks are discussed in Chapter 9.

Distillation can be carried out with an efficiency given by the von Neumann entropy $S(\rho)$, where ρ is the reduced statistical operator of a subsystem of AB [39]. This is a reversible process, in the sense that there is an asymptotic scheme in which the inverse conversion

$$|\Psi^-\rangle_{AB}^{\otimes k} \to |\Phi\rangle_{AB}^{\otimes n} \tag{6.74}$$

can be performed, again via CLOCC, with *equal* efficiency. The monotonicity condition (iii) of Section 6.7 implies that no entanglement distillation scheme can perform *better* than this asymptotic scheme.[25] Entanglement, like heat energy, cannot be increased by local operations on remote subsystems. These reversible transformations, consisting of only local operations that transform one entangled state into another, produce the analogue of the Carnot cycle. For *pure* states of a pair of qubits, both $D(\rho_{AB})$ and $E_f(\rho_{AB})$ are equal to the entropy $S(\rho_A)$ of the reduced statistical operator $\rho_A = \mathrm{tr}_B(\rho_{AB})$, that is,

$$E_f\big(P(|\Psi\rangle_{AB})\big) = D\big(P(|\Psi\rangle_{AB})\big) = S(\rho_A) \,, \tag{6.75}$$

where $|\Psi\rangle_{AB}$ is the pure state in question. This highly suggestive analogy has stimulated an investigation into the depth of the similarities between quantum information theory and thermodynamics.

6.13 The thermodynamic analogy

The analogy between entanglement theory and thermodynamics was first drawn by Sandu Popescu and Daniel Rohrlich, who pointed out that any process using collective local operations and classical communication (CLOCC) that preserves entanglement must be reversible [339]. In particular, they have provided an argument analogous to a traditional thermodynamic argument for the ideal efficiency of the Carnot cycle; condition (iii) of Section 6.7 can be viewed as the information-theoretic analogue of the second law of thermodynamics, because it imposes the condition that an increase of entanglement between systems in distinct laboratories cannot occur as a result of collective local operations on the systems and classical communication between laboratories. This thermodynamic analogy is further enabled by the introduction of a specific unit of entanglement via condition (iv) of Section 6.7, which has the

---

[25] The limit $n \to \infty$ is associated with the use of a standard unit of entanglement to describe the transformation process; taking this limit provides one with a well-defined ratio characterizing the conversion process of a whole number of states to a whole number of states, because the entanglement of formation may take any rational value. In the manipulation of n entangled pairs of particles in state $|\Psi\rangle_{AB}$, the optimal probability of obtaining k singlets tends to 1 when $k < D(P(|\psi\rangle_{AB}))$, in the infinite n limit; it is not possible to achieve the desired conversion for finite n. Potential problems arise from the use of this standard unit, however, which are discussed in Sect. 6.14, below.

consequence of establishing quantum entropy as *the* standard bipartite entanglement measure; the problem of finding a measure of entanglement of k pure states is thereby reduced to the problem of defining a measure of entanglement for n *singlet states* $|\Psi^-\rangle$.

A derivation from first principles of condition (iv) itself runs as follows [339]. The allowed local transformations are reversible only when the number of copies of a system becomes arbitrarily large. However, there is no way to define total entanglement for n infinite, as it would then clearly take an infinite, that is, *unphysical* value. It is therefore necessary to define entanglement intensively: the measure of entanglement for n singlets must be *proportional* to n. The entanglement of a collection of k systems in an arbitrary pure state, $|\Psi\rangle_{AB}$, then approaches that of n systems each in the singlet state $|\Psi^-\rangle$,

$$E(|\Psi\rangle_{AB}) = \lim_{n,k\to\infty} \left(\frac{n}{k}\right) E(|\Psi^-\rangle_{AB}) , \qquad (6.76)$$

providing the entropy of entanglement of the state [39]. Any such measure of pure-state entanglement is thus determined up to a constant factor. The constant term describing the entanglement of the singlet, so far left undetermined, is then taken to be *unity*, by convention.[26] This result provides constraints on entanglement manipulation that can be seen already to be similar to those involving heat in thermodynamics. It has been taken by some to be the starting point in the development of a full theory of *"entanglement thermodynamics"* analogous to traditional thermodynamics. However, it is not clear that any depth is added by pursuing a full-blown version of this analogy; because the entanglement measure chosen *is* an entropy, it perhaps would be more surprising if some sort of analogy could *not* be developed.

Under this analogy, entanglement plays the sort of role that heat energy does in traditional thermodynamics; the distillation of pure entangled states plays the role of *extracting* work from heat. Recall that the bound entanglement is given by

$$B(\rho) = E_f(\rho) - D(\rho) . \qquad (6.77)$$

This expression appears to be analogous to the Gibbs–Helmholtz equation of thermodynamics,

$$TS = U - A , \qquad (6.78)$$

where U is the internal energy and A the free energy, if TS is viewed as a "bound energy."

---

[26] However, there are objections to having to make such a choice, which are discussed in the following section.

When contemplating the analogy between entanglement theory and thermodynamics, it is helpful to consider the following simple formulation of the three basic laws of thermodynamics.

(1) Heat is a form of energy.
(2) It is impossible for any cyclic process to occur the sole effect of which is the extraction of heat from a reservoir and the performance of an equivalent amount of work.
(3) The entropy of a system approaches a constant value as the temperature approaches zero.

These three laws allow for the reversible transformation of work into heat and vice versa. One can then formulate the assumptions of thermodynamics as follows, "There is a form of energy (heat) that cannot be used to do work, that nonetheless can be used to store work though work can be stored in heat only if there is some heat to begin with, in which case work can be stored reversibly." [225].

Recall that, according to condition (iii) for well-defined entanglement measures given in Section 6.7, any good measure of entanglement, E, must satisfy

$$\rho \to \{\rho_i, p_i\} \Rightarrow E(\rho) \geq \sum_i p_i E(\rho_i) \qquad (6.79)$$

under LOCC. Consider then two parties, Alice and Bob, who initially share a collection of pairs of subsystems described by an n-fold product of the bipartite system state ρ, $\rho^{\otimes n} = \rho \otimes \cdots \otimes \rho$, where n is taken to be large, collectively and locally operate on the members of each shared pair, communicate using a classical information channel if they desire, and/or arrange their local subsystems into subensembles ρ_i represented with probabilities p_i. The above condition indicates that the average entanglement remaining at the end of such a CLOCC transformation cannot exceed the initial shared entanglement. The combination of the entangled-state distribution process and the entanglement distillation process that accumulates pure entangled states from mixed states and is seen to be analogous to the process of *cycling an engine* that obtains work from heat; bound entangled states, being those entangled states from which *no* pure entanglement can be distilled, are analogous to thermodynamic systems from which no work can be drawn and are seen as containing "fully disordered entanglement."

Accordingly, the following "laws of entanglement thermodynamics" have been suggested by analogy to the above traditional thermodynamic laws [225].

(1) The entanglement of formation is conserved.
(2) The disorder of entanglement can only increase.
(3) One cannot distill singlet states with perfect fidelity.

The "law" (1) corresponds to condition (ii) of Section 6.7. There is an analogy to reversible work extraction here, although in general one needs more entanglement (in the form of singlets) to create a state than can be drawn from it. In traditional thermodynamics, the second law dictates that any thermodynamical system has more energy than can be extracted from it, except when one of the reservoirs is at zero temperature. The same holds in this "thermodynamics of entanglement" where for a general mixed state ρ, $D(\rho) < E_f(\rho)$.

However, attempts to continue further with this treatment of entanglement in order to complete the analogy run into difficulties. The completion of such an analogy requires the completion of the correspondence between fundamental quantities in the two theories. Given that the role of entropy is played by $S(\rho)$, it is by no means clear *what* quantity is to play the role of temperature, T; one is required to find a well-defined "temperature of entanglement," $\bar{T}(\rho)$, for mixed states (when $S(\rho) > 0$) of the form

$$\bar{T}(\rho) = B(\rho)/S(\rho), \tag{6.80}$$

(*cf.* Eqs. 6.77–78) if the "entropy of entanglement" is to be taken to be $S(\rho)$, as is suggested by the fact that this results, for pure states, in the equality of the entanglement of formation and entanglement of distillation (*cf.* Eq. 6.76 and [111]). The temperature analogue is conspicuously absent from the above statement of "laws of entanglement thermodynamics." The lack of a well-defined such quantity brings this approach strongly into question because, for example, the third law of thermodynamics is expressed in terms of the behavior of entropy with respect to temperature.

6.14 Information and the foundations of physics

The superposition principle is a fundamental principle of quantum mechanics. In multipartite systems, this principle provides the entangled states, in which the most unusual quantum phenomena arise through extraordinary nonlocal correlations of physical properties. Physicists and philosophers have long suggested that by studying entanglement one might develop a deeper insight into the reality described by quantum mechanics. About this there can be little doubt, as witnessed by the history of results discussed in Chapters 2 and 3. More recently, as shown in this chapter, the quantitative study of entanglement by quantum information science has provided helpful and suggestive relationships between information in the possession of agents having the ability to perform local actions on quantum systems and to communicate with each other and thermodynamics. As just shown, some of these relationships suggest an analogy between entanglement and heat under which formal correspondences can be made between some thermodynamical quantities and entanglement measures, under certain specific conditions, through a particular novel use of quantum entropy functions.[27] The analogue of the thermodynamic limit is certainly important in the quantum information context, because the limit of an infinite collection of copies of a quantum states must be considered when quantifying the entanglements of formation and distillation, as we have seen. The relationship between quantum mechanics and information theory has led some to believe that information theory plays a special role in fully exposing the deepest aspects of physical reality. Some investigators have even suggested that information is more fundamental than *matter*, along the lines of John Wheeler's "it from bit" idea [452], that matter is reducible to information [471] or vice versa [269, 271].[28]

However, there are currently significant limitations to the information-resource theory of entanglement itself. In addition to the difficulty of completing "entanglement thermodynamics," the argument for the uniqueness of the quantum entanglement measure based on a *mutatis mutandis* argument may be seen to induce an unwarranted dependence on the choice of unit—the introduction of the Bell singlet state as providing an "e-bit" of entanglement—manifest in the *ratio problem* [314]: ratios of entanglement measures, such as the entanglement of formation or distillable entanglement of two different states, may depend on the *particular state* chosen as the basic unit of entanglement when the degree of entanglement is referenced to it. By contrast, the thermodynamic entropy *does have* a unique measure, as shown in the axiomatic approach of Giles [183]. Furthermore, the investigation of entanglement for multipartite systems reveals the existence of *different sorts of entanglement* not quantifiable in terms of a fundamental e-bit unit, as discuss in the next chapter. Moreover, it has been shown that no unique measure of entanglement exists in the case of *mixed* states [306]. These represent significant impediments to the reduction of quantum entanglement to information. Thus, although within the context of quantum information processing it clearly *is* possible to treat entanglement as an information-processing resource, it is by no means obvious that this approach is ultimately the *best* way of understanding entanglement itself in the broader physical context.

Nonetheless, given the benefits of viewing entanglement as a quantum resource, one may be under the impression that quantifying entanglement via entropy measures, involving condition (iv) and explicated in Sections 6.7 and 6.12, is the *only* good method of quantifying entanglement. However, another, related framework for quantifying entanglement has made significant progress where the information-theoretic approach has run into difficulties, namely, in the case of multipartite states. This second approach, outlined in the following

---

[27] See also [433], where it was shown that Giles's theory can be seen as encompassing both quantum and classical information-processing models due to similarities in mathematical structure.

[28] For a discussion of some of these ideas, see [421].

section, is more clearly rooted in traditional physical methodology in that it is based more on geometry and symmetry than on information theory.

6.15 The geometry of entanglement

Entanglement can be investigated from the geometric point of view, because the properties of quantum states and the classification of composite quantum states according to these properties relate to geometry as well as information. For example, measuring the degree of entanglement as the *distance* of the state from the nearest factorable state, that is,

$$E_G(|\Psi\rangle) = \frac{1}{2} \min || \, |\Psi\rangle - |\Xi\rangle \, ||^2, \tag{6.81}$$

where $|\Xi\rangle$ is a (normalized) product state in Hilbert space and the minimum is taken over the set of such normalized product states (*cf.* Eq. 6.23), provides the distance of the closest separable approximation proposed by Shimony [383]. The Hilbert-space angle $\phi \equiv \cos^{-1}\left(|\langle\Psi|\Xi\rangle|\right)$ is the natural distance between two state-vectors, and takes the state overlap to a distance function derivable from the Fubini–Study metric, which is a Riemannian metric on projective Hilbert-space [458]; see Section A.4. The above measure of entanglement is a very natural one due to its generality and direct relationship to the original definition of entanglement as nonfactorability unlike, say, the entanglement as measured by the von Neumann entropy of subsystem reduced states.

Any monotonically increasing function of $E_G(|\Psi\rangle)$ gives the same ordering of normalized vectors $|\Psi\rangle$, and serves as an equally acceptable such measure. In the case of pure states of two-qubit systems, at least six such measures can be found from conceptually distinct starting points that are monotonic functions of $E_G(|\Psi\rangle)$ [383]. One can find the nearest separable state to a given state by solving the corresponding nonlinear eigenvalue problem [448]. This quantity is defined independently of explicit locality considerations, something of value in light of the limitations of locality conditions discovered in the context of the use of Bell inequality violation for this purpose. Further geometrical treatments are discussed in the following chapter, where entanglement in larger multipartite states is explored.

The results of the geometric approach to quantum entanglement may improve our understanding of entanglement in quantum information processing and may also provide insight into the incompletely understood relationship between entanglement and quantum speedup, a question of fundamental importance to quantum information processing. Let us now consider how quantum entanglement is created in practice, before considering multipartite systems.

6.16 Creating entangled photons

Since Alain Aspect's famous tests of the Bell inequalities confirming the presence of nonlocal quantum phenomena, which used a two-photon source based on a double atomic-cascade transition, optical tests of quantum information processing principles have used increasingly efficient sources of entanglement based on spontaneous parametric down-conversion (SPDC) in second-order $\chi^{(2)}$ nonlinear crystals [16].[29] Bell-inequality violations and related phenomena have most commonly been demonstrated through the use of highly correlated pairs of photons generated by SPDC. Parametric down-conversion is an optical "three-wave mixing" process, in which an input light field "pump," with a frequency centered about a given value ω_P, induces oscillations in electrons within a dielectric medium, traditionally chosen to be a noncentrosymmetric nonlinear bulk crystal, which in turn radiates light at two lower frequencies; see Fig. 6.2. Because the electrons in the medium do not undergo state transitions during this process, it is referred to as a *parametric* process.

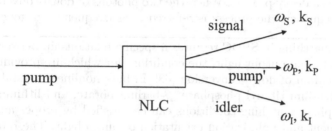

Fig. 6.2. Spontaneous parametric down-conversion (SPDC). Photons downconverted in a nonlinear crystal, NLC, from a pump laser beam, coming in at the left, emerge pairwise, to the right, each in one of two cones, one as "signal" and one as "idler," on opposite sides of the pump-beam direction with frequencies obeying phasematching conditions corresponding to the conservation of momentum.

An SPDC process is either of "type I" or of "type II," depending on whether the two photons of the down-conversion pair have identical or orthogonal polarizations, respectively. The two photons may leave the nonlinear medium either in the same direction as or in different directions from that of the pump beam, that is, collinearly and noncollinearly. Down-conversion photons are often accordingly described as collinear or noncollinear pairs, as well as being considered to be of type I or type II. In SPDC, conserved quantities of the electromagnetic field are preserved in the resulting photon pair due to the constraints of their respective conservation laws, as the electrons of the

---

[29] $\chi^{(2)}$ is the susceptibility tensor; see [206] for a general discussion of $\chi^{(2)}$ processes.

medium do not ultimately exchange energy or momentum with the fields.[30] In particular, at the quantum level, such down-conversion is of one photon into a *pair* of "daughter" photons, occurring spontaneously with a small probability that is proportional to the input light intensity due to stimulated emission induced by fluctuations of the vacuum field.[31]

The phases of the corresponding wave-functions must satisfy the relations

$$\omega_1 + \omega_2 = \omega_P, \qquad \mathbf{k}_1 + \mathbf{k}_2 = \mathbf{k}_P, \qquad (6.82)$$

the latter being known as *phasematching conditions*, on energies and momenta, respectively, where the $\mathbf{k}_i$ and ω_i are photon momenta and angular frequencies, respectively, and subscript P refers to pump photons. The down-conversion photons (here labeled $i = 1, 2$) are called the *signal* and the *idler* photon. In this process, energy and momentum are conserved *within* the down-conversion medium, where differences in indices of refraction allow for their conservation; as the down-converted photons leave the medium of down-conversion, they are refracted, transferring momentum to the medium. With strong pumping and phasematching conditions satisfied, down-conversion can be viewed as a decay process. When the two photons of down-conversion have the same frequency, the process is referred to as (frequency) *degenerate*.

> Phasematching in SPDC requires a specific relationship between input light and output light, the conditions for which are provided by the medium of down-conversion [63]. In bulk nonlinear crystal such as potassium titanyl phosphate, β-barium borate, and lithium niobate, phase matching conditions can be satisfied by proper angular alignment and polarization orientation of pump light. The range of wavelengths over which phasematching can be achieved can be limited, because the phase relationship between beams changes as light travels through nonlinear crystal but this can be counteracted by the natural birefringence of the crystal. Periodic inversion, or "*periodic poling*," of the $\chi^{(2)}$ nonlinearity in a crystal allows the power of down-conversion to continually increase, whereas it would otherwise lead to a cyclic growth and decay of down-conversion intensity over the interval of one coherence length. Such periodic poling provides "quasi-phasematching" and allows for significantly improved SPDC output [13, 167, 173]. Improvements of photon-pair generation efficiency of several orders of magnitude over traditional bulk crystal sources have been achieved by this method. For example, quasi-phase-matching in the ferroelectric crystal lithium niobate allows a nonlinear susceptibility tensor coefficient many times larger than the largest one that can be used for birefringent phasematching.

---

[30] Electrons can, however, be viewed as entering "virtual energy levels" within the medium during the down-conversion process.

[31] This is due to the very weak coupling among the three optical modes, even in media with very strong nonlinear susceptibilities.

In either type of SPDC, when the photons of a down-conversion pair are allowed different individual momenta or energies by the phasematching constraints, entanglement arises when the pair is prepared in such a way that its joint down-conversion possibilities are in principle indistinguishable; although the values of energy and momentum for each individual photon are indefinite, they are perfectly correlated. These photon pairs are created within a time window defined by the laser pump-field coherence time, τ. The ability to engineer such possibilities makes SPDC photon pairs especially useful in a range of practical applications [236]. In particular, singlet states produced in this way are a fundamental source of entanglement. The evolution of the three optical modes involved in SPDC can be described as

$$|\Psi(t)\rangle = \exp\left(-\frac{i}{\hbar}\int_0^t \hat{H}_I(t')dt'\right)|\Psi(0)\rangle \qquad (6.83)$$

in the interaction picture (about which see, for example, [359]). The interaction Hamiltonian is $H_I = \hbar g \hat{a}_1^\dagger \hat{a}_2^\dagger \hat{a}_p + \text{h.c.}$, where g is the interaction strength and $\hat{a}^\dagger(\hat{a})$ indicates a creation (annihilation) operator of an electromagnetic field mode; $|\Psi(0)\rangle$ is the initial state of the three-mode composite system and "h.c." denotes the Hermitian conjugate of the first term of the sum.[32]

Type-I down-conversion is often the preferred process for creating entangled photon pairs, because higher intensity down-conversion beams can generally be achieved under the corresponding phasematching conditions than in the type-II case [265]. The ideal photon pair as produced by type-I SPDC emerges in the state

$$|\Psi\rangle = |\text{vac}\rangle + \eta \int d\mathbf{k}_1 d\omega_1 \int d\mathbf{k}_2 d\omega_2$$
$$\Phi(\mathbf{k}_1, \mathbf{k}_2; \omega_1, \omega_2) \ \hat{a}_1^\dagger(\mathbf{k}_1, \omega_1)\hat{a}_2^\dagger(\mathbf{k}_2, \omega_2)|0\rangle|0\rangle \ , \qquad (6.84)$$

where $|\text{vac}\rangle$ is the vacuum state and $\Phi(\mathbf{k}_1, \mathbf{k}_2; \omega_1, \omega_2)$ describes the energy-momentum distribution of the resulting fields. The proportionality parameter, η, between the vacuum state (resulting when no down-conversion takes place) and the photon pair is the photon-pair creation efficiency; η is proportional to the interaction strength g and $\sqrt{\mu}$, and has a squared value much less than unity.

In this case, the two photons leave the nonlinear medium in the *same* polarization state, which is orthogonal to that of pump photons, and the two

---

[32] Note that the (optically) nonlinear interaction described here gives rise to a unitary transformation of the *composite system* of pump and down-conversion light beams. In the case of a planar continuous wave (c.w.) pump field, the initial state of the system is simply $|\Psi(0)\rangle = |0\rangle_1|0\rangle_2|\sqrt{\mu}\rangle_P$, where μ is the mean-photon number of the coherent state, $|\sqrt{\mu}\rangle_P$, describing the pump field that is an important parameter for producing effective single-pair production, and thus conditional single-photon states for quantum key distribution by "heralding" the production of one photon by the detection of the other.

down-conversion photons may be entangled in direction but not polarization, as in the state

$$|\Psi\rangle = \sum_{1,2} \delta(\omega_1 + \omega_2 - \omega_P)\delta(\mathbf{k}_1 + \mathbf{k}_2 - \mathbf{k}_P)|k_1\rangle|k_2\rangle , \qquad (6.85)$$

where $1, 2$, and P refer to signal, idler, and pump, respectively; the vacuum contribution has been left out. The down-conversion photons emerge within two broad cones, one corresponding to each photon, where in any given pair one photon appears within its cone on the opposite side of the direction defined by the pump beam direction from that of its "twin sister" photon. These two photons also generally differ in energy (color) in accordance with the phasematching conditions. Pairs of beams can then be spatially filtered to provide two spatial qubits, one for each beam, as in the double Mach–Zehnder arrangement for two-particle interferometry discussed in Section 3.6.

As an example of type-II SPDC, consider a situation involving a pulsed (noncontinuous wave) laser pump beam where, for simplicity, output beams are collinear; in such a case, the two-photon state can be written

$$|\Psi\rangle = \int d\omega_o \int d\omega_e\ \Phi(\omega_o,\omega_e)\ \hat{a}_o^\dagger(\omega_o)\hat{a}_e^\dagger(\omega_e)|0\rangle|0\rangle , \qquad (6.86)$$

(again neglecting the vacuum term) where o (ordinary) and e (extraordinary) indicate the orthogonal polarization states that define this type of SPDC. By taking the appropriate Schmidt decomposition

$$\Phi(\omega_o,\omega_e) = \sum_j a_j \xi_j(\omega_o)\chi_j(\omega_e) \qquad (6.87)$$

of this state, one can study the frequency entanglement of the output photon pair, where the amplitudes and eigenstates obey the integral equations

$$\int d\omega' K_o(\omega,\omega')\ \xi_j(\omega') = a_j^2 \xi(\omega) , \qquad (6.88)$$

$$\int d\omega' K_e(\omega,\omega')\ \chi_j(\omega') = a_j^2 \chi(\omega) , \qquad (6.89)$$

with kernels

$$K_o(\omega,\omega') = \int d\omega_e \Phi(\omega,\omega_e)\Phi^*(\omega',\omega_e) , \qquad (6.90)$$

$$K_e(\omega,\omega') = \int d\omega_o \Phi(\omega_o,\omega)\Phi^*(\omega_o,\omega') , \qquad (6.91)$$

providing single-photon spectral correlation functions.

Whenever more than one of the (without loss of generality, real) coefficients a_j are nonzero, the state is frequency-entangled. Numerical study of this state has shown an overwhelming majority of the state-vector components to lie in a Hilbert subspace of small dimension, so that it is capable of providing a physical qu-d-it for quantum information-processing applications. To study entanglement in this case, one can examine an *effective* Schmidt number by counting the number of nonnegligible contributions, a_i, and using the entropylike quantity

$$\bar{S} = \sum_{j=1}^{n} a_j^2 \log_2 a_j^2 , \tag{6.92}$$

which converges to the von Neumann entropy in the limit of infinite n [273].

Entangled multipartite systems

One unsatisfied desideratum for entanglement measures is that of full generality. There is no known single good entanglement measure applicable to all mixed states of systems with arbitrary numbers of subsystems. At present, the bipartite case is the only one in which definitive results may be said to have been obtained, by reference to the number of Bell states asymptotically interconvertible by local operations and classical communication to other states. The von Neumann entropy used in the previous chapter is a reliable measure only of bipartite entanglement. The *partial entropies*, defined as the number of Bell-state pairs convertible to subsystem states, can be *unequal* for distinct portions of a multipartite quantum system of more than two components. Because partial entropies are conserved by asymptotically reversible local operations and classical communication (LOCC) involved in the pertinent interconversions, they can therefore no longer be viewed as *absolute* entanglement measures beyond the bipartite case, in which there is only one way of partitioning the composite system [48]. This prevents the straightforward extension of the standard entanglement measure, the entanglement of formation, to the general multipartite case, as would be natural given its utility in characterizing bipartite entanglement. Schmidt number, a coarse measure, has been generalized to n-parties and then applied independently to various entanglement classes but, although it satisfies most of the conditions on entanglement monotones, it fails to satisfy condition (v) of Section 6.7 [148, 149].

In the case of composite systems distributed among several parties, some of the requirements on good entanglement measures are different from those for the special case of bipartite systems. Recall from the discussion in Section 2.6 that local operations include local unitary transformations (LUTs), the addition of ancillary particles and/or degrees of freedom, local measurements (including POVMs), and the discarding of parts of the system, as performed by any agent on the subsystem within its laboratory, and are described by CP maps. For a two-component partition of a composite system into a local subsystem in lab A, and the remainder of the total system in lab B, a general local operation (LO *without* communication) can be described by a superoperator transforming the total system statistical operator, namely,

$$\rho \rightarrow \mathcal{E}(\rho) = \sum_{i=1}^{m} \sum_{j=1}^{n} (A_i \otimes B_j)\rho(A_i \otimes B_j)^\dagger , \qquad (7.1)$$

where A_i and B_j are operators acting separately in the Hilbert spaces of A and B such that $\sum A_i^\dagger A_i = \mathbb{I}$ and $\sum B_j^\dagger B_j = \mathbb{I}$. When *communication* between laboratories, each possessing one part of the compound system, is allowed in addition to local unitary transformations, the transformations belong to the set of LOCC operations and both one-way and two-way classical communication may take place.

Those situations in which one laboratory, say A, may communicate measurement results to another laboratory, say B, known as *one-local operations*, are describable by superoperators of the form

$$\mathcal{E}(\rho) = \sum_{i,j}^{m,n} \sum_{k=1}^{p} (\mathbb{I} \otimes B_j^{(i)})(A_{kj} \otimes \mathbb{I})\rho(A_{kj} \otimes \mathbb{I})^\dagger(\mathbb{I} \otimes B_j^{(i)})^\dagger , \qquad (7.2)$$

where an operation B on lab B's local subsystem may depend on a measurement outcome k in lab A and $\sum_j A_{kj}^\dagger A_{kj} = \mathbb{I}$ for each value of k. *Two-local* operations are those allowing any sequence of such operations with classical communication from agents in *either* laboratory; the set of LOCC operations is that of all such operations by any pair among all the parties present in a given situation. If such transformations are performed on a number of *identical copies* of a system they are collective local operations and classical communications (CLOCC).[1] The generalization to multi-local transformations is straightforward.

In the case of operations on a *single copy* of a system, when a state transformation under LOCC always succeeds it is called *exact*. If the transformation can only be accomplished with some probability, it is a *stochastic* or *local filtering* (SLOCC) transformation. If a state transformation can only be accomplished in the presence of another state it is called *catalytic*, by analogy to situations in chemistry involving molecular reactions where the presence of the chemical catalyst is required for certain reactions to occur and the catalyst is intact after the reaction; a given such *catalyst state* can be used repeatedly in various transformations.[2] When catalyst states supplement a class of transformations, a "c" is appended to the acronym, as in the class of LOCCc transformations. When a number k of parties each possessing a local subsystem is involved in any of the above, the prefix k- can be prepended to the class designation.

---

[1] The entanglement distillation process shown in Fig. 9.1 is an example of CLOCC.
[2] For a discussion of catalytic processes, which will not be discussed further here, see [48].

In the special case of *tripartite* systems, entanglement for pure states of the available bipartite partitions can still be indirectly quantified using bipartite measures such as concurrence, by finding the residual tangle, τ_{ABC}; see Section 7.7 below. By considering the inter-convertability of three-qubit states under SLOCC transformations, one is able to isolate two equivalence classes of three-qubit pure states. However, for larger systems one has no choice but to seek new entanglement measures. The first three conditions and the fifth condition on entanglement measures E_X of Section 6.7 can be extended by replacing ρ_{AB} by $\rho_{AB...}$ and by removing the reference to singlet states in condition (iv) to arrive at an alternative form of the condition (iv'), specifically,

(iv') E_X should be additive for tensor products of independent states shared by the same parties.[3]

The resulting five conditions on good entanglement measures must then also be supplemented by a *sixth* condition, specifically,

(vi) E_X should be stable under the transferral of subsystems between parties.

> At the four-qubit level, a necessary condition for the existence of a reversible protocol for converting n copies of a four-qubit GHZ state,
>
> $$|GHZ_4\rangle = \frac{1}{\sqrt{2}}(|0000\rangle + |1111\rangle) , \qquad (7.3)$$
>
> into singlet states corresponding to the quantum entropy unit, is that the entropies of the initial and final states after transformation be the same. However, *no* combination of singlets shared among four parties has the same ratios of entropies as does the state $|GHZ_4\rangle$. Thus, it is impossible for the four-party GHZ state to be reversibly transformed into singlet states. This demonstrates the existence of a genuinely *new kind* of entanglement beyond the bipartite kind with the conditions imposed on entanglement measures above [48]. Indeed, even for the three-qubit GHZ state, which *is* entangled, any two reduced states shared by two of three parties have *no* bipartite entanglement. What is known is that any entangled state can be nonreversibly obtained from a sufficiently large number of copies of the singlet state.

In this chapter a number of results and relations are presented, many being essentially formal in nature, that can assist in navigating the states of multiple-qubit systems. Of the various sections that follow, only Section 7.8 is

---

[3] Condition (iv') is the *additivity conjecture* for multipartite systems.

necessary for a basic understanding of quantum information processing. The reader may wish to proceed immediately to that section on a first reading.

7.1 Stokes and correlation tensors

Symmetry has long played an important role in the investigation of quantum mechanics, as in all parts of physics. Not surprisingly, a geometrical approach also allows for progress in the classification of entangled states. The behavior of quantum states under local unitary transformations (LUTs) and stochastic local quantum operations and classical communication (SLOCC), both of which have associated symmetry properties, has helped illuminate the nature of entanglement in multipartite quantum systems; from the geometrical point of view, the associated group-invariant lengths are Euclidean and Minkowskian in character, as we show in Section 7.4, below. Recall that for two-qubit systems, in addition to being quantified by the von Neumann entropy of one of its single-qubit subsystems, entanglement can also be characterized in terms of symmetry-based quantities, concurrence and tangle, through the introduction of the two-qubit spin-flip operation, $\tilde{\rho} = \sigma_2^{\otimes 2} \rho^* \sigma_2^{\otimes 2}$, where ρ^* is the complex conjugate of the two-qubit statistical operator ρ and σ_2 is the Pauli operator performing universal single-qubit state-vector inversion [460]. As we show in Section 7.2 below, related geometrical quantities can be defined for the study of multiple-qubit states and have been used in attempts to provide absolute multipartite-entanglement measures [28, 240, 383, 447].[4]

A helpful method for grounding the study of multipartite entangled states that allows for their geometrical characterization is to connect them with straightforwardly measurable quantities. Any practical measure of multipartite entanglement must bear a clear relation to such measurable quantities, such as n-fold coincidence counting rates in multi-photon interferometry [236]. Vectors of correlations in multipartite systems can be constructed that capture their collective nature.[5] In particular, the Stokes parameters correspond to empirical counting rates and can be generalized to aid in this task. In particular, the N-qubit Stokes parameters generalizing the traditional (single-qubit) Stokes parameters can be used, which are given by

$$S_{i_1 \ldots i_N} = \text{tr}(\rho \, \sigma_{i_1} \otimes \cdots \otimes \sigma_{i_N}) \,, \qquad i_1, \ldots, i_N = 0, 1, 2, 3 \,, \qquad (7.4)$$

where $\sigma_\mu^2 = \mathbb{I}$, $\mu = 0, 1, 2, 3$, are the three Pauli matrices together with the identity $\sigma_0 = \mathbb{I}_2$, and $\frac{1}{2}\text{tr}(\sigma_\mu \sigma_\nu) = \delta_{\mu\nu}$ [240]. These generalized Stokes parameters are simply the expectation values of Pauli group elements.[6] These

---

[4] We show, however, that these measures have their own limitations.

[5] Quantum state tomography, for example, involves the measurement of these parameters for reconstruction of the quantum state ρ ; see Section 8.1.

[6] The elements of the Pauli group also play an important role in the description of quantum bit errors and their correction. Quantum error correction and the Pauli

directly observable parameters form an N-particle generalized Stokes tensor, $[S_{i_1...i_N}]$, the invariants of which are useful in characterizing states of multiple-qubit systems, just as in the case of bipartite systems discussed in Section 6.4. These quantities can be generalized so as to apply to qu-d-it systems [9].

One can then represent the statistical operator using the operators $\{\hat{\lambda}_i\}$ generating $SU(N)$ rotations, of which the Pauli matrices above are those for $N = 2$. In particular, the expectation values, $\lambda_i = \text{tr}(\rho\hat{\lambda}_i)$, of these generators form the $(N^2 - 1)$-dimensional *coherence vector*, $\boldsymbol{\lambda}$, in terms of which the statistical operator is

$$\rho = \frac{1}{N}\mathbb{I} + \frac{1}{2}\sum_{j=1}^{N^2-1} \lambda_j\hat{\lambda}_j , \qquad (7.5)$$

which is the *generalized Bloch-vector* form of the state.[7] Consider the standard-basis eigenvectors $\{|j\rangle\}$ for the Hilbert space of the system in question and the *transition operators*, $\hat{P}_{jk} \equiv |j\rangle\langle k|$, the $\hat{P}_{jj} = P(|j\rangle)$ being corresponding projectors. One then also has the vector of $N^2 - 1$ operators

$$\hat{\boldsymbol{\lambda}} = \{\hat{u}_{12}, \hat{u}_{23}, \ldots, \hat{v}_{12}, \hat{v}_{13}, \hat{v}_{23}, \ldots, \hat{w}_1, \hat{w}_2, \ldots, \hat{w}_{N-1}\} , \qquad (7.6)$$

where

$$\hat{u}_{jk} = \hat{P}_{jk} + \hat{P}_{kj} , \qquad (7.7)$$

$$\hat{v}_{jk} = i(\hat{P}_{jk} - \hat{P}_{kj}) , \qquad (7.8)$$

$$\hat{w}_l = -\sqrt{\frac{2}{l(l+1)}}(\hat{P}_{11} + \cdots + \hat{P}_{ll} - l\hat{P}_{l+1,l+1}) , \qquad (7.9)$$

with $1 \leq j < k \leq N$, $1 \leq l \leq N - 1$ (*cf.* Eqs. 1.19-21 for the case of $SU(2)$) [215]. In the case of the density matrices representing the statistical operator, one then has matrix elements $\rho_{jk} = \text{tr}(\rho\hat{P}_{jk})$ and corresponding expectation values for $\hat{u}_{jk}$, $\hat{v}_{jk}$, and $\hat{w}_l$ in terms of density matrix elements.[8]

It is also useful to construct *trace relations* involving the quantities

$$C(n, q) = \text{tr}(\rho^q) , \qquad (7.10)$$

where

$$C(n, 2) = \sum_{ij}^{n} \rho_{ij}\rho_{ji} \qquad (7.11)$$

$$C(n, 3) = \sum_{i,j,k}^{n} \rho_{ij}\rho_{jk}\rho_{ki} , \qquad (7.12)$$

---

group are discussed in Chapter 10, below. See also the discussion of the Pauli matrices in Sect. 1.3, which focuses on single-qubit Stokes parameters.

[7] In this section, the notation "^" is in some cases used to designate operators, to aid in distinguishing operators from scalars.

[8] These operators have been treated in detail [276] and applied to a range of situations, as discussed in [293].

and so on, which are invariant under unitarities and are related to state purity and entanglement in Section 7.6, below [215]. The $C(n, q)$ are particularly useful for quantum state spectroscopy; see also Section 8.3.

One can construct correlation tensors in the qu-d-it case, as follows. For bipartite decompositions, one has the tensor components

$$K_{ij} = \langle \hat{\lambda}_i^A \hat{\lambda}_j^B \rangle \, , \tag{7.13}$$

and correlations

$$M_{ij} = K_{ij} - \lambda_i^A \lambda_j^B \, , \tag{7.14}$$

which, again, are zero for product states; for entangled states some of the M_{ij} are nonzero (*cf.* Eq. 3.26–27), where A and B indicate subsystems. Cluster operators, useful for the study of entangled states of quantum networks, can also be constructed by taking the obvious tensor products of generators $\hat{\lambda}_i$, as in the case of the Pauli group, which are again expressible in terms of transition operators. Similarly, useful cluster sums can be formed by summing over products of expectation values λ_i, K_{ij}, and M_{ij}, and so on [293].

7.2 N-tangle

The bipartite tangle measure of entanglement, τ, which is equivalent to the square of the concurrence introduced in Section 6.10, can be generalized so as to apply to *any even number* of qubits. In particular, taking

$$|\tilde{\Psi}\rangle \equiv \sigma_2^{\otimes N} |\Psi^*\rangle \, , \tag{7.15}$$

where $|\Psi\rangle$ is a multiple-qubit state, one can define an N-tangle measure, τ_N, generalizing τ so as to apply to N-qubit states for those cases in which N is even [457].[9] In particular, one can take

$$\tau_N = |\langle \Psi | \tilde{\Psi} \rangle|^2 \, , \tag{7.16}$$

which is, therefore, a symmetry-based measure of entanglement as τ itself is. As an example, note that the four-qubit $|GHZ_4\rangle$ state can be seen by inspection to be unaffected by the global spin-flip operation $\sigma_2^{\otimes 4}$ and hence to have $\tau_4 = 1$. The N-tangle is the Lorentz-group invariant length for N-qubit states, and is further explored in Section 7.4, below [414]. For two-qubit pure states, the Lorentz-group invariant coincides with the tangle [240]. Relationships among entanglement, mixedness, and spin symmetry in multiple-qubit quantum states are found by exploiting these symmetry properties, as is shown in Section 7.4 below [10, 237].

---

[9] Note also that the σ_2 matrix performs the universal single-qubit inversion operation on pure states.

7.3 Generalized Schmidt decomposition

Multipartite extensions of the Schmidt decomposition of pure states introduced in Section 6.2 can also be found in special situations. For three parties, each in possession of a system described by a d-dimensional space, the combined pure state lies within in a d^3-dimensional Hilbert space and depends on $2(d^3 - 1)$ real parameters, whereas the transformations used to unitarily transform this state have only $3(d^2 - 1)$ independent real parameters. Thus, it is often impossible to obtain a Schmidt decomposition for a given pure state. However, the construction of a generalized Schmidt decomposition may proceed in *some* three-qubit systems [327]; the three-party states that are uniquely determined by two-party reduced statistical operators for pairings of component systems are just those admitting a three-particle Schmidt decomposition [285]. Furthermore, an N-partite pure state can be written in generalized Schmidt form if and only if *each* of its $N - 1$ partite reduced states, resulting from tracing out one party, is separable [418].

7.4 Lorentz-group isometries

As noted in Section 7.2 above, concurrence and N-tangle are naturally expressed in terms of spin-flip transformations. Recall from Section 6.10 that the generalized concurrence for bipartite pure states of any dimension is

$$C(|\Psi\rangle_{AB}) = \sqrt{2s_A s_B\left(1 - \mathrm{tr}(\rho_I^2)\right)} \,, \tag{7.17}$$

where ρ_I is the reduced statistical operator of either of the two subsystems, $I = A, B$ and the s's are just scaling factors [358]. This is simply the square root of a multiple of the mixedness $\mathcal{M}(\rho)$, as defined in Section 1.1, of the reduced state. The spin-flip operation generalized to *higher* dimensions is the *universal state inversion*, described by a superoperator $\mathcal{O}_d$ such that

   (i) $\mathcal{O}_d$ is an automorphism on Hermitian operators,
   (ii) $\mathcal{O}_d$ commutes with all unitarities, and
   (iii) The inner product $\left(\langle\Psi|_{AB}\mathcal{O}_{d_A} \otimes \mathcal{O}_{d_B}P(|\Psi\rangle_{AB})\right)|\Psi\rangle_{AB}$ is nonnegative for bipartite pure states $|\Psi\rangle_{AB}$, being zero for separable pure states.

The superoperators satisfying these conditions are the multiples of $(\mathbb{I} - \rho_i)$, where ρ_i is the reduced statistical operator of one subsystem of AB [357].

Complementarity relations involving this and more general invariants have been derived, as now shown [237]. First, note that the multi-local Lorentz-group invariant, that is, the $SL(2, \mathbb{C})^{\times N}$-invariant $S_{(N)}^2$ of the multiple-qubit Stokes tensor, is expressible in terms of the generalization of the spin-flip operation to any number of qubits (see box below), and for two-qubits coincides with the tangle:[10]

$$S_{(2)}^2\big(P(|\varPsi\rangle)\big) = \tau\big(P(|\varPsi\rangle)\big) \equiv C^2\big(P(|\varPsi\rangle)\big) \; . \qquad (7.18)$$

The $SL(2,\mathbb{C})^{\times N}$-invariant length, which can be compactly expressed as

$$S_{(N)}^2(\rho) = \mathrm{tr}(\rho\tilde{\rho}) \; , \qquad (7.19)$$

is expressible in terms of the generalization of the spin-flip operation to any number of qubits that takes

$$\rho \longrightarrow \tilde{\rho} \equiv \sigma_2^{\otimes N} \rho^* \sigma_2^{\otimes N} \; . \qquad (7.20)$$

This quantity is naturally related to the N-tangle, τ_N, a multipartite entanglement measure for even numbers of qubits [457]; for pure states, one has that

$$S_{(N)}^2\big(P(|\varPsi\rangle)\big) = |\langle\varPsi|\tilde{\varPsi}\rangle|^2 = \tau_N \; . \qquad (7.21)$$

τ_N and $|\tilde{\varPsi}\rangle$ are defined in Section 7.2. See also Eq. 6.30 and [414].

By considering the spin-flip symmetry measure,

$$I(\rho,\tilde{\rho}) \equiv 1 - D_{\mathrm{HS}}^2(\rho - \tilde{\rho}) \; , \qquad (7.22)$$

where $D_{\mathrm{HS}}(\rho - \rho')$ is the (renormalized) Hilbert–Schmidt distance

$$D_{\mathrm{HS}}(\rho - \rho') \equiv \sqrt{\frac{1}{2}\mathrm{tr}\big((\rho - \rho')^2\big)} \qquad (7.23)$$

in the space of statistical operators, which measures their distinguishability, the Lorentz-invariant length, the mixedness, and the spin-flip symmetry of multiple-qubit quantum states can be related. First, note that $S_N^2(\rho)$ and state purity are related by the square of the Hilbert–Schmidt distance between the state ρ and its spin-flipped counterpart $\tilde{\rho}$:

$$D_{\mathrm{HS}}^2(\rho - \tilde{\rho}) = \mathcal{P}(\rho) - S_N^2(\rho) \; , \qquad (7.24)$$

where $\mathcal{P}(\rho)$ is the purity of ρ, which is a Euclidean length in the real representation. One sees, then, that the following relation exists between the geometrical quantities of Lorentz-invariant length associated with multipartite-state entanglement, the Hilbert–Schmidt distance of ρ from $\tilde{\rho}$, and the state purity:

$$S_{(N)}^2(\rho) + D_{\mathrm{HS}}^2(\rho - \tilde{\rho}) = \mathcal{P}(\rho) \; . \qquad (7.25)$$

---

[10] Recall that the group $SO(3)$ acting locally on Stokes tensors, which corresponds to that of unitary transformations of statistical operators, is a subgroup of the Lorentz group, $SL(2,\mathbb{C})$; see Sect. 1.3.

Equivalently, one has the following simple general relation between multi-qubit Lorentz-group invariant and mixedness, $\mathcal{M}(\rho) = 1 - \mathcal{P}(\rho)$.

$$S^2_{(N)}(\rho) + \mathcal{M}(\rho) = I(\rho, \tilde{\rho}) \tag{7.26}$$

[237].[11] When the state ρ is spin-flip symmetric, $I(\rho, \tilde{\rho}) = 1$ so that this is a (restricted) complementarity relation. For pure states, for which $S^2_{(N)}$ is an entanglement measure, $\mathcal{P}(P(|\Psi\rangle)) = 1$, so that $\mathcal{M}(P(|\Psi\rangle)) = 0$, and this expression becomes the equivalence relation

$$S^2_{(N)}(|\Psi\rangle) = I(P(|\Psi\rangle), P(|\tilde{\Psi}\rangle)) . \tag{7.27}$$

Thus, for pure states and N even, entanglement so defined coincides with indistinguishability under the multi-local "spin-flip" transformation.

7.5 Entanglement classes

Consider now the question of quantum entangled-state classification for multipartite systems. A classification of entangled states can be obtained via the inherent transformational properties of states. A useful starting point is an ordering based on the accessibility of states from each other by local operations. For bipartite pure states under 2-LOCC, a partial ordering can be given based on majorization that is a *total* ordering. In particular, one state can be transformed into another by 2-LOCC if and only if the former is majorized by the latter (see Section 6.9). For more than two parties, however, it is no longer possible to find such a total ordering for all states. A generic multipartite state, ρ, can be converted into a state ρ' if and only if, for every ϵ, there is an integer m and a sequence of (deterministic) LOCC transformations L_n such that for any integer $n \geq m$

$$||L_n(\rho^{\otimes n}) - \rho'^{\otimes n}|| \leq \epsilon , \tag{7.28}$$

for n copies of the state, where $|| \cdot ||$ is the trace norm [306]. When these transformations are *reversible*, this allows for the identification of equivalence classes represented by a given state, ρ'.

One can also define classes of multipartite quantum states by accessibility through *SLOCC*. Recall that SLOCC transformations are local quantum operations together with classical communication that transform states with some finite *probability* of success, rather than with certainty. Two pure states are of the same class in this sense if the parties involved have a chance of successfully converting one state into another under SLOCC, that is, if $|\Psi'\rangle = M_1 \otimes M_2 \otimes \cdots \otimes M_N |\Psi\rangle$, where $M_i \in SL(d, \mathbb{C})$ is an invertible ILO acting on the d_i-dimensional Hilbert space of subsystem i [48].[12]

---

[11] The similarity of this expression to Eqs 3.28–29 is striking.

[12] Such classes of states pertain to the ability to perform quantum information-processing tasks with a given probability, such as in the KLM proposal for quantum computing (see Sect. 13.8).

Let us, therefore, turn to the problem of classifying multipartite states by finding equivalence classes under state transformations rather than via their utility as resources for various tasks. One can find equivalence classes under local unitary transformations (LUTs) of the statistical operator and equivalently under (local) rotations of the Stokes tensor, as compound states are equivalent in their nonlocal properties if they can be transformed into each other by such operations. Each group, $\mathcal{G}$, of transformations acts transitively on an orbit $\mathcal{O} = \mathcal{G}/\mathcal{S}$, where $\mathcal{S}$ is the stabilizer subgroup of the orbit and is a subgroup of $\mathcal{G}$.[13] This requirement is equivalent to invariance under the choice of local Hilbert space basis. Lower bounds on the number of parameters needed to describe equivalence classes have been provided that show the insufficiency of the total set of state descriptions of local systems for specifying the state description of the *compound* system they comprise.[14]

In particular, because states of N qubits are equivalent in entanglement when they lie on the *same orbit* under LUTs of the statistical operator, each such orbit corresponds to a single entanglement class with characteristic invariant quantities. The orbits have specific dimensionalities, $\dim\mathcal{O}$, given by the dimension of the stabilizer subgroups, $\dim\mathcal{S}$, of states on the orbit and the dimension $\dim\mathcal{G}$ of the group in question:

$$\dim\mathcal{O} = \dim\mathcal{G} - \dim\mathcal{S} , \qquad (7.29)$$

where for LUTs, $\mathcal{G}$, being local, has elements of the form $U_1 \otimes U_2 \otimes \cdots \otimes U_N$ so that each unitary transformation U_i acts on a Hilbert space corresponding to a component of the total system in the possession of a single party in its local laboratory. The dimension of the orbit is just the number of real parameters required to specific the location of a state in the orbit. The Hilbert space of pure states of N parties, each in possession of a single qubit is, as we have seen previously,

$$\mathcal{H}^{(N)} = \mathbb{C}^2 \otimes \mathbb{C}^2 \otimes \cdots \otimes \mathbb{C}^2 . \qquad (7.30)$$

Any pure state of the compound system is therefore described by $2(2^N - 1)$ real parameters, because there are 2^N *complex* parameters and so 2^{N+1} real parameters describing any state on this space, of which normalization reduces the number of real parameters by one, as does the freedom of global phase. The number of parameters describing a state thus grows *exponentially* with the number of components, N. Quantities invariant on an orbit thus specify nonlocal equivalence classes of states, as discussed in the next section.

In contrast to situations described by LUTs, in LOCC each agent can perform generalized measurements on its local subsystem and classically communicate measurement outcomes to other agents. The other agents can then choose their local transformations in way conditioned by these outcomes.[15]

---

[13] Consider $\mathcal{S}$ be the vector subspace kept fixed by a subgroup of elements of G_n, which is defined by Eq. 10.17, below; this subgroup is the *stabilizer*, S, of $\mathcal{S}$.

[14] The extra parameters are known as "hidden nonlocalities;" see, for example [247].

[15] Such a method is used, for example, in entanglement distillation; see Fig. 9.1.

More generally, one is interested in equivalence classes of states under SLOCC. One can seek equivalence classes of such multipartite states via the criterion of mutual accessibility via *invertible local operations* (ILOs). The number of state parameters that can be altered by a multiparty ILO grows linearly in the number of parties, being in particular $6N$.[16] It is difficult to find canonical states on the orbits of these multipartite states because the set of equivalence classes of multi-qubit states under SLOCC, in the space of orbits

$$
\frac{\mathcal{H}^{(N)}}{SL(2,\mathbb{C}) \times SL(2,\mathbb{C}) \times \cdots \times SL(2,\mathbb{C})} ,
\tag{7.31}
$$

depends on at least $[2(2^N - 1) - 6N]$ parameters [144]. For $N = 2, 3$ there is a finite number of equivalence classes, but there may be an *infinite* number for $N > 3$. The situation when one party possesses *more than one qubit* is worse, even in the case of three parties. In the case of two parties, there is a maximally entangled state from which all states may be accessed with certainty; in the case of three parties, there is generally *no* such state [284].[17]

7.6 Algebraic invariants of multipartite systems

The invariant lengths under the isometries corresponding to LUTs, LOCC, and SLOCC transformations providing multipartite-state equivalence classes have been explicitly considered as algebraic entities [275]. For bipartite systems, the situation is simple because the coefficients of the Schmidt decomposition form a complete set of LUT invariants. The next case of interest is that of LUT invariants for tripartite states. Consider invariants for three-qubit pure states

$$
|\Psi\rangle = \sum_{i,j,k=0}^{1} \alpha_{ijk} |ijk\rangle .
\tag{7.32}
$$

One obvious invariant, the invariant of degree two, is the *norm* of the state, the generalized Stokes parameter S_{000}, which can be written

$$
I_1 = \sum_{i,i',j,j',k,k'=0}^{1} \delta_{ii'}\delta_{jj'}\delta_{kk'}\alpha_{ijk}\alpha_{i'j'k'}^{*} = \sum_{i,j,k=0}^{1} \alpha_{ijk}\alpha_{ijk}^{*}
\tag{7.33}
$$

---

[16] A *local invertible operator* is an operator that can be written in tensor product form where each factor has a well-defined inverse and acts in a single-party Hilbert subspace. A single-qubit ILO described by a four-complex-component matrix is required to have a nonzero determinant scalable to unity because multiplication by a scalar does not affect accessibility, and depends only on six real parameters [144].

[17] Accessing a generic state in this case would require additional resources, such as shared singlet states.

(*cf.* Section 7.1 and [247]).

The *purities* of the statistical operators of three single qubits, obtained from $|\Psi\rangle$ by partial tracing out the remaining two systems, namely,

$$I_2 = \sum_{i,j,k,m,p,q=0}^{1} \alpha_{kij}\alpha_{mij}^{*}\alpha_{mpq}\alpha_{kpq}^{*} , \qquad (7.34)$$

$$I_3 = \sum_{i,j,k,m,p,q=0}^{1} \alpha_{ikj}\alpha_{imj}^{*}\alpha_{pmq}\alpha_{pkq}^{*} , \qquad (7.35)$$

$$I_4 = \sum_{i,j,k,m,p,q=0}^{1} \alpha_{ijk}\alpha_{ijm}^{*}\alpha_{pqm}\alpha_{pqk}^{*} , \qquad (7.36)$$

are LUT (and hence LOCC) invariants of degree four, which are sometimes also labeled J_i. Labeling the first qubit system A, the second B, and the third C, one has the following relations between these quantities and the single-qubit Minkowskian length

$$S_{(1)}^{2}(\rho_A) = 1 - I_2 , \quad S_{(1)}^{2}(\rho_B) = 1 - I_3 , \quad S_{(1)}^{2}(\rho_C) = 1 - I_4 , \qquad (7.37)$$

which are simply related to the concurrences obtained by bipartite decomposition of the corresponding three-qubit states; see below. The number of LUT invariants of a state then grows exponentially with the number of local parts. The LUT invariant of higher degree for three particles, known as the *Kempe invariant*, is

$$I_5 = \sum_{i,j,k,l,m,n,o,p,q=0}^{1} \alpha_{ijk}\alpha_{ilm}^{*}\alpha_{nlo}\alpha_{pjo}^{*}\alpha_{pqm}\alpha_{nqk}^{*} , \qquad (7.38)$$

which is not, in general, algebraically independent of I_2, I_3, and I_4 [247].

In the case of a general number of qubits, these states can be written $|\Psi\rangle = \alpha^{ijk}|ijk\ldots\rangle$, and the general polynomial can be written

$$F = \sum_{\text{indices }=0}^{1} a_{i_1 j_1 k_1 \ldots i_2 j_2 k_2 \ldots}^{i_r j_r k_r \ldots} \alpha^{i_1 j_1 k_1 \ldots} \alpha^{i_2 j_2 k_2 \ldots} \ldots \alpha_{i_r j_r k_r \ldots}^{*} \ldots , \qquad (7.39)$$

where the numbers of α and α^* terms are equal and all the indices are contracted between corresponding terms, by the coefficients $a_{i_1 j_1 k_1 \ldots i_2 j_2 k_2 \ldots}^{i_r j_r k_r \ldots}$ being products of Krönecker delta symbols each contracting an index, as is the case for the I_i above [275]. These allow one to fully distinguish the various orbits under LUTs. Similarly, in the context of SLOCC transformations, the polynomial invariants under $SL(2,\mathbb{C})^{\times N}$ of pure states, characterized by amplitudes $\alpha_{i,j,k,\ldots}$ in the computational basis $\{|ijk\ldots\rangle\}$, are

$$K_\sigma = \sum_{\text{indices}=0}^{1} \epsilon_{i_1 i_2}\epsilon_{j_1 j_2}\epsilon_{k_1 k_2} \cdots \epsilon_{i_{r-1} i_r}\epsilon_{j_{r-1} j_r}\epsilon_{k_{r-1} k_r} \cdots$$
$$\alpha^{i_{\sigma(1)} j_{\tau(1)} k_{\upsilon(1)} \cdots} \alpha^{i_{\sigma(2)} j_{\tau(2)} k_{\upsilon(2)} \cdots} \ldots \alpha^{i_{\sigma(r)} j_{\tau(r)} k_{\upsilon(r)} \cdots} \qquad (7.40)$$

where $\boldsymbol{\sigma} = (\sigma, \tau, \upsilon, \ldots)$, the $\sigma(i)$, $\tau(i)$, $\upsilon(i)$, and so on, are permutations over r elements, which correspond to the Lorentz-group invariant lengths $S_{(N)}^2$ that are squares of their moduli.[18]

All SLOCC invariants can then be written in terms of these basic polynomials. By examining state transformations beyond simple unitary operations, it has been shown that one can classify multipartite entangled states under Lorentz (SLOCC) transformations using the subset of filtering operations, as mentioned above. In particular, for three-party states it has been shown that there are nine different entanglement classes, which include the GHZ and W classes of entangled three-qubit states; see the following section, as well as [305]. In the case of *mixed* multipartite states, one considers the (squared) magnitude of K_{σ} and the complex state-vector coefficients are replaced by statistical operator elements.[19] In the bipartite case, the Schmidt decomposition always exists and provides the quantities invariant under local unitary operations. Furthermore, as mentioned above, it has been shown that pure states of *some* multipartite quantum systems are multi-separable: they provide, upon averaging over the state of any given party, separable (generally mixed) states if and only if they have a generalized Schmidt decomposition as does, for example, the GHZ state [418].

In the following chapter, we return to the examination of multiple-qubit entangled states that prove useful in the study of entanglement and for carrying out various quantum information-processing tasks. Now let us examine in detail entanglement properties and state classification in the cases of three- and four-qubit systems, and several other larger families of multiple-qubit states.

7.7 Three-qubit states and residual tangle

As mentioned in the introduction to this chapter, progress has been made in the quantification of multipartite entanglement for three-qubit states through the application of bipartite entanglement measures to their two-qubit subsystems. In particular, the residual genuinely three-party entanglement can be found by isolating it from the bipartite entanglement present in a three-qubit system. The *residual tangle* τ_{ABC} is a positive quantity for pure states,

$$\tau_{ABC} \equiv \tau_{A(BC)} - \tau_{AB} - \tau_{AC} \,, \tag{7.41}$$

---

[18] Here we have introduced the Levi–Civita symbol ϵ defined by the elements $\epsilon_{00} = 0 = -\epsilon_{11}$ and $\epsilon_{01} = 1 = -\epsilon_{10}$ and related to the σ_2 Pauli matrix by $\sigma_2 = -i\epsilon$. By contrast with the case of LOCC invariants, in the above one now contracts α terms with each other rather than α terms with their complex conjugates.

[19] It is worthwhile to consider Shimony's geometrical result regarding the equivalence of various entanglement measures in the case of bipartite pure states in this light; see [383], as well as Sect. 6.15.

that measures entanglement among the three components that does not arise from bipartite entanglement within the composite system. The residual tangle is invariant under permutations of the subsystems, as any good measure of inherently three-way entanglement must be.[20] In accordance with the above, the three-tangle for three subsystems A, B, C can be expressed in terms of two-qubit Lorentz-group invariant lengths, in particular,

$$\tau_{ABC} = \left| \sum_1^2 \epsilon_{i_1 i_3} \epsilon_{j_1 j_3} \epsilon_{k_1 k_4} \epsilon_{i_2 i_4} \epsilon_{j_2 j_4} \epsilon_{k_2 k_3} \right.$$
$$\left. \alpha^{i_{\sigma(1)} j_{\tau(1)} k_{\tau(1)}} \alpha^{i_{\sigma(2)} j_{\tau(2)} k_{\tau(2)}} \alpha^{i_{\sigma(3)} j_{\tau(3)} k_{\tau(3)}} \alpha^{i_{\sigma(4)} j_{\tau(4)} k_{\tau(4)}} \right|, \tag{7.42}$$

where, again, the $\sigma(i)$ and $\tau(i)$ are permutations.

The generic class of three-particle pure states can be written as

$$|\Psi_3\rangle = \lambda_1|000\rangle + \lambda_2 e^{i\theta}|101\rangle + \lambda_3|110\rangle + \lambda_4|111\rangle, \tag{7.43}$$

where the λ_i are real positive numbers such that $\sum_i \lambda_i^2 = 1$ and $\theta \in [0, \pi]$. These states include two classes of *separable* states, one that is fully separable into a product of single-party pure states (ABC), and one separable into a product of an entangled two-party pure (two-qubit) state and a single-party pure (qubit) state, ((AB)C, A(BC), B(AC)). These divisions are known as *two-splits*.[21] There are two locally inequivalent classes of *nonseparable*, hence genuinely tripartite-entangled pure states. One class is represented by a particularly useful such state, the Greenberger–Horne–Zeilinger (GHZ) state

$$|GHZ\rangle = \frac{1}{\sqrt{2}}(|000\rangle - |111\rangle), \tag{7.44}$$

which has been shown to violate the predictions of local realism ([55, 196, 195]) introduced in Chapter 3;[22] the generic state $|\Psi_3\rangle$ belongs to the GHZ class. The remaining class non-separable three-particle pure states is that represented by states of the form

---

[20] Thus, the apparent asymmetry of the above expression presents no difficulty.

[21] The general case of division of a composite system into n parts is referred to as an n-split, and illuminates the separability structure of larger compound-system states [143].

[22] The GHZ state is a eigenvector of all the Pauli group operators $\sigma_x \otimes \sigma_y \otimes \sigma_y$, $\sigma_y \otimes \sigma_x \otimes \sigma_y$, $\sigma_y \otimes \sigma_y \otimes \sigma_x$, with corresponding eigenvalue $+1$ and of the operator $\sigma_x \otimes \sigma_x \otimes \sigma_x$ with corresponding eigenvalue -1. With these four operators, a measurement of operators σ_x or σ_y on *any two* of the three qubits allows one to infer the outcome of the third. Local realism would then allow one to assign definite values to the local quantities $\sigma_x^{(i)}$ and $\sigma_y^{(i)}$, described by a function taking $\sigma_x^{(i)}$ and $\sigma_y^{(i)}$ each to the set $\{-1, +1\}$, where the superscript indicates the subsystem in question. There is therefore a violation of local realism: it is impossible to find a product of such local functions assigning the needed values, because the first three operators are assigned $(+1)$ but the fourth is assigned (-1).

$$|W\rangle = \lambda_1|001\rangle + \lambda_2|010\rangle + \lambda_3|100\rangle , \qquad (7.45)$$

and is a set of measure zero in the set of all pure states: a GHZ-class state as close as desired to any W-class state can be obtained by simply adding an additional term with a λ_4 as small as desired to it. Thus, the GHZ and W classes of states are representative of the two equivalence classes of three-particle *pure* states defined by interconvertibility under SLOCC transformations, the W-states being those such that $\tau_{ABC}(|W\rangle) = 0$ [436].

The three-qubit *mixed* states are similarly readily classified, as follows.

(i) S, the class of *separable* mixed states;

(ii) B, the class of *bi-separable* mixed states;

(iii) W, the class of states expressible as *convex combinations* of projectors onto the separable, bi-separable and pure W states;

(iv) GHZ, the *generic* class of three-qubit states.

These states are thus related as $S \subset B \subset W \subset GHZ$; states of later classes can be converted *stochastically* to states of preceding classes by the application of POVMs [4]. Furthermore, there exist methods for determining the class to which a given state belongs.

7.8 Three-qubit quantum logic gates

Before leaving the topic of three-qubit states, let us consider some quantum gates acting at the three-qubit level. Useful three-bit gates have been developed in the context of reversible computation, of which quantum gates are one sort of realization because they are carried out using unitary transformations, which are inherently reversible.

The *Toffoli gate* is one important three-qubit gate implementable in quantum computing that performs the following operation on computational values. $(x, y, z) \rightarrow (x, y, x \wedge y \oplus z)$, where $\oplus$ indicates the XOR operation and $\wedge$ the AND operation; see Section A.1. The truth tables of classical and quantum Toffoli gates, which are shown Figs. 3.6 and 7.1 respectively, are the same. That is, both gates change the third bit, z, conditionally on the first two being 1, and otherwise have no effect. A Toffoli gate is clearly its own inverse. Both the classical and quantum Toffoli gates are universal, in that one can construct a circuit computing any reversible function using *only* Toffoli gates; see Section 13.6. The unitary matrix representing the quantum Toffoli gate is given Section 3.8.

The quantum *Fredkin gate* is a three-qubit gate performing the following operation on three bits: $(x, y, z) \rightarrow (x, x \wedge z \oplus \neg x \wedge y, x \wedge y \oplus \neg x \wedge z)$, where $\neg$ indicates binary negation. The Fredkin gate has only *one* control input, whereas the Toffoli gate has *two* control inputs. It swaps the values of second and third bits if the first takes the value 0; see Fig. 7.1. Though the quantum Fredkin gate is reversible, its classical analogue is *not*.

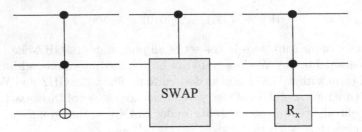

Fig. 7.1. The Toffoli, Fredkin, and (up to a phase) Deutsch gates.

The *Deutsch* (quantum) *gate*, which Deutsch designated $\mathbf{Q}$, is a universal quantum gate similar to the Toffoli gate, being again a controlled-controlled gate, with the operation on the third qubit being the combined phase shift/rotation operation $i\mathrm{R}_x(\theta)$ [129]. In the computational basis for three qubits, it performs the operation of switching the basis elements $|110\rangle$ and $|111\rangle$, leaving the others unchanged; see Fig. 7.1. More information regarding universal logic gates is provided in Section 13.6.

7.9 States of higher qubit number

Entangled states of more than three qubits are also important, particularly for quantum error correction, which is required for practical quantum information processing. For example, the smallest code states for arbitrary single-qubit errors are entangled five-qubit pure states; see Section 10.6.

In at least one sense, it is possible to generalize the entangled states of the Bell basis by retaining the symmetry of its elements under changes of scale from two to four and more qubits. The Bell gems are such a generalization that can be recursively defined [232].[23]

A *Bell gem*, $\mathcal{G}_d$, is a set of state-vectors of 2^N qubits lying in the $d = 2^{2^N}$-dimensional Hilbert space $\mathbb{C}^{2^{2^N}}$, of the form

$$\frac{1}{\sqrt{2}} \left(|i\rangle|i\rangle \pm |j\rangle|j\rangle \right) \tag{7.46}$$

$$\frac{1}{\sqrt{2}} \left(|i\rangle|j\rangle \pm |j\rangle|i\rangle \right) , \tag{7.47}$$

where $|i\rangle \neq |j\rangle$ are elements of a Bell gem $\mathcal{G}_{d'}$ of dimension $d' = 2^{2^{(N-1)}}$, $N \geq 2$, $N \in \mathbb{N}$, the simplest Bell gem, $\mathcal{G}_4$, being the Bell basis, namely,

---

[23] A generalization of the Bell state $|\Psi^-\rangle$ to N particles and N levels—the class of "supersinglet" states—has also been examined [97].

$$|\Phi^{\pm}\rangle = \frac{1}{\sqrt{2}}(|00\rangle \pm |11\rangle) , \tag{7.48}$$

$$|\Psi^{\pm}\rangle = \frac{1}{\sqrt{2}}(|01\rangle \pm |10\rangle) , \tag{7.49}$$

which is a basis for $\mathbb{C}^4$. The family of Bell gems has the following properties:

(i) The Bell gem $\mathcal{G}_{2^{2^N}}$ is an *orthonormal basis* for the 2^{2^N}-dimensional Hilbert space of state-vectors, $\mathcal{H}_{2^{2^N}} = (\mathbb{C}^2)^{\otimes 2^N}$, that is, of 2^N qubits.

(ii) The elements of $\mathcal{G}_{2^{2^N}}$ have *maximal 2^N-tangle*, τ_{2^N}.

The second-smallest Bell gem (after the Bell basis) is the four-qubit Bell gem, which has 16 elements, $|\mathbf{e}_i\rangle$, lying in $\mathcal{H}_{16} = (\mathbb{C}^2)^{\otimes 2^2}$:

$$\mathcal{G}_{16} = \{ \tfrac{1}{\sqrt{2}} (|\Phi^+\rangle|\Phi^+\rangle \pm |\Phi^-\rangle|\Phi^-\rangle), \tag{7.50}$$

$$\tfrac{1}{\sqrt{2}} (|\Psi^+\rangle|\Psi^+\rangle \pm |\Psi^-\rangle|\Psi^-\rangle), \tag{7.51}$$

$$\tfrac{1}{\sqrt{2}} (|\Phi^+\rangle|\Phi^-\rangle \pm |\Phi^-\rangle|\Phi^+\rangle), \tag{7.52}$$

$$\tfrac{1}{\sqrt{2}} (|\Phi^+\rangle|\Psi^+\rangle \pm |\Psi^+\rangle|\Phi^+\rangle), \tag{7.53}$$

$$\tfrac{1}{\sqrt{2}} (|\Phi^+\rangle|\Psi^-\rangle \pm |\Psi^-\rangle|\Phi^+\rangle), \tag{7.54}$$

$$\tfrac{1}{\sqrt{2}} (|\Phi^-\rangle|\Psi^+\rangle \pm |\Psi^+\rangle|\Phi^-\rangle), \tag{7.55}$$

$$\tfrac{1}{\sqrt{2}} (|\Phi^-\rangle|\Psi^-\rangle \pm |\Psi^-\rangle|\Phi^-\rangle), \tag{7.56}$$

$$\tfrac{1}{\sqrt{2}} (|\Psi^+\rangle|\Psi^-\rangle \pm |\Psi^-\rangle|\Psi^+\rangle)\} \tag{7.57}$$

[232]. The first four of these elements, $|\mathbf{e}_1\rangle, |\mathbf{e}_2\rangle, |\mathbf{e}_3\rangle$, and $|\mathbf{e}_4\rangle$, are the code states of the (extended) quantum erasure channel; see Chapter 10 and [39]. Furthermore, $|\mathbf{e}_2\rangle, |\mathbf{e}_3\rangle$, and $|\mathbf{e}_4\rangle$ are codes states of a one-error correcting detected-jump quantum code, as well as spanning a decoherence-free subspace in which universal four-qubit quantum computations can be carried out; see Chapter 13 and [8].

8

Quantum state and process estimation

In addition to having a conceptual understanding of the entanglement and other essential properties of quantum states, it is important to understand how states and essential functionals thereof can be *empirically* determined, particularly in a way that can be connected with formal results of the sort described in previous chapters. As mentioned at the outset, the state of a given quantum system cannot generally be discovered by simply measuring it once. For an unknown state, at least an *ensemble* must be measured for one to come to know an unknown state of a given quantum system. Quantum tomography is a general method for estimating ensemble averages for operators and states based on a complete set of quantum measurements.

Quantum state tomography allows one to find the *statistical operator* of a system: a state description for a quantum system requires the measurement of complementary properties of an ensemble in different, generally incompatible experimental arrangements, rather than merely compatible ones, by determining, for example, the full set of generalized Stokes parameters in the case of n-qubit systems. For present purposes, quantum state tomography and the associated method of quantum process tomography, which determines *transformations* of quantum states, allow one to characterize quantum sources and quantum information channels for applications such as quantum cryptography and quantum computing, which are described in later chapters. The basic elements of these estimation methods are discussed here.

In addition to the estimation of states and their transformations, it is also possible, and often necessary, to estimate quantum state functionals such as purity and entanglement. One method for doing this, which can be more efficient than the more general but often quite costly method of state tomography, is also described here.

8.1 Quantum state tomography

Given measurements on an ensemble of copies of a given quantum system, the state can be estimated by quantum state tomography. Historically, G. G. Stokes first introduced such a method, involving the four basic parameters that now bear his name and that are still commonly used to describe the polarization state of a light beam [406]. Such simple parameters also allow one to find the statistical operator describing a qubit ensemble, locating it in the Poincaré–Bloch sphere as described in Section 1.3 [241].[1] This procedure is now known as qubit-state tomography.

Qubit-state tomography can be readily extended to multiple-qubit systems, as well as to multiple-qu-d-it systems in which case it is referred to as *qu-d-it-state tomography* [420]. In general, $(d^2 - 1)$ parameters must be measured to reconstruct a state lying in a d-dimensional complex Hilbert space, as the global phase is not physical relevant. To find the state of a qubit, only three quantities need be found, corresponding to the Stokes parameters S_i ($i = 1, 2, 3$). The measurement of coincidence-count rates for multipartite systems correspond to generalized Stokes parameters and allow for the extension of this method to the tomography of multiple-qubit states. In particular, the statistical operator representing a quantum system state can, in principle, be found from a direct linear transformation of correlation data, corresponding to the generalized Stokes parameters [180, 459]. However, measurement errors and/or environmental noise may render ill-defined the operators constructed in this straightforward way, such as when the resulting matrices fail to be completely positive. Therefore, care must be taken to provide estimated states that are well defined. This generally requires additional measurements, as in the case of single qubits where one also measures the Stokes parameter S_0 [118]. A necessary and sufficient condition for the completeness of a set of tomographic measurement vectors (or *tomographic states*), is that the matrix of expectation values of the full set of Pauli-group operators, corresponding to measurement bases, be *nonsingular*.[2] This condition is the requirement for obtaining a well-defined density matrix from the data set of normalized coincidence-measurement outcomes.

Quantum state tomography of multiple-qubit systems can be carried out as follows. One first obtains a number of *identical copies* of the system in the unknown state ρ to be determined. One then measures the system properties using either a complete set of von Neumann measurements or a POVM [332]. The standard requirements for a matrix to represent a statistical operator are then kept in force during the construction of the matrix best representing ρ given the resulting data. A *likelihood functional*, $\mathcal{L}$, that describes the quality of the estimated density matrix can be used to produce such a matrix. One finds the optimal set of variables, for which the likelihood functional is

---

[1] Modern quantum tomography was first investigated in [277, 278, 346, 441].

[2] The Pauli group is defined in Sect. 10.4, below.

maximized, and arrives at a *best estimate* of the actual statistical operator describing the system. Traditionally, precise measurements are performed to find the values of the system properties, each being a measurement of an instance of the system state projecting onto a pure state; in the case of qubits, these measurements provide the generalized Stokes parameters described in the previous chapter, that is, the expectation values of Pauli-group elements.

Consider the case in which a complete set of measurements corresponding to the projectors $P(|\psi_i\rangle)$ is made, providing m distinct outcomes with relative frequencies f_i. The probabilities of measurement outcomes in the limit of an infinite sample size are provided by the Born rule, namely $p_i = \langle \psi_i | \rho | \psi_i \rangle$, such as are given by Eq. 1.22 for the case of a single qubit.[3] One then seeks the state ρ *most likely* to provide the observed *finite* number of measurement outcomes. A density matrix, ρ_{est}, for this state is one minimizing the Kullback–Leibler distance (*cf.* Eq. 4.8) between the relative frequencies provided by the data and the probabilities provided by the Born rule, considered as vectors [256].[4] Minimizing this distance is tantamount to finding the *maximum likelihood*, where the likelihood is given by the functional

$$\mathcal{L}(\rho) = \prod_i p_i^{f_i} \, . \tag{8.1}$$

Treating this as a linear positive (LP) problem, one can make use of the *expectation maximization* (EM) *algorithm*. One writes

$$p_i = \sum_j r_j h_{ji} \, , \tag{8.2}$$

where the r_j are the components of the vector $\mathbf{r}$ used to describe the system during a given step in the solution of this problem and $[h_{ji}]$ is a positive kernel. The EM algorithm iterates the value of such a vector, which at step n is given by

$$r_j^{(n)} = r_j^{(n-1)} \sum_i \frac{h_{ji} f_i}{p_i(\mathbf{r}^{(n-1)})} \, , \tag{8.3}$$

beginning from a first step with an initially chosen positive $\mathbf{r}^{(1)} \equiv \{r_j^{(1)}\}$. In the basis $\{|\lambda_i\rangle\}$ in which the density matrix ρ is diagonal, one has

$$\rho = \sum_i \lambda_i P(|\lambda_i\rangle) \, , \tag{8.4}$$

the λ_i being its eigenvalues. The Born rule then provides a well-defined LP problem.

---

[3] The Born rule is Postulate II of quantum mechanics; see Appendix B.

[4] In general, measurements can be made using a set of measurements of *nonorthogonal* states. However, uncertainties for a given number of measurements will increase as progressively fewer orthogonal basis elements are measured, so that a given accuracy requires that increasingly larger data sets be collected.

In particular, one addresses the LP problem provided by the (first-order) variational contribution

$$\delta \ln \mathcal{L} = \sum_j \delta r_j (\langle \phi_j | R | \phi_j \rangle - 1) + i\theta \, \text{tr}(G) , \tag{8.5}$$

where the operator R is given by

$$R = \sum_i \left(\frac{f_i}{p_i} \right) P(|\psi_i\rangle) , \tag{8.6}$$

the operator $G \equiv i[\rho, R]$ is the generator of a unitary transformation allowing updated eigenvalues to be determined, and θ is a very small angle of rotation corresponding to this unitary operation. Success in attaining the desired global maximum values is guaranteed by the convexity of $\mathcal{L}$ [350].

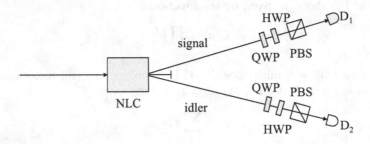

Fig. 8.1. An apparatus for quantum state tomography on two-qubit joint-polarization states of photon pairs produced by spontaneous parametric down-conversion (about which, see Section 6.16). The PBSs alone enable measurements of the projections involving σ_3; invoking the half-wave plates (HWP) as well enables measurements involving σ_1; using the PBS, entirely deflecting one linearly polarized component (*cf.* Fig. 2.2), together with the quarter-wave plates (QWP) enables measurements at detectors D_1 and D_2 involving σ_2 (*cf.* Section 1.3 and, for example, [11, 302, 454]).

As an example, consider now *single-qubit-state* tomography in particular. The most straightforward tomographic approach is to measure three orthogonal components of the Pauli operators, σ_i, providing the single-particle Stokes parameters, S_i. Determining these three parameters involves the measurement and estimation of *six* quantities, one pair for each of the Pauli operators, on an ensemble of qubits; when measuring *each* of the σ_i, two measurement outcomes can occur. In the case of the measurement of the photon polarization qubits, for example, one performs *ellipsometry* in a linear-optical apparatus constructed of beam-splitters, polarizing beam-splitters, and polarization rotators feeding photodetectors providing photon-counting statistics, providing a set of six counting bin values, n_{ij} ($j = 1, 2$), corresponding to the operators σ_i above (*cf.* Eqs. 1.19–22) and the states of the three bases, the computational, diagonal, and circular bases [376]. These values are then normalized by dividing each by the sum of the total number of counts, $\bar{N} = \sum_{ij} n_{ij}$, providing the needed likelihoods

$$p_{ij} = \frac{n_{ij}}{\bar{N}} \, , \tag{8.7}$$

and thus the p_i; for example, see [226].

Two-qubit state tomography can be performed by doubling the above-described apparatus and constructing a probability vector composed of coincidence count rates corresponding to the two-photon generalized Stokes parameters to obtain an invertible square matrix from these coincidence measurements for the possible pairings of single-qubit states; see Fig. 8.1 [3]. For larger numbers of qubits one similarly extends the measurement apparatus to include the corresponding number of duplicates of the basic apparatus.

8.2 Quantum process tomography

Quantum process tomography involves the reconstruction of the *process* $\mathcal{E}(\rho)$ describing the transformation of a quantum system, such as occurs in the transmission of a quantum system through a quantum channel.[5] It is convenient to consider the *process matrix*, M, representing this superoperator acting on the vector of independent coefficients of the nonnormalized density matrix ρ_{input}, such that

$$\rho_{\mathrm{output}} = \mathrm{M}\rho_{\mathrm{input}} \, , \tag{8.8}$$

constructed using the multi-particle Stokes parameters for a complete set of measurements,

$$\rho \doteq \Big(\rho_{11}, \rho_{22}, \dots, \mathrm{Re}(\rho_{12}), \mathrm{Im}(\rho_{12}), \mathrm{Re}(\rho_{13}), \dots, \mathrm{Im}(\rho_{n-1,n}) \Big)^{\mathrm{T}} \, . \tag{8.9}$$

Though, formally, the matrix M can be thought of as being obtained by inversion, errors in actual measurement can give rise to ill-defined resulting "density matrices." Thus, as mentioned in the previous section, it is a better strategy to estimate M using *maximum-likelihood estimation* to find the appropriate completely positive superoperator. The same optical elements from a wing of the apparatus above can be used for the initial state-tomography step in the case of two-qubit systems, for example; see Fig. 8.1 and [302, 340].

In particular, to carry out *quantum process tomography* on a quantum channel, one prepares quantum systems in *different states* $\{\rho^{(i)}\}$, transforms

---

[5] Quantum channels are discussed in detail the following chapter.

them by the process of interest, and performs quantum-state tomographic measurements of the resulting states. It is convenient to consider the decomposition elements, E_j, of the operator-sum representation of $\mathcal{E}(\rho)$, where $E_j = \sum_k e_{jk} \bar{E}_k$ (where, for example, in the single-qubit case, $\bar{E}_k = \sigma_k$, with $k = 0, 1, 2, 3$ [315]), so that we also have

$$\mathcal{E}(\rho) = \sum_{k,l} \bar{E}_k \rho \bar{E}_l^\dagger M_{kl} , \qquad (8.10)$$

where the M_{kl} are the elements of the process matrix. Thus,

$$M_{mn} = \sum_k e_{km}(e_{kn})^* . \qquad (8.11)$$

Then

$$\mathcal{E}(\rho_j) = \sum_k c_{jk} \rho_k \qquad (8.12)$$

and, writing $\bar{E}_n \rho_j \bar{E}_p^\dagger = \sum_k m_{jk}^{np} \rho_k$, we have

$$M_{np} = \sum_{j,k} (m^{-1})_{jk}^{np} c_{jk} , \qquad (8.13)$$

forming the desired matrix representation of the process.

8.3 Direct estimation methods

In addition to the estimation of states and the processes describing their transformation, it is possible and often desirable to estimate other simpler quantities discussed in previous chapters, such as purity and analytically computable entanglement measures. Although it is possible to perform tomography to obtain a state as well as information about its dynamics from which these can be evaluated in many cases, it is valuable and more efficient to have a method of *directly* estimating these properties. One may perform some of these estimates using LOCC alone [156].

Quantum interferometry provides such a method. To see this, consider a Mach–Zehnder interferometer including coupling to an ancilla by a controlled unitary operation; see Fig. 8.2. The usual interferogram resulting from variation of the phase shift in such an interferometer is sensitive to this coupling. Consider the expectation value of the corresponding unitary operator,

$$v e^{i\alpha} = \operatorname{tr}(U\rho) , \qquad (8.14)$$

where the real parameters v and α are (an ideal) visibility and (a Patcharatnam) phaseshift, respectively, that depend on ρ. The form of this expression

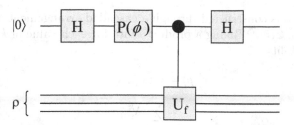

Fig. 8.2. Quantum circuit for the direct estimation of quantum state functionals of a system in state ρ [156].

shows that the observed effect on the interferogram of the controlled-U operation is simply the expectation value of the corresponding operator, U. This can be exploited to perform more efficient estimates of important quantities.

For example, to directly estimate the state purity, $\mathcal{P}(\rho)$, one can take U to be the SWAP operation, namely,

$$V = \sum_{ij} |ji\rangle\langle ij|$$

$$= \sum_{ij} \mathcal{T}(|i\rangle\langle j|) \otimes \mathrm{Id}_n(|i\rangle\langle j|) \,,$$

where $\mathcal{T}$ is the transposition operator, noting that

$$V(|\phi\rangle|\psi\rangle) \to |\psi\rangle|\phi\rangle \,, \quad V\big(P(|\phi\rangle|\psi\rangle)\big) \to P(|\psi\rangle|\phi\rangle) \tag{8.15}$$

for all single-party states $|\psi\rangle$ possessed by Alice and Bob; see Section 2.6. Now, if the input to this interferometer is two *identical* subsystems, described by a separable statistical operator $\rho_A \otimes \rho_B = \rho^{\otimes 2}$, a measurement of the resulting interferogram provides a measurement of state overlap, which in this case is

$$v(\rho) = \mathrm{tr}\big(V(\rho \otimes \rho)\big) \tag{8.16}$$

$$= \mathrm{tr}(\rho^2) \tag{8.17}$$

$$= \mathcal{P}(\rho) \,, \tag{8.18}$$

from which the Rényi entropy can be directly obtained as well.[6]

Replacing the SWAP operation by its generalization, the *shift* operation $V^{(i)}$ defined by

$$V^{(i)}(|\psi_1\rangle|\psi_2\rangle \ldots |\psi_i\rangle) = |\psi_i\rangle|\psi_1\rangle \ldots |\psi_{i-1}\rangle \,, \tag{8.19}$$

and considering the fully separable input state of k copies of ρ, $\rho^{\otimes k}$, the above construction provides an estimate of the *eigenvalue spectrum* of ρ [248]. This

---

[6] Such a measurement has been explicitly carried out [73, 240].

shift operation is constructed through a cascaded arrangement of $i-1$ SWAP gates, V [156, 275].[7] Taking a product state for each value of k provides one with the visibility

$$v'(\rho) = \mathrm{tr}\big(V^{(k)}(\rho^{\otimes k})\big) \tag{8.20}$$

$$= \sum_{i=1}^{n} \lambda_i^k , \tag{8.21}$$

where the λ_i are the eigenvalues of ρ. Thus, estimation of the statistical-operator spectral elements can be performed.

This direct-estimation method also allows one to estimate the SLOCC-invariant length $S_{(N)}^2$ corresponding to the multipartite-entanglement measure τ_N in the case of pure states; see Sections 7.2, 7.4 and [240]. For example, in the case of $N = 2$, taking $\rho_A = P(|\Psi_{12}\rangle)$ and $\rho_B = P(|\tilde{\Psi}_{12}\rangle)$ provides one with the invariant

$$S_{(2)}^2(\rho) = \mathrm{tr}(\rho\tilde{\rho}) = \tau(\rho) , \tag{8.22}$$

namely, the tangle; this accords with what we have just seen in that the concurrence, of which the tangle is the square, is a function of the eigenvalue spectrum only.

---

[7] Note, however, that the shift operation $V^{(k)}$ is *not* implementable as a LOCC operation, but must be implemented by a global network, as described in [307]. Note also that, though the eigenvalue spectrum may be found in this way, the full density matrix is *not* provided, as it is in the more laborious procedure required for quantum state tomography.

9

Quantum communication

The communication of information between two spatially separated parties requires a directed resource, such as a bit or qubit, that is naturally constrained by the speed of light and, in the case of the quantum information, is subject to the constraints of the no-cloning theorem. Several specifically quantum-mechanical protocols are discussed in this chapter that illustrate different uses of the combination of resources available to support communication and information processing when quantum resources are available. It is important to note that, despite the value of entanglement for communication, classical communication cannot be simulated by the resource of shared quantum entanglement alone in an attempt to circumvent this speed-of-light constraint, due to the undirected nature of entanglement.[1]

In the direct transmission of a quantum system, communication resource requirements are satisfied via a single channel. In quantum teleportation, resource requirements are satisfied by two distinct systems, one passing through a classical channel as classical information and one passing through a quantum channel, in order to transmit *quantum* information. A third task considered here, quantum dense coding, uses previously shared entanglement and classical communication to double, in a specific sense, the capacity of a quantum channel to transmit classical information.

Entanglement swapping and entanglement purification are quantum resource distribution tasks that are discussed in this chapter as well. Quantum cryptography, which uses both a quantum channel and a classical channel to distribute correlate keys securely, is discussed later in Chapter 12; see Fig. 12.1. We begin now by considering what quantum channels and their communication capacities are.

---

[1] For more on the "peaceful coexistence" of quantum mechanics and relativity, see Ch. 8 of [384].

9.1 Quantum channels

A *quantum channel* is a means for transmitting quantum information. Physically, a quantum channel may be viewed as a medium of transmission, such as an optical fiber, together with an ensemble of quantum systems, such as photons, prepared by the sender in quantum states ρ_i, $i = 1, 2, \ldots, n$, carrying n symbols with corresponding probabilities p_i. As with classical channels, quantum channels that transmit information without introducing errors are called *noiseless*; those that introduce errors are called *noisy*.[2] However, quantum channels are fundamentally different in character from classical channels, just as qubits differ from bits. For example, due to the inevitable, generally irreversible, interaction of a transmitted quantum system with the environment of a realistic quantum channel, the input quantum states *per se* will not be retrievable from output states themselves by unitary transformations alone. Nonetheless, quantum signal theory allows one to use encoding–decoding to *improve* signal fidelity.

To understand a quantum channel through which a pure quantum state $|\psi\rangle$ can be sent, it is again helpful to consider the *fidelity*

$$F(|\psi\rangle, \rho') = \langle \psi | \rho' | \psi \rangle \,, \tag{9.1}$$

where now ρ' is the state of the system after transmission, as a measure of channel faithfulness; a quantum channel is *faithful* if this expectation value goes to unity in the appropriate information-processing limit. It is also useful to consider the effect of quantum channels in the operations formalism discussed in Chapter 2, wherein the final state of a statistical operator after the effect of a channel is given by

$$\rho' = \sum_i K_i P(|\psi\rangle) K_i^\dagger \,, \tag{9.2}$$

$\{K_i\}$ being an operator decomposition of the CPTP map $\mathcal{E}(\rho)$ describing the channel.[3]

Quantum channels are most often taken to be stationary and memoryless, so as to have the same effect on every block of qubits they may transmit. The CPTP map describing a channel is sometimes referred to as the *superscattering operator* and is analogous to the Markov matrix describing the probabilities of outputs in terms of inputs in the description of classical channels. Any one of the above K_i has the effect of projecting a pure state onto a pure state, whereas the collective effect of the operator *sum* is typically to take pure states to *mixed* states.

---

[2] Quantum channels can be characterized empirically using quantum process tomography, described in the previous chapter.

[3] Operator decompositions were previously discussed in Sect. 2.6.

As noted in Section 2.6, the K_i are generally not unique. In particular, any pair of decompositions $\{K_i\}$ and $\{\bar{K}_j\}$, of r and s elements respectively, for the same operation are related as

$$K_i = \sum_{j=1}^{s} f_{ij} \bar{K}_j , \qquad (9.3)$$

where $[f_{ij}]$ is the matrix representing a maximal partial isometry between the vector spaces $\mathcal{H}^{(r)}$ and $\mathcal{H}^{(s)}$. A *partial isometry* is an operation more general than a unitary transformation, such that $VV^\dagger = P$ where P is a projector. In the event that $r \neq s$, V is nonunitary; a *maximal* partial isometry is a V for which either $VV^\dagger$ or $V^\dagger V$ is the identity. The freedom of choice of decomposition can be viewed in this context as the freedom of choice of the basis for the environment of the channel [27].

If a quantum channel is noiseless and distortion-free, then it is described by the identity operator $\mathbb{I}$ and fully preserves the quantum coherence of the input state. At the other extreme, the completely decohering channel destroys *all* off-diagonal elements of the statistical operators input to it, so that $\rho \rightarrow \rho' = \sum \rho_{ii} P(|\psi_i\rangle)$.[4] The completely decohering channel can transmit *classical* information perfectly, but will destroy the coherence properties essential for transmitting genuinely quantum information.[5] Interestingly, the ratio of the entanglement-assisted classical capacity to the unassisted classical capacity generally *increases* with the amount of noise in the quantum channel, even when the quantum capacities go to zero. Let us now precisely define a number of quantum channel capacities and further consider the relationship between various pairs of them.

9.2 Quantum channel capacities

An information source for a quantum channel provides an ensemble of quantum states; a given state that a quantum source is capable of producing can be broadcast over a quantum channel using a number of copies of the state. In the case of two parties sharing such a channel, subsystems of a larger quantum system can also be prepared in an entangled joint state ρ_{AB}, shared by sender Alice and receiver Bob, the partial trace of which over either subsystem, A or B, is described by ρ_I, $I = A, B$.[6] These two situations reflect the situations in quantum key distribution using the BB84 and E91 protocols, respectively.

---

[4] This expression corresponds to a transition between states given in Eqs. 11.48–49.
[5] Specific examples of quantum channels are given below in Sect. 9.6.
[6] Note, however, the existence of constraints on broadcasting, such as the "no-broadcasting" theorem, described in Sect. 11.1.

In general, n qubits can be *encoded* by a process described by a super-operator $\mathcal{E}(\rho)$ to m inputs for a quantum transmission channel, which may be affected by noise process $\mathcal{N}$, and then *decoded* back to n qubits from m outputs of the channel in a process described by another superoperator, $\mathcal{D}(\rho)$. Though $\mathcal{E}(\rho)$ and $\mathcal{D}(\rho)$ serve opposite purposes, they are *not* necessarily operator-inverses of each other. Various operations can then be performed locally to prepare various joint states shared by the parties at the ends of the channel.

Any quantum channel can be attributed at least three basic types of transmission capacity: a classical capacity, C, an (unassisted) quantum capacity, Q, and an entanglement-assisted classical capacity, C_E [49, 50]. C is simply the capacity of a channel to transmit classical information using quantum systems. Q is the capacity for transmitting intact quantum states. C_E is the capacity for transmitting classical information making use of quantum entanglement resources, as in the case of quantum dense coding which is discussed in detail in Section 9.8, below.

The (asymptotic) *classical capacity* of a quantum channel is given by

$$C(\mathcal{N}) = \lim_{\epsilon \to 0} \limsup_{n \to \infty} \left\{ \frac{n}{m} \;\middle|\; \exists_{m,\mathcal{E},\mathcal{D}} \forall_{|\psi\rangle \in \{|0\rangle,|1\rangle\}^{\times n}} \right.$$
$$\left. \langle\psi|\mathcal{D}\mathcal{N}^{\otimes m}\mathcal{E}\big(P(|\psi\rangle)\big)|\psi\rangle > 1 - \epsilon \right\}, \quad (9.4)$$

the maximum asymptotic rate at which *bits* can be transmitted with arbitrarily good reliability using elements of the computational basis. It is the optimal asymptotic (classical) mutual information per channel use, where possibly entangled input quantum states are mapped back to classical data by possibly collective measurement during decoding.[7] For cases where only *one* use is made of the channel, it provides the *one-shot classical capacity*, $C_1(\mathcal{N})$.

The (protected subspace definition of the) *unassisted quantum channel capacity* is

$$Q(\mathcal{N}) = \lim_{\epsilon \to 0} \limsup_{n \to \infty} \left\{ \frac{n}{m} \;\middle|\; \exists_{m,\mathcal{E},\mathcal{D}} \forall_{|\psi\rangle \in H^{2^n}} \right.$$
$$\left. \langle\psi|\mathcal{D}\mathcal{N}^{\otimes m}\mathcal{E}\big(P(|\psi\rangle)\big)|\psi\rangle > 1 - \epsilon \right\}. \quad (9.5)$$

$Q(\mathcal{N})$ is bounded from below by $C(\mathcal{N})$, because if a quantum channel faithfully transmits a generic qubit state then it can always at *least* transmit a computational basis state, $|0\rangle$ or $|1\rangle$. $Q(\mathcal{N})$ is known to be a nonadditive quantity, in that it can surpass the maximum value of the coherent information that can be sent by a single channel use. Forward classical communication

---

[7] Note the role of the channel fidelity in the context of encoding-decoding in this and the following definition.

cannot increase the quantum capacity of a channel [47].[8] However, if a classical back-channel from Bob to Alice is *also* available, allowing for two-way communication, an increase in channel capacity is then possible; the quantum capacity in that case will have a potentially higher value $Q_2(\mathcal{N}) \geq Q(\mathcal{N})$ through multiple adaptive uses of communication [46]. $Q_2(\mathcal{N})$ is known as the *classically assisted quantum capacity.*

The *entanglement-assisted classical capacity*, $C_E(\mathcal{N})$, is defined similarly to $Q(\mathcal{N})$ but instead for an interactive protocol that makes use of the quantum channel, shared entanglement, and unlimited classical communication between source and destination laboratories, instead of a quantum encoding-decoding scheme. This capacity can be written

$$C_E(\mathcal{N}) = \sup_{P(|\psi\rangle)} \left\{ S(\rho) + S(\mathcal{N}(\rho)) - S((\mathcal{N} \otimes \mathbb{I})P(|\psi\rangle)) \right\}, \qquad (9.6)$$

where ρ is the state obtained by taking the partial trace of $P(|\psi\rangle)$ over system B.[9] The value of $Q(\mathcal{N})$ bounds $C_E(\mathcal{N})$ from below; for example, see Section 9.8 below.

Various further capacities exist for quantum channels when assisted by entangled states shared by senders and receivers [46, 88]. As mentioned in the introduction to this chapter, entanglement alone, being undirected, cannot serve to transmit information from A to B but is capable of *improving* the capacity of a quantum channel, as in quantum dense coding. In particular, for a channel described by the identity map the entanglement-assisted classical capacity is *twice* that of the unassisted classical capacity when the dense coding protocol is used; see Section 9.8, below.

9.3 Holevo's theorem

Let us now consider the limitations on optimal communication using a *noiseless* quantum channel and data compression.[10] The *Holevo theorem*, which provides the *Holevo bound*, which was originally conjectured by Gordon [191] and stated without proof by Levitin [279, 280, 281], describes the fundamental limit on the amount of *classical* information from a sender that is accessible to a receiver in terms of the entropy of an ensemble of quantum systems decomposable as signal states $\{p_i, \rho_i\}$.

The unassisted quantum channel capacity given by Eq. 9.5 can also be written in terms of entropies as

---

[8] Note that the definition of $Q(\mathcal{N})$ involves consideration of *all* n qubit states, rather than only the *two computational basis states* of each bit, as in the definition of $C(\mathcal{N})$.

[9] It is worthwhile to compare this expression with that of Eq. 4.18.

[10] Quantum data compression is specifically addressed in Sect. 10.8.

$$Q(\mathcal{N}) = \max_{p_i} \left(S\left(\sum_i p_i \rho_i \right) - \sum_i p_i S(\rho_i) \right) , \qquad (9.7)$$

the quantity optimized being referred to as the *Holevo information*

$$\chi \equiv S\left(\sum_i p_i \rho_i \right) - \sum_i p_i S(\rho_i) , \qquad (9.8)$$

which has a form that accounts for the fact that the information present in the ensemble is *reduced* from that given by the von Neumann entropy as its components become increasingly impure. This quantity can also be written in terms of quantum relative entropy as

$$\chi = \sum_i p_i S(\rho_i \| \rho_{\text{avg}}) , $$

where $\rho_{\text{avg}} = \sum_i p_i \rho_i$.

This can be understood in terms of *information accessible to the receiver*, as follows. Consider a message received as such an ensemble. The optimal value of the mutual information $I(A : B)$ between sender Alice's input, A, and receiver Bob's measurement result, B, is bounded by the Holevo information:

$$I(A : B) \leq \chi , \qquad (9.9)$$

with Bob making measurements providing outcomes m with probabilities q_m resulting in a post-measurement ensemble $\{p_{i|m}, \rho_{i|m}\}$ and mutual information

$$I(i : m) = H(\{p_i\}) - \sum_m q_m H(\{p_{i|m}\}) , \qquad (9.10)$$

which Bob desires to *maximize* over all possible measurement strategies, H being the Shannon entropy.[11] In this way, Bob arrives at the maximum accessible information $I(A : B) = \max I(i : m)$. If the signal states are *pure states*, then $I(A : B) \leq S(\rho)$, the bound being achieved if and only if the encoding states ρ_i commute and Bob measures in the basis where they are represented by diagonal matrices. This upper bound is not generally a very strong one, however.

A tighter bound holds when the receiver's measurements are *not complete* and must be described by a POVM. Take the elements of such a POVM to be $\{E_m\}$ and the corresponding measurement outcomes to be indexed by m. The tighter, *Schumacher–Westmoreland–Wootters bound* is then given by

$$I(A : B) \leq \chi - \sum P_m \chi_m , \qquad (9.11)$$

---

[11] The Holevo information has the properties of additivity and monotonicity inherited from the von Neumann entropy, asymptotic continuity, and of being bounded by the Hartley entropy, $\log_2 \dim \rho$.

where p_m is the probability of measurement outcome m and χ_m is the Holevo information for the state after a measurement with outcome m.[12] The bound on the information is thus reduced by the amount of information that can still be obtained from the system *after* this measurement as well as the Holevo bound on that information [371].

The Holevo bound implies that $I(A : B) \leq S(\rho) \leq \log_2 d$, where d is the dimensionality of the Hilbert space of the encoding system, indicating that the amount of information encodable in a system is bounded by d, which corresponds to the number of orthogonal states available. In particular, one sees that the greatest amount of information that can be encoded in a qubit is *one bit*.

9.4 Discrimination of quantum states

The use of quantum channels to send information has specific advantages over the use of classical channels. For example, the use of qubits to encode bits allows one to perform cryptographic key distribution (QKD) without trusted couriers, and so to provide cryptographic security based on fundamental physical principles, something that has never been achieved classically. The ability to distinguish possible signals in communication is central to the transmission of information of any kind, quantum or classical. The problem of distinguishing quantum states is essential for QKD, and is relevant to many issues in the foundations of quantum theory as well.

For successful quantum communication in general, one needs to be able to distinguish the different states in which a quantum system can be prepared. Unless a measured collection of systems constitutes an ensemble of orthogonal subensembles, a given signal state cannot be perfectly discriminated from other possible signal states, so one must find an *optimal strategy* for discriminating between these possibilities. Indeed, the ability to perfectly discriminate all members of a set of *nonorthogonal* states would contradict the "no-cloning theorem," which is a simple corollary of basic quantum postulates; see Section 9.5 below. This fact is the basis of QKD protocols, which are based on the use of states from conjugate nonorthogonal bases. Furthermore, the behavior of any QKD strategy must be consistent with Holevo's theorem. The problem of determining the state of an individual qubit prepared in one of two, not necessarily orthogonal states $|p\rangle$ and $|q\rangle$ was explicitly considered in the 1980s as an issue in the foundations of quantum mechanics [135, 230, 328]. This issue has since been approached in several ways.

One approach considers what is now known as the problem of *hypothesis testing* or *ambiguous state discrimination* [211, 218, 466]. It is to find the procedure that yields *on the average* a maximum number of correct classifications

---

[12] The Schumacher–Westmoreland–Wootters bound is sometimes also called the *Holevo–Schumacher–Westmoreland bound*.

of state in an ensemble of such cases, assuming that for each member of the ensemble a *definite* classification is made. A different approach is to consider what is known as the problem of *unambiguous state discrimination*. It is to find the procedure that enables one, in a maximum number of cases, to infer *with certainty* whether the system was prepared in $|p\rangle$ or $|q\rangle$, leaving a minimum number of cases *unclassified*.

The primary interest has been in solving the problem of unambiguous state discrimination, which pertains directly to QKD, and which here we address first [152, 146]. The solution to the unambiguous state-discrimination problem in the simple case where one assumes that half of the ensemble is prepared in the state $|p\rangle$ and half in the state $|q\rangle$, as is ideally the case in, for example, BB84 QKD, was found through the following evaluation of the maximum probability of correct classification and minimum probability of no decisions.

$$P = 1 - |\langle p|q\rangle| \tag{9.12}$$
$$1 - P = |\langle p|q\rangle| , \tag{9.13}$$

the former expression, now known as the *Ivanovic–Dieks–Peres* (IDP) *limit*, is the probability of *correct* classification, the latter expression being the probability of making *no* classification [230]. The overlap $|\langle p|q\rangle|$ of the two states $|p\rangle$ and $|q\rangle$ is a measure of the degree to which they *cannot* be distinguished.

To better understand this result, consider preparing an *ancillary system* in an initial state $|s_0\rangle$ in addition to the given qubit and inducing a unitary state evolution on the resulting *composite* system:

$$|p\rangle|s_0\rangle \to a|p_1\rangle|s_1\rangle + b|p_2\rangle|s_2\rangle , \tag{9.14}$$
$$|q\rangle|s_0\rangle \to c|q_1\rangle|s_1\rangle + d|q_2\rangle|s_2\rangle , \tag{9.15}$$

the state-vectors $|s_1\rangle, |s_2\rangle, |p_1\rangle, |p_2\rangle, |q_1\rangle, |q_2\rangle$ being normalized and such that

$$\langle p_1|q_1\rangle = 0 , \tag{9.16}$$
$$\langle s_1|s_2\rangle = 0 , \tag{9.17}$$

$|q_2\rangle$ being identical to $|p_2\rangle$ except for a (unphysical) phase factor [135, 328]. This process provides one with the ability to make a measurement on the ancilla that distinguishes $|s_1\rangle$ from $|s_2\rangle$, something typically done by a QKD eavesdropper. When the state is $|s_1\rangle$, then measurement outcomes p_1 and q_1 determine whether the state of the qubit is $|p\rangle$ or $|q\rangle$. When the state is instead $|s_2\rangle$, that of the qubit *must be* $|p_2\rangle$ (or, equivalently, $|q_2\rangle$) and the question of whether one has state $|p\rangle$ or state $|q\rangle$ is left undetermined.

This approach was extended in the mid-1990s to the consideration of situations in which an *arbitrary* proportion r of systems of the ensemble are initially prepared in $|p\rangle$, with the remaining amount, $s = 1 - r$, being prepared in $|q\rangle$ (where, without loss of generality, one may take $r \geq s$) rather than making the simplifying assumption that $r = s = \frac{1}{2}$ [238]. Again, one considers the unitary evolutions above, but then seeks to optimize the probability

$$P = r|a|^2 + s|c|^2 \, , \qquad (9.18)$$

where $|\langle p|q \rangle| = |b| \, |d| \, |\langle p_2|q_2 \rangle|$ and so $|b| \, |d| \geq |\langle p|q \rangle|$. This is achieved when $|b|^2 = \max\{|\langle p|q \rangle|\sqrt{s/r}, |\langle p|q \rangle|^2\}$ and $|p_2 \rangle = |q_2 \rangle$. There are then two possibilities: one in which the above maximum is achieved by the former quantity and one in which it is achieved by the latter quantity. One then finds

$$P = 1 - 2\sqrt{rs}|\langle p|q \rangle| \qquad (9.19)$$

in the former case and in the latter case

$$P = r\left(1 - |\langle p|q \rangle|^2\right) \, , \qquad (9.20)$$

known as the *Jaeger–Shimony bound* [103, 238]. This solution is the one of interest in QKD because, for example, eavesdroppers seek to exploit more realistic situations where the QKD system is working *imperfectly*, in that signal states will not always be sent with the identical probabilities called for by the QKD protocol in use, so the quantum key distribution system must be described by r as well as by the nonorthogonal states designated in the ideal protocol.

Consider a system of interest attributed a Hilbert space of greater dimension than that of a qubit, which is the case for the qubits used in real-world QKD, where qubits are encoded in photons. In such situations a similar result is found; the above method of solution can still be followed in such cases *without* the need for an external ancilla, because additional degrees of freedom can serve the same function as an ancilla (*cf.* 6.86–92). Such a procedure might be followed by a QKD eavesdropper wishing, for example, to exploit other degrees of freedom in the photon that have may been neglected by QKD system designers or users. Because of the greater dimension of such a Hilbert space in this case, $|p \rangle$ and $|q \rangle$ can be written

$$|p \rangle = a|p_1 \rangle + b|p_2 \rangle, \qquad (9.21)$$
$$|q \rangle = c|q_1 \rangle + d|p_2 \rangle, \qquad (9.22)$$

where $|p_1 \rangle, |q_1 \rangle$ and $|p_2 \rangle$ are orthonormal, b and d satisfy the same conditions as above, and a and c are *real* numbers. Then $|q \rangle$ can be expressed in terms of $|p \rangle$ and a normalized state $|p^\perp \rangle$ orthogonal to it:

$$|q \rangle = e^{i\theta} n|p \rangle + (1 - n^2)^{1/2}|p^\perp \rangle, \qquad (9.23)$$

where $n = |\langle p|q \rangle|$. This produces the same results as the problem of determining a qubit state described above, but without the need of introducing an ancillary system [238].

A different approach can be used to solve the *hypothesis testing* problem not addressed above, that of find the maximal correct classification *on average*, with a determinate classification being made for *every* member of the ensemble. This can be done in the case of a general preparation by seeking a binary scheme that measures a projection operator O on the Hilbert space of the system and classifies for every measurement the measured state as either $|p\rangle$ or $|q\rangle$, depending on the measurement outcome. Label the projector eigenvalues as 1 and 0, where in measurements with the outcome 1, $|p\rangle$ is selected and in those with the outcome 0, $|q\rangle$ is selected. The probability of a correct selection is then

$$P = r\langle p|O|p\rangle + s\big(1 - \langle q|O|q\rangle\big) \ . \tag{9.24}$$

One considers $|q\rangle$ as a superposition of orthonormal vectors $|p\rangle$ and $|p^\perp\rangle$, as in Eq. 9.23, then finds the expectation values of this projection operator O for the states $|p\rangle$ and $|q\rangle$ and solves the corresponding optimization problem to find the appropriate projection, arriving at

$$P = \frac{1}{2}\left(1 + \sqrt{1 - 4rs|\langle p|q\rangle|^2}\right) , \tag{9.25}$$

in accordance with the *Helstrom bound* [211, 238]. This result has been applied to the problem of providing an error-free optimum quantum receiver for binary pure quantum signals [21].[13]

9.5 The no-cloning theorem

A problem related to the above results regarding quantum state discrimination is that of *copying* quantum states. It is impossible to make perfect copies of an unknown state of a quantum system by unitary operations. Were unknown quantum states able to be perfectly so cloned, a large number of perfect copies could be made and used to distinguish quantum states to whatever precision desired, contradicting the Holevo bound, for example. Another simple argument that such cloning cannot be performed by a unitary operation is that such an operation would then allow for the *simultaneous* measurement of two properties represented by noncommuting operators, which is precluded by the basic principles of quantum mechanics: it would enable the measurement of two such properties via the measurement of a different one of the two in *each* of the identical copies. The *no-cloning* theorem is the direct mathematical demonstration of the impossibility of cloning an unknown quantum state by a unitary operation.

---

[13] Related problems, including that of discriminating a larger number of states as well as *mixed states* in the form of *state comparison*—the determination of whether two systems are described by the same state—and *state filtering*—discriminating a pure state from a set of pure states—have also been considered since 2000; see, for example, [53, 104, 157, 158, 356, 409, 410].

A simple proof of the theorem is the following. Consider a unitary operator U that could perform both of the following transformations on two *different* *nonorthogonal vectors* $|\psi\rangle, |\phi\rangle$:

$$|a\rangle|\psi\rangle \rightarrow |\psi\rangle|\psi\rangle , \tag{9.26}$$

$$|a\rangle|\phi\rangle \rightarrow |\phi\rangle|\phi\rangle , \tag{9.27}$$

resulting in perfect copies of two such two unknown vectors $|\psi\rangle$ and $|\phi\rangle$ made from a given quantum state $|a\rangle$. This transformation would then give

$$\langle\psi|\phi\rangle = \langle\psi|\langle a|a\rangle|\phi\rangle = c \tag{9.28}$$

$$\rightarrow \langle\psi|\langle\psi|\phi\rangle|\phi\rangle = \langle\psi|\phi\rangle\langle\phi|\psi\rangle = (\langle\psi|\phi\rangle)^2 = c^2 ; \tag{9.29}$$

but this is possible *only if* $c = 0$ or $c = 1$, implying that $|\psi\rangle$ and $|\phi\rangle$ are either identical or orthogonal, contradicting our initial assumption. Thus, no unitary process can make identical copies of a general, *unknown* quantum state via such a process. At the same time, this calculation *is* compatible with the important task of copying of states from a *known orthogonal basis*.

The impossibility of a universal cloning procedure strongly distinguishes quantum information from classical information, and has broad practical implications such as lending security to quantum key distribution; see Chapter 12. Given this result, the practical problem of interest becomes that of optimal universal *copying*, which is addressed in Section 11.2.

9.6 Basic quantum channels

Let us now consider the effects of several nontrivial quantum channels on individual qubits, to better understand what quantum channels are like in practical, rather than noiseless, circumstances.

The *depolarizing channel* has been extensively studied in the context of the polarization encoding of quantum information. It can be viewed as taking a system to a fully mixed state with a probability, p, known as the strength of depolarization, or leaving it unchanged with a probability $q = 1 - p$, and so is described by

$$\mathcal{E}(\rho) = p\frac{1}{2}\mathbb{I} + (1 - p)\rho , \tag{9.30}$$

with corresponding decomposition operators

$$E_0 = \sqrt{1 - (3/4)p}\sigma_0, \ E_i = (1/2)\sqrt{p}\sigma_i,$$

where $i = 1, 2, 3$. The effect of this channel on the set of states initially described by the entire Poincaré–Bloch sphere (*i.e.* the pure states) is to uniformly reduce its radius by multiplying it by q in the Stokes parameter representation, in accordance with Eqs. 1.24–25. This channel is the quantum analogue of the classical binary symmetric channel discussed in Section 4.1.

Pauli channels. The *phase-flip channel* is described by the map

$$\mathcal{E}(\rho) = p(\sigma_3 \rho \sigma_3) + (1 - p)\rho , \qquad (9.31)$$

which can be described by the two decomposition operators

$$E_0 = \sqrt{1 - p}\sigma_0 \text{ and } E_1 = \sqrt{p}\sigma_3 ,$$

where σ_3 is the Pauli operator corresponding to the single-qubit gate describing phase-flipping. This channel has the effect on a set of states initially described by the entire Poincaré–Bloch sphere of reducing its width in the equatorial plane by multiplying it by a factor of $1 - 2p$. This channel also acts as a *phase-damping channel* channel as the operation elements are related by unitary transformations (*cf.* the box below Eq. 2.31). Under it, quantum information can be lost without energy being lost.

The descriptions and effects of the *bit-flip* and *bit+phase-flip* channels are analogous to that of the phase-flip channel, with the Pauli operators σ_1 and σ_2, respectively, taking the place of the σ_3 in the above, producing contractions of the Poincaré–Bloch sphere occurring along the corresponding orthogonal directions. Analogous basis vectors are similarly left unaffected by these channels. Decomposition operators for these are thus

$$E_0' = \sqrt{1 - p}\sigma_0, E_1' = \sqrt{p}\sigma_1, \text{ and } E_0'' = \sqrt{1 - p}\sigma_0, E_1'' = \sqrt{p}\sigma_2,$$

respectively. A qubit pure-state $|\psi\rangle$ that undergoes an *arbitrary* "error" by coupling to an environment, taken to be in some initial state $|\bar{E}\rangle$, evolves unitarily together with the environment into the entangled state

$$|\psi\rangle|\bar{E}\rangle = (a_0|0\rangle + a_1|1\rangle)|\bar{E}\rangle \rightarrow \quad (a_0|0\rangle + a_1|1\rangle)|\bar{E}_0\rangle \qquad (9.32)$$
$$+ (a_1|0\rangle + a_0|1\rangle)|\bar{E}_1\rangle \qquad (9.33)$$
$$+ (a_0|0\rangle - a_1|1\rangle)|\bar{E}_2\rangle \qquad (9.34)$$
$$+ (a_1|0\rangle - a_0|1\rangle)|\bar{E}_3\rangle , \qquad (9.35)$$

where the $|\bar{E}_\mu\rangle$ are not necessarily orthogonal states of the environment. Each of the above summands is the result of a distinct one of the above Pauli "errors"; see [401] and Sections 10.2–4.

The *amplitude-damping channel* produces the asymmetrical "decay" of one computational-basis state to the other, for example $|1\rangle$ to $|0\rangle$ as assumed below, with probability p. It can be described by the two decomposition operators

$$E_0 = \begin{pmatrix} 0 & \sqrt{p} \\ 0 & 0 \end{pmatrix} , \qquad (9.36)$$

$$E_1 = \begin{pmatrix} 1 & 0 \\ 0 & \sqrt{1 - p} \end{pmatrix} . \qquad (9.37)$$

Its effect on the states of the Poincaré–Bloch sphere is to produce an ellipsoidal set of states with a height multiplied by a scaling factor of $q = 1 - p$, contracting it upward (as in our choice of decay direction here, or downward in the alternative choice) rather than uniformly contracting it as in the cases above, due to its asymmetrical character, and a width multiplied by $\sqrt{q}$, contracting it uniformly inward as well.

In the context of quantum information processing, the most significant effect on a system is decoherence, which can take at least two forms, decay and dephasing, and is discussed in detail in the following chapter.[14] A number of specific values of the capacities introduced in Section 9.2 for some of the above-described channels can be found in [51].

9.7 The GHJW theorem

The GHJW theorem points out the fundamental nature of mixed states. By performing different measurements on individual qu-d-its of a system in a bipartite pure state $|\Psi\rangle_{AB}$, decompositions of a single-system ensemble can be produced that differ but are described by the same statistical operator.

Consider two different decompositions of the same statistical state of the first system

$$\rho_A = \sum_i |a_i|^2 P(|\psi_i\rangle) , \tag{9.38}$$

$$\rho_A = \sum_i |a_i'|^2 P(|\psi_i'\rangle) , \tag{9.39}$$

via Hilbert-space bases $\{|\psi_i\rangle\}$, $\{|\psi_i'\rangle\}$. These have differing purifications describing the total bipartite system that can be written

$$|\Psi\rangle = \sum_i a_i |\psi_i\rangle |\chi_i\rangle , \tag{9.40}$$

$$|\Psi'\rangle = \sum_i a_i' |\psi_i'\rangle |\chi_i'\rangle , \tag{9.41}$$

where $\{|\chi_i\rangle\}$ and $\{|\chi_i'\rangle\}$ are orthonormal bases for the Hilbert space of the second system. By measuring the *second* qu-d-it, B, in these bases, two different ensembles are therefore obtained that are described by the *same* reduced statistical operator ρ_A. The purifications $|\Psi\rangle$ and $|\Psi'\rangle$ are related to each other by a unitary transformation of the state of the second system acting only on the space of the second system, that is, of the form $\mathbb{I} \otimes U$. Thus, *either* of the two ensemble descriptions is obtainable by a measurement on the second system alone. Likewise, one can consider different decompositions for the reduced

---

[14] For detailed examples of quantum channels producing decoherence effects via dephasing, see [465].

state of the second system and find bases for the first system and purifications giving rise to that statistical operator by measurements of the first system. This result, known as the *GHJW theorem*, was shown by Gisin, Hughston, Jozsa, and Wootters [187, 229], is similar to a result originally obtained by Schrödinger [249, 298], and has been extended by Cassinelli *et al.* [99].[15]

This result can also be seen to describe "quantum erasure" at its most general, in that it describes the effect of the choice of measurement basis or, equivalently, choice of measurement apparatus: the information obtained by a measurement of system B, when classically communicated to A, results in a change of the description of the subsystem at A. It shows that any finite ensemble of bipartite quantum states can be remotely prepared by two agents in distant laboratories through local operations and classical communication (LOCC). A range of experiments demonstrating quantum erasure have been carried out (*cf.* [213]).

9.8 Quantum dense coding

A quantum communication scheme that provides insight into the value of entanglement for facilitating communication is *quantum dense coding*, first proposed by Charles Bennett and Stephen Wiesner [52]. Without shared entanglement, the transmission of a single qubit between Alice and Bob can only communicate *one bit* of information, as per Holevo's theorem. Because sending two units of information requires twice the resources needed to send a single unit of information, Holevo's theorem appears to require *two* qubits to be physically transmitted to send two bits of information when using a quantum channel. A property of systems in Bell states is that local operations on one qubit of the pertinent pair enable transformations between any one of the Bell states and any other. Quantum dense coding is a means of using a Bell state already shared by Alice and Bob that exploits this property to achieve the transmission of *two bits* of information by directly transmitting only *one* qubit. Quantum dense coding demonstrates that the addition of shared entanglement resources can enable Alice and Bob, in effect, to enhance the capacity of a shared quantum channel "beyond" the Holevo limit, as mentioned in Section 9.2 above.

A standard implementation of quantum dense coding proceeds as follows.

(i) Alice and Bob are provided with a shared pair of qubits in one of the states of the Bell basis, say, the singlet state $|\Psi^-\rangle$.

(ii) Alice performs on her qubit either the identity, basis-state flip, phase flip or basis-state+phase flip transformation, placing the full two-qubit system in the one of the four Bell states of her choice, then sends it to Bob.

(iii) Bob performs a Bell-state measurement on the pair of qubits now completely in his possession, providing him with *two* bits of information.

---

[15] This theorem is not to be confused with the identically named factorization theorem in differential geometry.

Thus, only *one* physical qubit is transmitted from Alice to Bob, in step (ii); step (i) can be carried out by a well-characterized entanglement source in the possession of *neither* Alice *nor* Bob. The choice of Alice is effectively one among the four available Bell states encoding a two-bit message that is obtained by Bob through his Bell-state measurement (*cf.* box and Fig. 3.5).

Because quantum communication is generally realized using photons, it is useful to explicitly consider Bell-state analysis in that context.

The *Bell-state measurement* (*Bell-state analysis*) used in the dense-coding protocol and elsewhere can be performed on a pair of photons that are polarization-encoded and, being bosons, must be described by a *total* wavefunction that is symmetric under interchange of particles. Because the total wavefunction of the photon pair includes both polarization *and* spatial parts, it can be written in the form

$$|\Psi\rangle = |\Sigma\rangle_{\text{space}} |\Pi\rangle_{\text{polarization}} . \qquad (9.42)$$

The spatial part can be consider to be specified by the *beam* occupation of the pair after encountering a (nonpolarizing) beam-splitter. This spatial part itself, $|\Sigma\rangle_{\text{space}}$, can be either symmetric *or* antisymmetric, as can the polarization part, $|\Pi\rangle_{\text{polarization}}$. In the polarization-Bell-state basis, there are four vectors each of which must be correlated with the spatial part in order to provide the required overall state symmetry. Only in the case of the (anti-symmetric) *singlet* state $|\Psi^-\rangle_{\text{polarization}}$ can the spatial part be anti-symmetric. The remainder of the polarization-Bell-basis states (the triplet states) are symmetric. In the former case, both photons must emerge anti-symmetrically from the beam-splitter, *i.e.*, in the *same* beam. This case will thus not appear in measurements conditioned by a positive simultaneous *coincidence* of single photons in different beams; this allows the polarization singlet, in which *both* spatial and polarization parts are anti-symmetric, to be distinguished from the remaining three cases by using two detectors, one in each beam.

One then notes that, although the remaining states are all exchange-symmetric polarization states, they can still differ in *polarization correlations*. Only the state $|\Psi^+\rangle$ will have its photon polarizations *anti-correlated*. Therefore, by also performing *polarization selection* using polarizers in the two output beams, one performs a *partial* polarization-Bell-state analysis using only linear optical elements (a beam-splitter and a pair of polarizers) into a total of *three sets* of polarization Bell state: $|\Psi^-\rangle$, $|\Psi^+\rangle$, and $\{|\Phi^+\rangle, |\Phi^-\rangle\}$. The remaining two states of the third set can then finally be distinguished using a measurement of polarization with appropriate *nonlinear* optics. For a detail discussion of Bell-state discrimination, see [472].

9.9 Quantum teleportation

Consider now the task of transmitting an unknown *qubit state*, $|\psi\rangle$.[16] This can be done provided, as in quantum dense coding, Alice and Bob have first come to share a Bell state (of two additional qubits), using the technique of *quantum teleportation*. To accomplish this task, Alice first measures, in the Bell basis, the state of the two qubits in her laboratory, namely, the qubit the state of which is to be transmitted and one qubit of the pair which she jointly shares with Bob, and communicates the result of this measurement to Bob as two *bits*. Based on the values of these two bits of classical information, when received, Bob performs one of four operations on the *second qubit* of the pair the state of which is shared by Alice and Bob, in order finally to obtain the desired state $|\psi\rangle$ of his qubit [42, 69, 71].

Let us consider this process in greater detail. Take the unknown state to be transmitted to be

$$|\psi\rangle = a_0|0\rangle + a_1|1\rangle \qquad (9.43)$$

and the shared initial Bell state to be $|\Phi^+\rangle = \frac{1}{\sqrt{2}}(|00\rangle + |11\rangle)$. Take the first of the shared qubits to be in the possession of Alice and the second to be in the possession of Bob. The three qubits involved in teleportation thus begin in the total, three-qubit state

$$|\Psi\rangle = |\psi\rangle|\Phi^+\rangle = \frac{1}{\sqrt{2}}(a_0|0\rangle + a_1|1\rangle)(|00\rangle + |11\rangle) . \qquad (9.44)$$

By an algebraic rearrangement, this state can be written as a sum of four terms wherein each term has the first two qubits—those in the lab of Alice—in a *different* Bell state, namely,

$$\begin{aligned}
|\Psi\rangle = \frac{1}{2}\big[\ &(|00\rangle + |11\rangle)\,(a_0|0\rangle + a_1|1\rangle) \\
+\ &(|00\rangle - |11\rangle)\,(a_0|0\rangle - a_1|1\rangle) \\
+\ &(|10\rangle + |01\rangle)\,(a_1|0\rangle + a_0|1\rangle) \\
+\ &(|10\rangle - |01\rangle)\,(a_1|0\rangle - a_0|1\rangle)\big] .
\end{aligned}$$

When Alice performs a Bell-state measurement on her qubits, she thus effectively obtains a two-bit result, corresponding to the particular one of the four Bell states her two qubits ended up in after her measurement. It is *these* two bits that she then sends to Bob, allowing him to perform the required local unitary transformation on his qubit, in order for it to end up in the unknown state $|\psi\rangle$, as described at the bottom of the box below.

---

[16] This task should be distinguished from the task of transmitting a *physical* qubit system itself, as it is only *state* "teleportation."

It is worthwhile here to consider in some detail how the Bell-basis measurement of the quantum state teleportation protocol can be carried out by Bell-state analysis described by a simple quantum circuit. In particular, this allows one to see how a two-bit outcome can be obtained by Alice more explicitly than by mere reference to the fact that she obtains a measurement result that is one among four alternatives.

To obtain two bits, Alice can transform her two qubits during the "pre-measurement" portion of this analysis, into the four two-qubit computational-basis states $|00\rangle, |01\rangle, |10\rangle, |11\rangle$ that correspond explicitly to the two-bit result of her measurement *before* a quantum state projection is effected. In particular, by performing a C-NOT operation on the two qubits in her laboratory (using the qubit the state of which is to be teleported as the *control* qubit) followed by a one-qubit Hadamard transformation on this unknown (control) qubit state, the four Bell-states are transformed into computational-basis states in a deterministic way. (Together, these two operations of the Bell-state measurement constitute running *backward* through the circuit for Bell-state creation shown in Fig 3.5.)

After these two unitary transformations, a correlated total *three-qubit* state results that is an evenly weighted linear combination of four states, each of which is of a form where the first two qubits are in a computational-basis state corresponding to a two-qubit eigenvalue and the *third* qubit is in a computational-basis superposition state with weights arising from that of the unknown state $|\psi\rangle$ being teleported, namely,

$$|\Psi\rangle = \frac{1}{2} \big[\, |00\rangle \, (a_0|0\rangle + a_1|1\rangle) \tag{9.45}$$
$$+ \, |01\rangle \, (a_1|0\rangle + a_0|1\rangle)$$
$$+ \, |10\rangle \, (a_0|0\rangle - a_1|1\rangle)$$
$$+ \, |11\rangle \, (-a_1|0\rangle + a_0|1\rangle) \big] \, .$$

Writing the three-qubit state in this way explicitly exhibits the perfect correlation between the (eigen)values of Alice's possible measurement outcomes and the possible quantum states of Bob's single qubit. When the Bell-state analysis is complete, the corresponding measurement projection produces an explicitly two-bit outcome and only one of the four above addends remains. By communicating the two-bit eigenvalue of her measurement result, Alice provides Bob with the required information as to the local unitary operation (described by a Pauli operator σ_μ) he must perform in order to recover the still unknown state $|\psi\rangle = a_0|0\rangle + a_1|1\rangle$ in the qubit in his laboratory, namely, the appropriate combination of (potentially trivial) bit-flip and (potentially trivial) phase-flip: for the bit-string 00, $\mu = 0$, for 01, $\mu = 1$, for 10, $\mu = 3$, and for 11, $\mu = 2$.

It is important to note that neither Alice *nor* Bob comes to learn the state $|\psi\rangle$ as a result of the teleportation process itself; the state is merely transferred from the first qubit to the third qubit. Alice learns *nothing* about the unknown qubit state itself by performing the Bell state measurement, in that she has come to know nothing about the (complex) amplitudes a_0 and a_1. Even Bob must still measure the qubit in his laboratory after teleportation in order to come to learn anything about its state.[17] Quantum teleportation of qubits and more complicated quantum systems has been carried out experimentally, for example, in systems described by continuous variables [177].

9.10 Entanglement "swapping"

Another quantum information-processing task closely related to quantum teleportation is the redistribution of entanglement. For example, two particles that are initially entangled with respective partner particles but not with each other can become entangled with each other if an appropriate measurement is made on a pair of qubits, one from each pair.[18] This is accomplished by a procedure known as the *entanglement-swapping protocol* [70, 320, 474].

For specificity, let us examine this process in an optical context. Consider two photon pairs simultaneously emerging in Bell-singlet states from two sources, each pair created by spontaneous parametric down-conversion (as discussed in Section 6.16) in a separate nonlinear crystal; call the photons from one source 1 and 2 and those from the other source 3 and 4. Performing a Bell-state measurement on two photons, *one from each source*, provides a projection of the state of the remaining two photons, also from different sources, onto a Bell state, changing the pairing of photons that are entangled, in that sense transferring the entanglement. The initial state of the pair of photon pairs, rewritten in the following way, exhibits the pertinent correlations.

$$|\Xi\rangle = \frac{1}{2}\big(|0\rangle|1\rangle - |1\rangle|0\rangle\big)_{12}\big(|0\rangle|1\rangle - |1\rangle|0\rangle\big)_{34} \tag{9.46}$$

$$= \frac{1}{2}\big(|\Psi^+\rangle_{14}|\Psi^+\rangle_{23} + |\Psi^-\rangle_{14}|\Psi^-\rangle_{23} + |\Phi^+\rangle_{14}|\Phi^+\rangle_{23} + |\Phi^-\rangle_{14}|\Phi^-\rangle_{23}\big)\,.$$

A Bell-state measurement of the joint state of photons 2 and 3, for example, will leave photons 1 and 4 in the same Bell state as photons 2 and 3, accomplishing the task of redistributing the entanglement among the photons.

---

[17] Even *then* Bob could not come precisely to learn the state $|\psi\rangle$ by a single measurement unless he were told the *basis* in which it had initially been prepared.

[18] As in the case of quantum teleportation, this activity was first experimentally carried out by members of the research groups of Francesco de Martini in Italy and Anton Zeilinger in Austria.

9.11 Entanglement "purification"

As the above protocols demonstrate, entanglement resources allow one to perform various uniquely quantum communication and information-processing tasks. Yet another important such task can be accomplished, namely, the consolidation of quantum resources shared between laboratories. If a source of entanglement is imperfect or the quantum states involved are imperfectly transmitted due to the quantum channel being noisy, resulting in, for example, a product of q-basis states with $q \neq 0, \frac{1}{2}, 1$ (see the box above Eq. 6.23), the shared collective state may be distilled via LOCC to a more valuable collective state by agents in two laboratories. This process is referred to as *entanglement purification* [43].

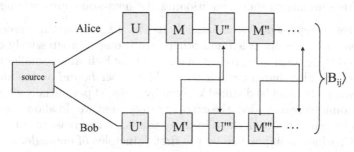

Fig. 9.1. The entanglement-distillation process. The Us are local unitary transformations. The Ms are local measurements. The lines represent classical communication between laboratories. $|B_{ij}\rangle$ represents the factors of the result [39].

A simple example of an entanglement-purification protocol (EPP), sometimes also referred to as an *entanglement-concentration* or *entanglement-distillation* protocol, uses local quantum operations and classical communication between two parties (2-LOCC) on an initially shared state of a *number of copies* of a bipartite quantum system that may be mixed or pure but nonmaximally entangled [43, 47]. After performing a number of sets of prescribed operations in stages, a smaller number of pure, highly entangled bipartite states of these systems can come to be shared, *e.g.* Bell states $|B_{ij}\rangle$.

Recall that the quantity of distillable entanglement can be identified with the number of Bell-basis states that can be so obtained. A quantum state $\rho \in \mathcal{H}$ is considered *distillable* if there is a number $k \in \mathbb{N}$ and state $|\Psi\rangle \in \mathcal{H}_s \subset (\mathcal{H}_A \otimes \mathcal{H}_B)^{\otimes k}$, where $\mathcal{H}_s$ is a 2×2-dimensional subspace, such that

$$\langle \Psi | (\rho^{T_A})^{\otimes k} | \Psi \rangle < 0 . \tag{9.47}$$

The problem under consideration then is the conversion, by 2-LOCC, of a number n of joint states ρ_{AB} constituting the collective state $(\rho_{AB})^{\otimes n}$, into a smaller number, nR, of pure fully entangled states, say $P(|\Phi_{AB}^+\rangle)^{\otimes nR}$. To carry out this task, one introduces a third, *purifying* system, E.

After purification, the state of all three systems will thus be

$$P(|\Phi_{AB}^+\rangle)^{\otimes nR} \otimes \rho_E .\qquad(9.48)$$

where ρ_E is the final state of the purifying system. In general, it is difficult to distill out of a mixed state *exactly* the number of Bell states required for its production. One-way communication may also be insufficient for entanglement distillation. In an EPP protocol, the two parties, Alice and Bob, begin with a bipartite state of n entangled pairs in the state ρ_M. The protocols proceed by the repeated application of the following actions by the two parties.

(i) LUTs U are performed on local subsystems.

(ii) Measurements M are performed on subsets of local subsystems.

(iii) Measurement results conditioning the next stage are exchanged.

This process is represented schematically in Fig. 9.1. During this process, some shared particles are discarded and others are brought progressively closer to the desired state, such as a product of a number of Bell states smaller than the number of initially shared particle pairs. The *upper bound* on the amount of entanglement that can be distilled is given by $E_f(\rho_M)$ per initially shared pair; if this amount were exceeded, the extra entanglement would allow *more* mixed states to be generated than those initially shared, an increase of entanglement by LOCC, which is impossible by the first principles of entanglement theory (*cf.* Sections 6.6–7).

For example, the *Schmidt projection method* for entanglement concentration proceeds from an initial set of n shared pure entangled pairs of qubits described by a product of entangled pure states

$$|\Psi\rangle = \prod_{i=1}^{n} \left(\cos\theta|\alpha_1(i)\beta_1(i)\rangle + \sin\theta|\alpha_2(i)\beta_2(i)\rangle \right) ,\qquad(9.49)$$

where the α_j and β_j label Schmidt-basis states of two qubits, that when expanded binomially has n^2 terms and $n+1$ distinct coefficients [39]. When one of the two agents sharing these pairs makes a precise measurement with outcome k described by a projection onto a state $|\Psi_k\rangle$ in one of the corresponding $n+1$ orthogonal subspaces of dimension $2\binom{n}{k}$ corresponding to the distinct coefficients, where $k = 0,\ldots,n$, they are left with a shared maximally entangled state in a known subspace of the initial 2^{2n}-dimensional state space, each of the two agents being in possession of a system state lying in a $\binom{n}{k}$-dimensional subspace.

If a particular such state, labeled by k, is desired, it is obtained with probability

$$p_k = \binom{n}{k} \left(\cos^2\theta \right)^{n-k} \left(\sin^2\theta \right)^k .\qquad(9.50)$$

The corresponding residual state is maximally entangled and can be transformed into the form of a product of a smaller number of Bell states by again making use of the Schmidt decomposition. This method is efficient when n is large, and works for any $n > 2$. A less efficient process, known as the *Procrustean method* can also be used that will work with as little as *one* shared entangled pair by making use of bivalent POVMs rather than precise measurements [39]. The above procedure can be straightforwardly extended to the case of entangled states of qu-d-its.

One reason entanglement purification may be required is that the quantum channel used to share entanglement is *noisy*. In that case, entanglement purification improves the fidelity of a subset of the entangled states agents intend to distribute. In this sense, entanglement purification can be viewed as a form of quantum error correction, about which more is provided in the following chapter. Furthermore, because quantum error-correction methods do not traditionally allow for classical communications, whereas entanglement purification protocols *may*, EPPs can be used for error correction in situations where quantum error-correction codes are insufficient.

For example, if Alice and Bob have a very noisy quantum channel between them and Alice desires to send an unknown qubit to Bob, she may not be able to do so using an error correction code but can do so by sharing Bell states with Bob through this noisy quantum channel, purifying them to a smaller number of pairs after transmission, and using the resulting purified pairs to *teleport* the qubit to Bob with the aid of their classical channel. This illustrates the fact, mentioned in Section 9.2, that the quantum capacity assisted by two-way classical communication, Q_2, can exceed the unassisted quantum capacity, Q.

9.12 Quantum data compression

In order to efficiently store quantum states produced by a known source of quantum systems in m possibly nonorthogonal states $|\psi_j\rangle$, which is represented by $\rho = \sum_{j=1}^{m} p_j P(|\psi_j\rangle)$ when each occurs with a corresponding probability p_j, a quantum version of data compression can be used. If an n-symbol string of such states is sent, it will occur with probability $\prod_{k=1}^{n} p_k$. An ensemble of possible n-symbol messages $\{\bar{p}_i, P(|\Psi_i\rangle)\}$ will be represented by $\bar{\rho} = \rho^{\otimes n}$. In general, such an ensemble will contain redundancies, which quantum data compression effectively eliminates.

In order to effect such compression, one can use a quantum code that allows one to move messages into states on a Hilbert subspace *smaller* than the $2^{n \log_2 m}$-dimensional one corresponding to $\bar{\rho}$. This can be realized by collecting a large number of copies of the system at the source and encoding their joint state into a smaller system that is transmitted through a quantum channel; the compressed state can be decoded finally by the receiver into a system of the same kind as the original system, in a state sufficiently close to the original one.[19] Keeping in mind the classical noiseless coding theorem given in Section 4.6, one may view the source as essentially *classical* in nature and described by a Shannon entropy $H(\{\lambda_{i_j}\}) = S(\rho)$, sending messages as strings of eigenstates of ρ with probabilities equal to the product $\lambda_{i_1}\lambda_{i_2}\cdots\lambda_{i_n}$ of their eigenvalues by considering the orthonormal basis in which the statistical operator ρ is *diagonal*. As in classical coding theory, it is helpful for this to consider a "typical sequence." Its analogue in quantum coding takes the form of the *typical quantum subspace* associated with $\bar{\rho}$, in the sense now described.

Consider a (discrete) quantum source capable of producing qu-d-its, considered without loss of generality as $d \log_2 d$ qubits, together with an operation to carry out quantum data compression on the chosen message. Again, as might be expected by analogy to the classical data-compression results, $S(\rho)$ qubits per source state is the minimum number required [367]. Nontrivial compression can be carried out provided the message is *nonrandom*, that is, provided $\rho \neq \frac{1}{2}\mathbb{I}$. The von Neumann entropy value indeed describes the limit of compression of quantum information in an ensemble described by the source state ρ, regardless of the manner in which the quantum information is generated. n qubits can therefore be encoded in at best $nS(\rho)$ qubits.

In particular, consider independent identically distributed (i.i.d.) states sequentially produced by a quantum source and characterized by Shannon entropy $H(\{\lambda\}) = S(\rho) \leq \log_2 d$. One may make use of the following *typical subspace theorem* (Schumacher [367]): Given a real $\epsilon > 0$, for any $\delta > 0$ and sufficiently large n there exists a projector $P^{(n)}$ onto an at most $2^{n[S(\rho)+\epsilon]}$-dimensional subspace, such that $\text{tr}(\bar{\rho}P^{(n)}) > 1 - \delta$, that commutes with $\bar{\rho}$. Under the same conditions, the dimension $\text{tr}(P^{(n)})$ of this "typical subspace" is bounded:

$$(1 - \delta)2^{n[S(\rho)-\epsilon]} \leq \text{tr}(P^{(n)}) \leq 2^{n[S(\rho)+\epsilon]} , \qquad (9.51)$$

that is, the signal strings converge to states on the typical subspace as n becomes large. For a projector $\bar{P}^{(n)}$ onto any subspace of at most 2^{nR} dimensions, for a fixed $R < S(\rho)$, for any $\delta > 0$ and sufficiently large n, one has

$$\text{tr}(\bar{\rho}\bar{P}^{(n)}(\rho)) \leq \delta . \qquad (9.52)$$

One also has the *quantum noiseless coding theorem* (Schumacher): Given an i.i.d. quantum source described by ρ, there exists a reliable compression scheme with compression rate $R > S(\rho)$ for it, and for $R < S(\rho)$ there exists no such reliable scheme.[20] It therefore suffices in practice to code the typical

---

[19] When the states characterizing the subensembles are orthogonal, the problem reduces to a classical one. When the states are nonorthogonal, however, classical compression methods will damage them. Furthermore, the quantum encoder must not contain any memory of them or they will not be recoverable during decoding (*cf.* comments at the end of Sect. 9.9).

[20] In addition to Schumacher's original proofs cited above, one can find alternative proofs of these theorems in [315].

subspace and to ignore its complement. Jozsa and Schumacher accordingly provided the following scheme for *quantum data compression* [245]. Given a source, characterized by ρ, and large M-symbol blocks, $|\Psi^{(M)}\rangle$, states of the typical subspace of dimension $2^{M[S(\rho)+\epsilon]}$ are transformed by a unitary operation U to the subspace of the first $M[S(\rho)+\epsilon]$ qubits of a corresponding sequence of $M \log_2 d$ qubits, where d is the dimension of the states produced by the source. Placing in the state $|0\rangle$ the remaining $R = M \log_2 d - M[S(\rho)+\epsilon]$ qubits of the full sequence of qubits results in a (generally mixed) state $\rho^{(M)}$. In a transmission making use of such data compression, sender Alice carries out the above process and then sends only the *first* $M[S(\rho) + \epsilon]$ qubits. The receiver Bob then decompresses them, by first *adding* R null qubits in $|0\rangle$ so as to recover $\rho^{(M)}$, and then performing the transformation U^{-1}. The result is that Bob obtains the signal block with an average fidelity greater than $1 - \delta$.

9.13 Quantum communication complexity

Given the procedures of entanglement purification and quantum data compression, a straightforward modification of the schema of classical communication allows one to use quantum resources such as entangled qubit pairs to obtain substantial improvements in the efficiency of communication. Quantum communication complexity theory is the general study of computational tasks and their efficiency in a context where quantum channels as well as classical communication channels are available to agents [262, 464]. One can consider communication that remains largely classical but where an unlimited supply of quantum *entanglement* resources are available to the parties that have the ability to perform local operations [109]. For example, rather than communicating using only bits, parties communicate making use of entangled qubits as in dense coding, as described above. Yet more complex situations involve communication using qubits as well as previously shared entangled systems.

As already shown in this chapter, quantum resources allow a number of tasks to be accomplished more efficiently than possible when only the transmission of classical bits is allowed and quantum entanglement between physical systems is not present; without quantum entanglement, laboratories must communicate with each other to come to possess nonlocal correlations, whereas with shared entanglement present no communication is necessary to produce such correlations, which can then be used to carry out tasks.

Indeed, quantum nonlocality demonstrations can be understood as communication complexity problems in which shared quantum entanglement reduces the amount of communication required. In particular, a two-party quantum communication protocol has been shown to operate more efficiently than the corresponding classical protocol for determining a binary function $f(x, y)$ of two input bits x_0 and y_0 received by an Alice and a Bob, respectively, who can thereafter no longer communicate. Such a function can be determined provided the two agents output two bits a and b, respectively, such that $a \oplus b = x_0 \wedge y_0$ with as high a probability as possible, where $\oplus$ indicates XOR (addition mod 2) and $\wedge$ indicates AND.[21]

Assume an Alice and a Bob initially to have random two-bit strings $a = a_1 a_2$ and $b = b_1 b_2$, respectively, with bits satisfying the conditions $a_i \oplus b_j = 0$ for all i, j, except when $i = j = 1$ in which case $a_1 \oplus b_1 = 1$. All four conditions cannot be simultaneously satisfied deterministically, but any three of them *can* be; they can thus be satisfied by classical random variables with probability at best $\frac{3}{4}$. Assume the a_i and b_j to be randomly distributed independently of inputs x_0 and y_0. There is a quantum protocol in which Alice sends a single bit to Bob, which he uses to compute the function $f(a, b) = a_1 \oplus b_1 \oplus (a_2 \oplus b_2)$ but must sometimes *fail* to accomplish this task; Bob can compute $f(a, b)$ if he can obtain the proper information about Alice's string (either a_1 or the binary sum of her two bits), which can be done with probability $p(1-p) = \cos^2(\pi/8) > \frac{3}{4}$ by performing measurements on a Bell state and performing the appropriate pair of rotations on Alice's and Bob's qubits [85].

To see this, consider Alice and Bob each to possess one of two qubits initially in the shared Bell state $|\Phi^-\rangle_{AB}$ and then to perform the following actions, with the string $x_0 y_0$ as their input. They wish to determine the function $f(x, y) = x_1 \oplus y_1 \oplus (x_0 \wedge y_0)$. If $x_0 = 0$, then Alice applies a $-\frac{\pi}{16}$ rotation to her qubit in the appropriate plane; otherwise, she applies a rotation of $\frac{3\pi}{16}$. She then measures the qubit, obtaining the outcome bit a. Bob proceeds equivalently with his qubit, obtaining his outcome bit b. These actions produce a two-qubit superposition of states $|\Phi^-\rangle_{AB}$ and $|\Psi^+\rangle_{AB}$ wherein the probability amplitude of the former is $\cos(\theta_A + \theta_B)$ and that of the latter is $\sin(\theta_A + \theta_B)$, where the rotational angles θ_X ($X = A, B$) are those actually performed. Thus, the probability that $a \oplus b = 0$, that is, the two qubits end up in their original state is $p(a \oplus b) = \cos^2(\theta_A + \theta_B)$, where the rotation angles actually executed are here labeled according to the party implementing them and $p(a \oplus b)$ is the probability of producing the desired outputs. In all cases, $\theta_A + \theta_B = \frac{\pi}{8}$. Alice then sends $a \oplus x_1$ to Bob, who sends $b \oplus y_1$ to Alice, so that both parties can *individually* determine $f(a, b)$, because they can at that point determine the bit $(a \oplus x_1) \oplus (b \oplus y_1) = x_1 \oplus y_1 \oplus (a \oplus b)$, which equals the $f(x, y)$ given above with probability $\cos^2(\frac{\pi}{8})$, which exceeds $\frac{3}{4}$.

Investigations of quantum communication complexity have been made experimentally [84] and extended to multiparty situations [86]. Consider n parties in possession of partial input data for some n-variable function. The communication complexity of the function is the minimum number of classical bits that must be broadcast for every party to come to know the value of the function. Interestingly, it has been found that there exists a broad class of quantum communication complexity protocols that can improve the efficiency of solution of communication complexity problems beyond what is possible classically if and only if a Bell-type inequality for *qutrits* is violated [84].

---

[21] This function has also been used to probe quantum mechanics *itself* [102].

Quantum decoherence and its mitigation

One of the greatest challenges in quantum information science is the development of methods for efficiently maintaining the uniquely quantum properties of states during the performance of quantum information-processing tasks. In particular, quantum information-processing hardware designs are faced with two competing requirements, that of strong interactions between their internal components for the performance of controlled quantum gates and that of isolating qubits from the environment in which the hardware operates.[1] *Quantum decoherence* is the loss of coherence within a quantum state behind the second of these requirements, and can take the form of dephasing or the loss of state population described by a nonunitary evolution of local system states, due to unwanted interaction with the environment.[2] This interaction generally necessitates the correction of errors it induces, which are highly detrimental to quantum information-processing tasks.

One of the benefits of information processing with few qubits, say the one or two qubits of quantum key distribution with photons, is that the coupling of qubit and environment can be kept small for photons. On the other hand, effective deterministic quantum computing generally requires *many* qubits and the ability to implement conditional quantum gates between different, intersecting sets of qubits. In the case of multi-photon states, the lack of direct photon–photon coupling renders *deterministic* quantum computation ineffective, but probabilistic approaches remain viable. In the case of more strongly interacting particles, their coupling with the environment tends to be strong as well, leading to increased decoherence. A number of techniques for avoiding and mitigating decoherence have been developed that are discussed here, along with principles and methods of quantum error correction that are vital to the further development of quantum information processing.

---

[1] See, for example, [429].

[2] A more general theory of decoherence was laid out by Roland Omnès in [317].

10.1 Quantum decoherence

Quantum decoherence can result from an environment's interacting with individual qubits in such a way that an independent random average phase shift $\langle \phi_i \rangle$ $(i = 0, 1)$ is given to each basis state, and thus a phase shift relative to its complement, that is, the environment may induce the following transformation on these states

$$|0\rangle \to e^{i\langle \phi_0 \rangle}|0\rangle , \tag{10.1}$$

$$|1\rangle \to e^{i\langle \phi_1 \rangle}|1\rangle , \tag{10.2}$$

resulting in an observable *relative phase difference* (*dephasing*) between computational basis states. Decoherence may also transform qubits in such a way that the population of qubit states is changed in a manner describable by a nonunitary evolution of the statistical operator of each qubit.[3] Initially pure states are then transformed to mixed states in which off-diagonal terms have simply vanished due to *decay* (*i.e.*, amplitude damping).[4] More precisely, the decoherence of the state of an ensemble of pure qubit states $|\psi\rangle_Q = a_0|0\rangle_Q + a_1|1\rangle_Q$ may occur in a quantum channel so that

$$P(|\psi\rangle_Q) \to \rho'_Q = \begin{pmatrix} |a_0|^2 & a_0 a_1^* e^{-\gamma(t)} \\ a_0^* a_1 e^{-\gamma(t)} & |a_1|^2 \end{pmatrix} , \tag{10.3}$$

where $\gamma(t)$ is a positive, time-dependent real parameter characterizing coupling to the environment and the statistical operator is kept normalized. For example, in a d-level system, the dephasing due to coupling with the environment has the following effect in the case of a generic, possible mixed, initial qubit state.

$$\rho_Q \to \rho'_Q = \frac{\epsilon}{d}\mathbb{I} + (1 - \epsilon)\rho_Q , \tag{10.4}$$

where ϵ is the strength of depolarization. Such behavior has been carefully studied experimentally in the important case of the qubit ($d = 2$), as described in Chapter 8 [334].

Decoherence can also be viewed as the unitary evolution of a compound system consisting of the qubit and its environment under which the system and environment become entangled. This is just the situation faced by *Schrödinger's cat* when it becomes entangled with its environment [365]. Consider two pertinent environmental states $|\bar{E}_0\rangle$ and $|\bar{E}_1\rangle$ that are not necessarily orthogonal, in particular, that are such that $\langle \bar{E}_0|\bar{E}_1\rangle = e^{-\gamma(t)}$.[5] The pure

---

[3] See [317] for the presentation of a general theory of decoherence.

[4] See also Sect. 9.6, where the joint evolution of a qubit and environment is described. For a detailed treatment of decoherence effects described as quantum channels and a discussion of the effects on state entanglement, see [465].

[5] The bar notation for the $|\bar{E}_i\rangle$ here, as before, serves to distinguish these states from decomposition operators for CPTP maps used to describe the effects of quantum channels.

time-dependent statistical operator for the *total system* of qubit and its environment can be written

$$\rho_{Q+\bar{E}}(t) = |a_0|^2 |Q_0\bar{E}_0\rangle\langle Q_0\bar{E}_0| + a_0 a_1^* |Q_0\bar{E}_0\rangle\langle Q_1\bar{E}_1| \qquad (10.5)$$
$$+ a_0^* a_1 |Q_1\bar{E}_1\rangle\langle Q_0\bar{E}_0| + |a_1|^2 |Q_1\bar{E}_1\rangle\langle Q_1\bar{E}_1| .$$

The reduced statistical operator of the qubit provides its individual description, namely,

$$\rho_Q(t) = \text{tr}_{\bar{E}}\rho_{Q+\bar{E}}(t) \qquad (10.6)$$
$$= \langle \bar{E}_0 | \rho_{Q+\bar{E}}(t) | \bar{E}_0\rangle + \langle \bar{E}' | \rho_{Q+\bar{E}}(t) | \bar{E}'\rangle , \qquad (10.7)$$

where $|\bar{E}'\rangle = (|\bar{E}_1\rangle - e^{-\gamma(t)}|\bar{E}_0\rangle)/\sqrt{1 - e^{-2\gamma(t)}}$ is a state effectively orthogonal to $|\bar{E}_0\rangle$ at time t. As the qubit and its environment continue to interact over the decoherence time, an increasingly mixed ensemble of computational-basis states with *no* mutual coherence ultimately results, which is the limiting behavior of the initially pure qubit state described by Eq. 10.3, namely,

$$\rho_Q(t) \to \rho_Q' = |a_0|^2 P(|0\rangle_Q) + |a_1|^2 P(|1\rangle_Q) . \qquad (10.8)$$

Decoherence occurs rapidly in quantum computing systems, which by nature are generally far more complex than, say, those required to implement quantum key distribution.[6]

Quantum information-processing systems must be engineered and their states encoded so as to minimize such unwanted interaction with the environment and to sufficiently slow the rate of decoherence that all needed quantum logic operations may be well performed. Random errors accumulate as steps in a random walk do, that is, with a probability that accumulates *linearly* in the number of operations leading to them, say, the number of quantum gates effected. Three primary methods for mitigating and counteracting the effect of decoherence in quantum information-processing systems are decoherence-free subspace methods, quantum error-correction methods, and dynamical decoupling methods. The first two of these are considered below.

10.2 Decoherence and mixtures

It might be thought that the need to describe quantum systems by statistical operators arises simply from the influence of *noise* on quantum systems, that is, that all such states could be described by some pure state, being contaminated by random external influences. However, noise is generally *local* in character, as in the description above, and is certainly so in cases where one is dealing with photons separating from each other at the speed of light.

---

[6] The behavior of such a large number of qubits under decoherence is also described in Sect. 1.7.

This severely constrains the ability of noise alone to give rise to mixed states. Mixed states that *can* be produced by local noise influences alone are referred to as *locally contaminated* (LC) and those in which local measurements and classical communication may occur are considered states produced by local contamination and classical communication (LCCC). This lends further support to the view that mixed states are fundamental in nature.

Let us consider in some detail the question of which statistical operators are accessible from pure states under such local processes. For N-partite systems with qu-d-it subsystems, the initial pure states are described by $2(d^N-1)$ real parameters, whereas the number of real numbers necessary for describing a general mixed state of the system is larger, namely $d^{2N} - 1$, as discussed in Chapter 7. That not all statistical operators are accessible from initially pure states via noise effects can be seen by noting that the number of real parameters characterizing state contamination is only *linear* in the number of parties, N, so that the number of parameters describing the pure states and their contamination is *less* than the number describing the statistical states.

In particular, note that the most general transformation of *each* of N qu-d-its $|i\rangle$ interacting locally with its environment is

$$|i\rangle|0\rangle_{\bar{E}} \rightarrow \sum_j |j\rangle|\bar{E}_{ij}\rangle_{\bar{E}} \ , \tag{10.9}$$

where $|\bar{E}_{ij}\rangle_{\bar{E}}$ are (not necessarily orthonormal) environmental states specified by d^4 parameters and constrained by d^2 conditions arising from the orthonormality of the qu-d-it states, which are therefore described by $(d^4 - d^2)$ parameters. Thus, the number of parameters governing the interaction of all N qu-d-its with their local environments together with the number of real parameters describing the initial pure states is $2(d^N - 1) + N(d^4 - d^2)$, whereas the statistical operators of the qu-d-its may require up to $(d^{2N} - 1)$ parameters to be fully described. As a result, it is not possible to account for the appearance of all statistical operators from initially pure states by local environmental effects alone for systems composed of more than two qu-d-its [212].

10.3 Decoherence-free subspaces

In the decoherence-free subspace method of decoherence mitigation, one makes use of a specially chosen type of subspace, a *decoherence-free subspace* (DFS), of the full Hilbert space of the information-processing system that by construction is protected from the influence of noise from its environment. This subspace is spanned by logical-qubit states of several physical qubits [468, 469]. This approach to decoherence mitigation typically assumes that the qubits under consideration are subject to decoherence of a sort in which the environmental disturbances are *identical* for all qubits, with each qubit interacting with the environment in a similar but uncorrelated way, sometimes called *collective decoherence*. The logical states (sometimes called *error-avoiding quantum code states*) of decoherence-free subspaces accordingly possess a symmetry, for example, under permutation of physical components. This symmetry is associated with the fact that the logical states of DFSs are often maximally entangled states.

To see how a properly chosen encoding subspace can help mitigate decoherence, consider the encoding of a logical qubit into the Bell states $|\Psi^{\pm}\rangle$, by the mapping

$$|0\rangle \mapsto |0_L\rangle = |\Psi^-\rangle \, , \tag{10.10}$$

$$|1\rangle \mapsto |1_L\rangle = |\Psi^+\rangle \, . \tag{10.11}$$

A dephasing-noise environment such as the first one described above in Section 10.1 will have no influence on encoded states in the subspace spanned by $|\Psi^{\pm}\rangle$. It will have the effect

$$|\Psi^-\rangle \to e^{i\langle\phi_0\rangle}|0\rangle e^{i\langle\phi_1\rangle}|1\rangle - e^{i\langle\phi_1\rangle}|1\rangle e^{i\langle\phi_0\rangle}|0\rangle \tag{10.12}$$

$$= e^{i(\langle\phi_0\rangle+\langle\phi_1\rangle)}\big(|0\rangle|1\rangle - |1\rangle|0\rangle\big) \tag{10.13}$$

$$= e^{i(\langle\phi_0\rangle+\langle\phi_1\rangle)}|\Psi^-\rangle \tag{10.14}$$

in the case of the Bell state $|\Psi^-\rangle$, the resulting global phase factor being unobservable; a similar global phase will clearly also result in the case of such noise on the Bell state $|\Psi^+\rangle$, and therefore on linear combinations of the two. This sort of noise will accordingly have no computationally relevant effect on the logical states of this decoherence-free subspace, because $|0\rangle_L \to e^{i(\langle\phi_0\rangle+\langle\phi_1\rangle)}|0\rangle_L$ and $|1\rangle_L \to e^{i(\langle\phi_0\rangle+\langle\phi_1\rangle)}|1\rangle_L$; the logical states *remain orthogonal* despite the noise, as can be seen by taking their inner product.[7]

10.4 Quantum coding, error detection, and correction

Bit errors in classical systems can be corrected, for example, through the use of a simple *repetition code*, wherein each bit is encoded into a *logical bit* consisting of several duplicates of itself: $i \mapsto i_L = ii \ldots i$, $i = 1, 2$. In this way, provided errors are independent and unlikely, by checking the values of the encoding bits and maintaining each at the dominant bit value, one may render single-bit errors incapable of flipping the value of the logical bit, which is similarly decoded by the receiver using such a *majority vote*.[8] More generally, linear codes are special subgroups of the full error group consisting of errors that take codewords to codewords and, as such, are undetectable.

---

[7] This situation has been well investigated experimentally by the groups of Paul Kwiat and Aephraim Steinberg; for example, see [264, 302].

[8] The classical case is described in Sect. 4.6.

A noisy quantum channel can similarly cause the value of a transmitted bit to flip value with a given probability, p, analogously to the situation in the classical binary symmetric channel (see Fig. 4.2). A quantum majority vote procedure can also be carried out. For example, for three qubits, one can use the encoding

$$|i\rangle \mapsto |iii\rangle \qquad (10.15)$$

for this purpose. A simple quantum circuit consisting of a C-NOT gate controlling each of two ancillae initially in the state $|0\rangle$ suffices for the construction of these codestates. Although an essential component of any quantum error-correction scheme is the encoding of quantum information, quantum states are often susceptible to a *broader* range of errors that may render unrecoverable superpositions of states encoded in this simple way.[9]

In the quantum context, a majority vote procedure analogous to the classical one encounters the difficulty that one must perform a *measurement* on all the "voter" qubits in order to correct errors, a process that can also destroy quantum coherence. Unlike classical bit errors, qubit errors are not discrete but rather have a continuous character just as have qubit states themselves. Thus, a qubit can enter any of an *infinite* number of possible states, rather than just two, as a result of the presence of environmental noise. Because of this continuous character, the quantum fidelity

$$F\big(P(|\psi\rangle), \rho\big) = \langle\psi|\rho|\psi\rangle \qquad (10.16)$$

is a helpful tool to supplement the simple discrete accounting of errors. It is also important to note that the unitary operations required by quantum information processing may also be imperfectly executed. A corrected or processed qubit must have nearly full fidelity, $F = 1$, with its desired state. In fault-tolerant quantum information processing, a corresponding *amplitude of failure* is also often introduced [258].

The quantum parallelism of quantum computing uses high-visibility multiqubit interference to operate efficiently, which requires coherence to be maintained throughout computation. To avoid the difficulties potentially posed by decoherence, quantum error correction uses entangled states and partial-state measurements to extract only that information associated with errors without disturbing vital state coherence. If errors on distinct qubits occur independently and the probability p of an error to occur on any given qubit

---

[9] Note that here, because the basis is known and the states involved are *orthogonal*, this does not violate the no-cloning theorem. Only *bits* are actually copied in this process. Also note that by contrast with quantum data compression, which seeks to eliminate redundancy, quantum coding *introduces* it.

is sufficiently small, it is a good approximation to ignore the possibility of more than n total errors occurring, because more than n errors will occur with a very small probability, of the order $O(p^{n+1})$. Therefore, one generally considers the most common situations, which deal with a limited number of errors.

Similarly to the logical states of decoherence-free subspaces discussed in the previous section, quantum error-correcting code (QECC) states are typically also entangled states. QECCs can be understood as functioning, in essence, by storing information in the quantum correlations among different components of the composite system realizing the code. If a portion of a system in a code state is influenced by its environment but remains well correlated with other portions of the system, then the encoded information still remains in these correlations and remains salvageable from them through error recovery procedures.

The continuous nature of quantum errors does not present an insurmountable problem because quantum errors can be reduced to a few *types* of error. Just as the classical binary symmetric channel produces a number of classical bit-flip errors $\mathbf{e} \in GF(2)$ with a given maximum Hamming weight, quantum *Pauli channels* are those where single-qubit errors and products thereof produce *at most* a given number of errors. The two fundamental types of qubit error are the bit-flip error and sign-flip error. A *bit-flip error* on a single qubit is described by the operation of the Pauli operator σ_1 on it. *Sign-flip errors* (also known as *phase-flip errors*) are similarly described by the operation of σ_3 on the qubit at which they occur, as mentioned in the discussion in Section 9.6 on the quantum channels inducing them.[10] The errors arising from a number of these basic errors in Pauli channels are thus naturally described by the *Pauli error group*, G_n, for n qubits:

$$G_n = \left\{ e_1 \otimes e_2 \otimes \cdots \otimes e_n | e_i \in G, i \in \{1, \ldots, n\} \right\}, \qquad (10.17)$$

where G is the single-qubit Pauli group, $\{\pm\sigma_\mu, \pm i\sigma_\mu\}$ $(\mu = 0, 1, 2, 3)$, which is G_1.[11] Thus, the multiple-qubit error group consisting of such tensor products is seen to be a subset of the group $U(2^n)$ of unitary operators on $(\mathbb{C}^2)^{\otimes n}$. An *error weight* consisting of the sum of the number of bit-flip errors plus the number of phase-flip errors is attributed to a given error-group element

---

[10] Both types of qubit error are thus describable by the corresponding gates discussed in Sect. 1.4.

[11] See Sect. 9.6 for a discussion of quantum channels inducing the errors comprising G, the Pauli channels. Note that G is a finite group of order 16, with a center consisting of diagonal matrices with a related "effective" error group G_{eff} being G modulo this center, which is a cyclic group of order 4. The effective error group is thus an Abelian two-group of order 4 that is isomorphic to $\mathbb{Z}_2^2$. The algebra of the Pauli matrices is such that any of them can be constructed from at most two of the remaining three.

$e \in G_n$. These errors are correctable, as are linear combinations of them, because G_n is a basis for the space of $2^n \times 2^n$ matrices.

Consider now a single-qubit state of the generic form

$$|\psi\rangle = a_0|0\rangle + a_1|1\rangle \tag{10.18}$$

to be mapped into a linear combination of *quantum-code states* (logical states) $|0_L\rangle$, $|1_L\rangle$ lying in a higher-dimensional Hilbert space of a larger system including additional, ancillary qubits. In the DFS example presented in the previous section, only a single ancilla was used to support this. More robust logical states for error correction involve a greater number of ancillae, and hence larger spaces. Specifically, the state of each qubit Q_i of a k-element quantum register is encoded in a logical state of an element $\bar{C}_i$ of a larger n-element register by a mapping of the form

$$(a_0|0\rangle_{Q_i} + a_1|1\rangle_{Q_i})|00\ldots0\rangle \mapsto a_0|0_L\rangle_{\bar{C}_i} + a_1|1_L\rangle_{\bar{C}_i} , \tag{10.19}$$

so that k logical qubits are present in an integral number $n \geq k$ of physical qubits.[12]

Geometrically, QECCs are subspaces such that any error in a small number of qubits moves the state in a direction *perpendicular* to them, allowing the error to be corrected by reference to this change. The (logical) computational basis $\{|0_L\rangle, |1_L\rangle\}^{\otimes k}$ is chosen such that errors induced by the environment also take each basis state to a subspace that preserves the required quantum coherence and leave the computational register and the environment in a joint tensor product state. States susceptible to decoherence are initially encoded by a unitary operation into the corresponding *error-correction codespace* within a larger subspace of the full encoding system of original bits plus the ancillae. The ancillae evolve into mutually orthogonal states dependent on interaction with the environmental noise, so that error correction is possible based on the results of measurements of these ancillae, which provide information as to the specific errors induced, in a procedure known as *error extraction*: *error detection* is performed, whenever necessary, using a quantum measurement that projects the computational state onto a superposition of projectors each derived from a specific code states by introducing the corresponding error in it. The original state can be then recovered by a unitary transformation into the codespace that depends on the result of the detection known as the *error syndrome*, which corresponds to a specific error.

Consider a projector P onto the codespace. To each error syndrome there will correspond one of a set of orthogonal subspaces. *Error correction* is carried out via a (trace-preserving) recovery operation $\mathcal{R}(\rho)$. In the statistical-operator description, a proper error-correction recovery process corrects the error process $\mathcal{E}(\rho)$ in the sense that

---

[12] Note that even in cases of repetition coding this process does not involve the cloning of qubits because the *local* physical qubit states obtained by partial tracing out all others will not in general be the same as that of the initial qubit.

$$\mathcal{R}\big(\mathcal{E}(\rho)\big) = \rho \,, \tag{10.20}$$

where ρ is the statistical operator of the system being corrected. Such an operation exists for a given error if and only if

$$Pe_i^\dagger e_j P = a_{ij} P \,, \tag{10.21}$$

where the e_i, e_j are the errors correctable by $\mathcal{R}(\rho)$ and a_{ij} is a complex scalar.

For a multiple-qubit pure state $|\Psi_0\rangle$ lying in the codespace of a nonde-generate error-correcting code, C, whereby k qubits are encoded in a n-qubit space, a quantum codespace is a 2^k-dimensional Hilbert subspace such that

$$\langle\Psi_0|e_{\nu_1}^{\alpha_1} e_{\nu_2}^{\alpha_2} \dots e_{\nu_n}^{\alpha_n}|\Psi_0'\rangle = 0 \,, \tag{10.22}$$

where $1 \le n \le 2K$, for any two, possibly identical states $|\Psi_0\rangle$, $|\Psi_0'\rangle$ in the codespace and any product of Pauli matrices of up to $2n$ factors in the error group acting on different qubits, K being the *number of errors* that the code can correct, where the action of $(\sigma_1)_{\nu_i}$, $(\sigma_2)_{\nu_i}$ or $(\sigma_3)_{\nu_i}$ on any of the n qubits is considered to be a single error, the ν_i indicating the *error locations* and the α_i indicating the *multiplicities* of the errors. A good error-correcting code will have a coding rate and an error rate that tend to nonzero values as n becomes large.

In the face of known types of error, one thus wishes to preserve a 2^k-dimensional subspace, $\mathcal{S}$, within the full 2^n-dimensional space using the code C, which is then called an $[n, k]$ *quantum code*. The $(n - k)$ ancillae are used as primary memory during the error correction process. Let A be a family of *interaction operators* $e_a = \langle\mu_a|U|\chi_0\rangle$, where $\{|\mu_a\rangle\}$ is a Hilbert-space basis for the environment the initial state of which is $|\chi_0\rangle$. The necessary and sufficient conditions for correcting individual errors resulting from G_k are together known as the *Knill–Laflamme conditions*:

$$\langle 0_L|e_a^\dagger e_b|1_L\rangle = 0 \,, \tag{10.23}$$
$$\langle 0_L|e_a^\dagger e_b|0_L\rangle = \langle 1_L|e_a^\dagger e_b|1_L\rangle \,, \tag{10.24}$$

[47, 155, 256]. The first condition requires that the encoded states remain orthogonal under any correctable error; the second condition requires that the lengths and inner product of the encoded states remain equal under any correctable error. The error behavior in a network of quantum gates can be represented by a sum of all networks that represent all possible errors, known as the *error expansion* of the network [258, 259].

10.5 The nine-qubit Shor code

Let us now consider some specific error-correction codes used to implement the above general methods. As mentioned previously, classical bit-flip errors have obvious quantum analogues, where the value of a qubit is flipped in the computational basis. Such bit flips can be detected and corrected in direct analogy to the manner in which they are detected and corrected for classical bits, by a majority vote method. However, in particular, the phase-flip error, described by the action of the Pauli operator σ_3, and the bit+phase-flip error induced by the product of both errors $-i\sigma_2$, have *no* classical counterparts. Nonetheless, recalling from their discussion in Section 1.4 that the bit-flip and phase-flip operations interchange roles in the diagonal basis $\{|\nearrow\rangle, |\searrow\rangle\}$, one sees that phase-flip errors can still be corrected if one uses an encoding taking computational-basis states into diagonal-basis states.

In order to be able to handle both sorts of error at one time, one can, for example, make use of the following binary linear code and then apply it repetitively. Taking the *dual* of the classical Reed–Müller code $RM(1,3)$, namely $C = \{(0,0,0),(1,1,1)\}$ discussed in Section 4.4, one can use the following two (three-qubit) GHZ-type entangled states

$$| \Uparrow \rangle = \frac{1}{\sqrt{2}}\big(|000\rangle + |111\rangle\big), \tag{10.25}$$

$$| \Downarrow \rangle = \frac{1}{\sqrt{2}}\big(|000\rangle - |111\rangle\big) \tag{10.26}$$

as logical qubits.[13] These logical qubits inherit the one-bit-flip-error correction property of the classical code used in its construction [402]. Using these logical qubits and an *additional* repetition coding step, one has

$$|0\rangle_L = | \Uparrow \rangle| \Uparrow \rangle| \Uparrow \rangle, \tag{10.27}$$

$$|1\rangle_L = | \Downarrow \rangle| \Downarrow \rangle| \Downarrow \rangle. \tag{10.28}$$

The resulting code is the *nine-bit Shor code* which makes use of nine physical qubits to encode *each* logical qubit [391].[14]

Recall, from the discussion in Section 10.1, that the decoherence process can be understood as a process in a larger space whereby an environment described by the pure state $|\bar{E}_0\rangle$ becomes entangled with computational basis state-vectors of one of the qubits, that is,

$$|\bar{E}_0\rangle|0\rangle \rightarrow |a_0\rangle|0\rangle + |a_1\rangle|1\rangle, \tag{10.29}$$

$$|\bar{E}_0\rangle|1\rangle \rightarrow |a_2\rangle|0\rangle + |a_3\rangle|1\rangle. \tag{10.30}$$

Thus, for example, as a result of the entangling of the environment with the first physical qubit of the three-qubit state $| \Uparrow \rangle$, the following evolution of the overall environment and system state is

---

[13] An rth order Reed-Müller code $RM(r,m)$ consists of the binary strings of length $n = 2m$ associated with the Boolean polynomials of degree at most r.

[14] This code is degenerate in the sense that different errors in the total space have an *identical* effect in the codespace.

$$|\bar{E}_0\rangle|\Uparrow\rangle \rightarrow \frac{1}{\sqrt{2}}((|a_0\rangle|0\rangle + |a_1\rangle|1\rangle)|00\rangle + (|a_2\rangle|0\rangle + |a_3\rangle|1\rangle)|11\rangle), \quad (10.31)$$

and similarly for the three-qubit state $|\Downarrow\rangle$,

$$|\bar{E}_0\rangle|\Downarrow\rangle \rightarrow \frac{1}{\sqrt{2}}((|a_0\rangle|0\rangle + |a_1\rangle|1\rangle)|00\rangle - (|a_2\rangle|0\rangle + |a_3\rangle|1\rangle)|11\rangle). \quad (10.32)$$

These two expressions can be rearranged as a product of superposition states of the environmental final states $|a_i\rangle$ and three-qubit GHZ-type computational basis superposition states $|ijk\rangle \pm |lmn\rangle$, with the result that the same environmental states are found to be entangled with orthogonal states in the encoding basis $\{|\Uparrow\rangle, |\Downarrow\rangle\}$. Then, by examining the states of the remaining six qubits, one can determine the original states. Finally, ancillary qubits can be introduced to allow error extraction, providing an error syndrome describing *which* of the three encoded qubits was in fact the one that suffered decoherence and whether there was phase flip between the two elements of the three-qubit code GHZ superposition state [391]. With this information, the error can be corrected.

10.6 Stabilizer codes

In applying quantum error-correction methods, it is often more convenient to work with a set of *operators* than with a set of state-vectors to find specific implementations.[15] The *stabilizer-code formalism* is useful for finding codespaces in this way. The Pauli group G_n described above is the basic mathematical structure underlying this formalism. Take S to be the vector subspace kept fixed by a subgroup of elements of G_n; this subgroup is the *stabilizer*, S, of $\mathcal{S}$. Given the generators of a subgroup, the subspace it stabilizes can be efficiently found.

For example, the *Steane code* is given by the logical states

$$|0_L\rangle = \frac{1}{\sqrt{8}}(\; |0000000\rangle + |1010101\rangle$$
$$+|0110011\rangle + |1100110\rangle$$
$$+|0001111\rangle + |1011010\rangle$$
$$+|0111100\rangle + |1101001\rangle \;) , \quad (10.33)$$

$$|1_L\rangle = \frac{1}{\sqrt{8}}(\; |1111111\rangle + |0101010\rangle$$
$$+|1001100\rangle + |0011001\rangle$$
$$+|1110000\rangle + |0100101\rangle$$
$$+|1000011\rangle + |0010110\rangle \;) , \quad (10.34)$$

---

[15] A method for simplifying the description of QECCs and their construction based on orthogonal geometry can also be found in [98].

of seven physical qubits [402].[16] The six generators, s_i, of the stabilizer subgroup corresponding to this code are the operators

$$\sigma_0 \otimes \sigma_0 \otimes \sigma_0 \otimes \sigma_1 \otimes \sigma_1 \otimes \sigma_1 \otimes \sigma_1 \qquad (10.35)$$

$$\sigma_0 \otimes \sigma_1 \otimes \sigma_1 \otimes \sigma_0 \otimes \sigma_0 \otimes \sigma_1 \otimes \sigma_1 \qquad (10.36)$$

$$\sigma_1 \otimes \sigma_0 \otimes \sigma_1 \otimes \sigma_0 \otimes \sigma_1 \otimes \sigma_0 \otimes \sigma_1 \qquad (10.37)$$

$$\sigma_0 \otimes \sigma_0 \otimes \sigma_0 \otimes \sigma_3 \otimes \sigma_3 \otimes \sigma_3 \otimes \sigma_3 \qquad (10.38)$$

$$\sigma_0 \otimes \sigma_3 \otimes \sigma_3 \otimes \sigma_0 \otimes \sigma_0 \otimes \sigma_3 \otimes \sigma_3 \qquad (10.39)$$

$$\sigma_3 \otimes \sigma_0 \otimes \sigma_3 \otimes \sigma_0 \otimes \sigma_3 \otimes \sigma_0 \otimes \sigma_3 \; . \qquad (10.40)$$

Consider the set of unitary operators leaving G_n unchanged; this set of operators is the *normalizer* $N(G_n)$. The set of errors $e \in G_n$ for which $eg = ge$ for all $g \in S$ is the *centralizer*, $Z(S)$. Encoding, decoding, error detection and recovery for stabilizer codes require only gates in the normalizer, which is generated by the tensor products of identity, C-NOT, Hadamard, and phase gates. The conditions for a stabilizer code to be an error-correcting code sufficient for the correction of an error set $\{e_i\}$ are that $e_i^\dagger e_j \notin N(S) \setminus S$, for all i, j.

The *smallest* code that allows for the correction of *arbitrary* errors within a single-qubit subspace is the *five-qubit* code

$$\begin{aligned}
|0_L\rangle = \frac{1}{4} \big(& |00000\rangle + |10010\rangle + |01001\rangle + |10100\rangle \\
& + |01010\rangle - |11011\rangle - |00110\rangle - |11000\rangle \\
& - |11101\rangle - |00011\rangle - |11110\rangle - |01111\rangle \\
& - |10001\rangle - |01100\rangle - |10111\rangle + |00101\rangle \big) \\
|1_L\rangle = \frac{1}{4} \big(& |11111\rangle + |01101\rangle + |10110\rangle + |01011\rangle \\
& + |10101\rangle - |00100\rangle - |11001\rangle - |00111\rangle \\
& - |00010\rangle - |11100\rangle - |00001\rangle - |01111\rangle \\
& - |01110\rangle - |10011\rangle - |01000\rangle + |11010\rangle \big) \; ,
\end{aligned}$$

which has the stabilizer

$$\sigma_1 \otimes \sigma_3 \otimes \sigma_3 \otimes \sigma_1 \otimes \sigma_0 \qquad (10.41)$$

$$\sigma_0 \otimes \sigma_1 \otimes \sigma_3 \otimes \sigma_3 \otimes \sigma_1 \qquad (10.42)$$

$$\sigma_1 \otimes \sigma_0 \otimes \sigma_1 \otimes \sigma_3 \otimes \sigma_3 \qquad (10.43)$$

$$\sigma_3 \otimes \sigma_1 \otimes \sigma_0 \otimes \sigma_1 \otimes \sigma_3 \qquad (10.44)$$

$$\sigma_3 \otimes \sigma_3 \otimes \sigma_3 \otimes \sigma_3 \otimes \sigma_3 \qquad (10.45)$$

$$\sigma_1 \otimes \sigma_1 \otimes \sigma_1 \otimes \sigma_1 \otimes \sigma_1 \qquad (10.46)$$

---

[16] For an extended pedagogical discussion of this code, its encoding and syndrome quantum circuits and relationship to the corresponding classical Hamming code, see [343].

[266]. The entanglement inherent in these code states allows one to combat the unwanted entanglement of qubits with their environment. These code states and those of the seven-qubit Steane code, like the GHZ state, can be shown to contradict local realism; see Footnote 22 of Chapter 7 and [139].

10.7 Concatenation of quantum codes

In quantum concatenated coding, quantum codes are *combined* so that data are encoded in some $[n, k, d]$ code, as categorized analogously to classical coding discussed in Section 4.5, where n describes the size of the codespace, k is the number of bits encoded, and d is the Hamming distance of the code. Each qubit in a block of the first code is encoded an *additional* time, in an $[n_1, 1, d_1]$ code. Qubits forming blocks in the second code can be further encoded using an $[n_2, 1, d_2]$ code, and so on, to a desired number of levels, l. After encoding is complete, one has an $[nn_1 n_2 \cdots n_{l-1}, k, dd_1 d_2 \cdots d_{l-1}]$ code. To find the error syndrome for a concatenated code, one first finds the error syndrome for the $[n_{l-1}, 1, d_{l-1}]$ code at the first level of code, for all of the blocks of n_{l-1} qubits. One then finds the error syndrome for the $[n_{l-2}, 1, d_{l-2}]$ code at the second level of code, and so on, for all l levels of code, each level being measured *in parallel*. One thus finds the error syndrome for the overall code in a total number of steps obtained by summing those required at each level of code. A simplification often made is to assume that the operations at level j are essentially similar to the operations at level $j + 1$, as in the above example. This coding method allows frequently occurring errors to be preferentially corrected.

In the evolution of an encoded state, the effect of errors is reduced by such concatenation, simultaneously allowing for error correction at various levels. For sufficiently low basic-error rates, arbitrarily long computations can be performed with arbitrarily low error-rates by implementing a sufficient number of concatenation levels, allowing one to accomplish fault-tolerant quantum computation, providing a computational accuracy threshold; for example, see [345]. The nine-bit Shor code discussed in Section 10.5 is an example of the use of the quantum dual to the Reed–Müller code as the inner code, where GHZ-type states were used as the inner code the logical qubits of which were then encoded with the three-logical-qubit repetition code as the outer code. As mentioned previously, one can similarly view some of the states of the Bell gem $\mathcal{G}_{16}$ of Eq. 7.63 as the result of the encoding $|0(1)\rangle \mapsto |\Psi^{\pm}\rangle = |0(1)\rangle_L$, repeated once [235].

Quantum broadcasting, copying, and deleting

Even with a good set of tools for maintaining the coherence properties of quantum states, there exist fundamental limitations on the set of tasks that can be carried out in quantum communication and quantum information processing. In this brief chapter, we consider some of these limitations and methods of working within them to approximate quantum tasks that are desired but cannot be performed perfectly. For example, perfect quantum deleting cannot be performed due to the linearity of quantum mechanics. Similarly, the no-cloning theorem discussed in Section 9.5, which is also based on the superposition principle, precludes the *exact* copying of an unknown quantum-information bearing pure state. Particularly important for quantum communication are limitations arising in the context of *state broadcasting*, that is, the distribution of the same quantum information to a number of parties.

Consider a quantum copier, a quantum machine for producing quantum states that approximate as closely as possible a given original state or set of states with the smallest possible effect on originals. One finds that approximate copying of unknown quantum states *is* possible: copies so produced approximate the state of the measured system to the degree there is overlap between the original and projected states. The goal in such situations is thus to find the precise bounds imposed by fundamental quantum principles.

11.1 Quantum broadcasting

Quantum state broadcasting is the provision of locally identical quantum states to each of a number of spacelike-separated parties. Consider a quantum system in an arbitrary unknown (invertible) mixed state, ρ. The *no-broadcasting theorem* is that the broadcasting of one such input state to two identical output copies is impossible [26]: it is impossible to find a transformation of a mixed state ρ to the state of a composite system ρ_{AB}, consisting of the original and a copy, such that the partial traces of ρ_{AB} over one system A and another system B, respectively, are *both* equal to ρ, where ρ is, say, an element of a set of two *arbitrarily chosen* states $\{\rho_0, \rho_1\}$. Such a transformation would broadcast this original single quantum state onto two systems that can be considered separately. A more powerful process would be that of *cloning* quantum states, wherein

$$\rho \otimes \tau \to \rho \otimes \rho \,, \tag{11.1}$$

where τ is a specified standard state of a system similar to the one the state of which is to be broadcast.

For pure states, quantum state broadcasting and quantum cloning are identical processes; deterministic state broadcasting is *impossible* for pure states. Nonetheless, this does not preclude *superbroadcasting*, wherein one begins the broadcasting process with *more than one* instance of the input state forming a product state. In particular, it has been shown that it is possible both to broadcast quantum states in this way and, when beginning with at least four input copies of a state, to *purify* the broadcast state in the process [119]. For mixed states, a distinction exists between the cloning of states and the broadcasting of states, in that there are ways to specify nonseparable composite system states giving rise to identical subsystem-state descriptions through partial tracing, which are the states locally accessible to agents. Proving the impossibility of broadcasting arbitrary mixed states is more difficult than proving a no-cloning theorem for mixed states in that, unlike in the case of pure states, for mixed states the latter is insufficient for a no-broadcasting result: there are many ways of broadcasting a mixed state without the result of this broadcasting being a state that is of *product* form, as in Eq. 11.1.

Although the broadcasting and cloning of a *single copy* of an arbitrary quantum state is impossible, it is quite possible to clone *known* quantum states. One can in that case make use of classical information from precise measurements of a quantum state in order to prepare copies of the eigenstate onto which the system state is thereby projected. However, the general problem of practical interest in quantum broadcasting is that of performing optimal imperfect broadcasting of *unknown states*. Let us now consider quantum copying in a more general sense.

11.2 Quantum copying

The basic problem of quantum copying is the following: given an unknown state, ρ, find a device that will produce a number of copies of this state, collectively described by $\rho^{\otimes n}$, in either a deterministic or in a probabilistic way. This is essentially what one requires of a classical copier, but with quantum states being copied. One provides the copier with a number of blanks and receives a particular number of copies when it is run.

In quantum information theory, by contrast to classical information theory, there is a distinction between the (unitary quantum) copying and the (unitary quantum) *transposition* of information from one quantum system, M, to another, M': in the quantum case, perfect copying (quantum cloning) requires also that the state of M be *unchanged* by the process; transposition does *not* require that the original remain intact but rather requires that the state of the original be brought to the null or blank state. Quantum teleportation is an example of quantum state transposition. In the simplest case, quantum states can be approximately or statistically copied in a process described by a unitary evolution together with a quantum measurement. In particular, given two nonorthogonal states of a quantum system, $|\phi_1\rangle$ and $|\phi_2\rangle$, there exists a total (nonunitary) process involving both a unitary transformation *and* a measurement such that

$$|\phi_i\rangle|\Sigma\rangle \rightarrow |\phi_i\rangle|\phi_i\rangle \, , \qquad (11.2)$$

for $i = 1, 2$; see, for example [142].

In general, there exists, for any unknown state chosen from a set $\mathsf{A} = \{|i\rangle\}$ ($i = 1, 2, \ldots, k$) a copying machine that produces, by executing a unitary evolution U, a linear superposition of multiple clones together with possible failure copies [322]. Given states $|i\rangle \in \mathsf{A}$ belonging to Hilbert space $\mathcal{H}_A = \mathbb{C}^{\otimes N_A}$ of the primary system A, a state consisting of a number M of "blank" states (each of dimension N_A) lying in the second Hilbert space $\mathcal{H}_B = \mathbb{C}^{\otimes N_B}$ associated with an ancillary system B, and yet another state in a third, N_C-dimensional space $\mathcal{H}_C = \mathbb{C}^{\otimes N_c}$ of the subsystem C used to measure the *number* of copies produced, the copying process results in a composite system (ABC) state of the form

$$U(|i\rangle|\Sigma\rangle|P\rangle) = \sum_{n=1}^{M} \sqrt{p_n^{(i)}} |i\rangle^{\otimes(n+1)} |0\rangle^{\otimes(M-n)} |n\rangle$$

$$+ \sum_{l=M+1}^{N_c} \sqrt{f_l^{(i)}} |\Psi_l\rangle_{AB} |l\rangle \, , \qquad (11.3)$$

where $p_n^{(i)}$ ($i = 1, 2, \ldots, M$) is the probability with which n copies of the original input state can be produced, and $f_l^{(i)}$ is the *failure rate* of the machine, for the ith input state. In the above expression, we have consider as "blanks" the states $|\Sigma\rangle = |0\rangle^{\otimes M}$ and have taken the state $|P\rangle \in \mathcal{H}_C$ to be the initial state of the copy-number indicator. The resulting output is found upon measurement.

A *universal quantum copier* is a copier that outputs two identical copies of the original with a quality that is *independent of* the specifics of the input state. The total quantum system involved in universal quantum copying consists of three parts: the *original*, the *blank system* onto which the state is to be copied, and the *copier*. In the copying process, one desires to maximize the *fidelity* of copying, that is, to minimize the difference between output states, say as measured by the (single-copy) fidelities $f_i = \langle \psi | \rho_i | \psi \rangle$, where ρ_i is the reduced statistical operator of the ith copy.

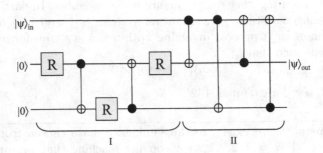

Fig. 11.1. Quantum circuit description of a universal quantum circuit realizing a universal quantum cloning machine (*cf.* [95]). I indicates the preparation stage and II the copying stage. The R are single-qubit rotations.

A simple model of a single-qubit *universal quantum cloning machine* (UQCM) that outputs two copies of a qubit state $|\psi\rangle$ involves the two-step pre-measurement procedure illustrated in Fig. 11.1. In this model, a four-dimensional ancilla is prepared in a known blank state $|\Theta\rangle$ and coupled to the qubit to be cloned, as shown in the left side of the figure. Then, a unitary operation is performed on this combined system, resulting in an output state, $|\Psi_f\rangle$, containing two clones of the input state in the sense that the reduced statistical operator for each of the resulting systems is

$$\rho = (1 - \eta)\frac{1}{2}\mathbb{I} + \eta P(|\psi\rangle) \,, \tag{11.4}$$

where η characterizes the quality of the copies.[1] The optimal UQCM is that carrying out cloning with the greatest fidelity

$$\max F\big(P(|\psi\rangle), \rho\big) = \max \langle \psi | \rho | \psi \rangle \,, \tag{11.5}$$

which is the probability that the copy produced is found upon measurement to be in the desired state; in this case, $F_{\max} = \frac{1}{2} + \frac{\eta}{2}$.

---

[1] Compare this with the depolarizing channel of Sect. 9.6.

The unitary transformation realizing this process transforms the product basis including the computational basis states of the original qubit so that

$$|0\rangle_A|\Theta\rangle_{BC} \to \sqrt{\frac{2}{3}}|00\rangle_{AB}|0\rangle_C + \sqrt{\frac{1}{3}}|\Psi^+\rangle_{AB}|1\rangle_C , \qquad (11.6)$$

$$|1\rangle_A|\Theta\rangle_{BC} \to \sqrt{\frac{2}{3}}|11\rangle_{AB}|1\rangle_C + \sqrt{\frac{1}{3}}|\Psi^+\rangle_{AB}|0\rangle_C . \qquad (11.7)$$

This UQCM makes two copies of a single qubit with a fidelity $F_{\max} = \frac{5}{6} < 1$, that is, has $\eta = \frac{2}{3}$ [96].[2]

One can go on further to define *symmetric* quantum copying machines as those producing all copies with equal fidelity, and *asymmetric* copying machines as those in which the individual copy fidelities f_i may differ. A copying machine is said to be *optimal* if the fidelities of the copies are maximal.[3]

11.3 Quantum deleting

Just as perfect deterministic quantum copying cannot occur, perfect reversible deletion of nonorthogonal quantum states cannot be performed. Consider two copies of an unknown qubit in a pure state, $|\psi\rangle$. The *quantum no-deleting theorem* states that it is impossible to delete one copy of such a state: no linear transformation exists from $\mathcal{H}_A \otimes \mathcal{H}_B \otimes \mathcal{H}_C$ to itself such that $|\psi\rangle|\psi\rangle|C\rangle \mapsto |\psi\rangle|B\rangle|C'\rangle$, where $|\psi\rangle$ is the state to be deleted, $|B\rangle$ is a blank state, and $|C'\rangle$ is a state that is *independent of* $|\psi\rangle$; the only linear transformation capable of performing a transformation of the above form is one that violates this final requirement by *swapping* the unknown state $|\psi\rangle$ with the ancilla state $|C\rangle$; that is, that has $|C\rangle = |\psi\rangle$.

To see this, consider any two different nonorthogonal qubit states, $|\phi\rangle$ and $|\phi'\rangle$. Given two copies of $|\phi\rangle$, consider the deletion of one copy by sending it to the null computational basis state $|0\rangle$ by a unitary transformation of the two states together with that of the environment

$$|\phi\rangle|\phi\rangle|\bar{E}_0\rangle \to |\phi\rangle|0\rangle|\bar{E}\rangle , \qquad (11.8)$$

as well as using this *same* transformation to delete one of two copies of the second nonorthogonal state $|\phi'\rangle$,

$$|\phi'\rangle|\phi'\rangle|\bar{E}_0\rangle \to |\phi'\rangle|0\rangle|\bar{E}'\rangle . \qquad (11.9)$$

---

[2] Universal copying via the specific physical process of parametric down-conversion has been carried out using type-II phase-matched parametric down-conversion producing polarization-entangled two-photon singlet-state output; see [393].

[3] A comprehensive review of quantum cloning and its relation to quantum cryptography can be found in [362].

As in the case of cloning, a unitary transformation accomplishing *both* tasks is impossible: such a transformation is identical to the state swapping transformation and has the effect that the environment enters *two different* states, which allows for the recovery of the states by inversion of the unitary transformation.

11.4 Landauer's principle

The no-cloning and no-deleting theorems have suggested to some a fundamental conservation of quantum information, somewhat different from that discussed in Chapter 6. This is becausequbits cannot be erased by unitary transformations, as we saw in the previous section, and, like classical bits, are subject to *Landauer's principle*, laid down by Rolf Landauer.[4] This principle is that bit erasure dissipates a minimum energy of $kT \ln 2$ into the environment. Quantum error-correction processes can be viewed as being similar to Maxwell's demon, in that they both act to prevent the increase of entropy of encoded states. In particular, the storage of the results of *syndrome measurements* of error correction in a finite memory come with a thermodynamic cost due to the need for their inevitable erasure as *classical bits*, even though in the correction of each error the system state is unitarily returned to its original state.

Very recent results show that it is impossible to *partially erase* quantum information, even when irreversible (nonunitary) processes are present, if partial erasure is taken to correspond to a *reduction of the size of the parameter space* of the quantum state encoding the quantum information in question. In particular, such partial erasure reduces the dimension of the parameter domain of a qubit without leaving its state entangled with other systems, say by restricting a qubit to a great circle of the Poincaré–Bloch sphere, where *partial erasure* is taken to be the CPTP map of all pure states $|\psi_i\rangle$ $(i = 1, \ldots, n)$ with real parameters π_i from a Hilbert space $\mathcal{H}^{(n)}$ to pure states in a smaller, m-dimensional Hilbert subspace $\mathcal{H}^{(m)}$ under a constraint $\kappa_i(\pi_i)$. It has been shown that one can erase *complete* information with an associated cost, as mentioned above, but not such *partial* information: no physically allowed operation can partially erase a pair of qu-*d*-its that are nonorthogonal [323]. Moreover, no qu-*d*-it can be partially erased by any irreversible operation.

---

[4] Landauer's principle dictates that the erasure of information is irreversible, with a "energy cost" of $kT \ln 2$ per bit in an environment of temperature T [270, 272].

12

Quantum key distribution

The most regularly performed nontrivial task using quantum information to date has been quantum key distribution. *Quantum key distribution* (QKD) is a method of securely distributing cryptographic key material for subsequent cryptographic use. In particular, it is the sharing of random classical bit strings using quantum states. Its use of a set of nonorthogonal quantum states then requires this key material to be considered *quantum* information.[1] The quantum encoding of cryptographic keys for distribution is valuable because the no-cloning theorem and the superposition principle governing quantum states confer a uniquely powerful form of information security during transmission of key bits. For maximal security, it can be followed by one-time pad message encryption, which is the only cryptographic method that has been proven to be unbreakable once a random key has been securely shared.

12.1 Cryptography and cryptosystems

Cryptography is a method of providing information security by preventing unintended recipients from coming to know private information by *encryption*, a special application of coding. *Cryptanalysis* is the methodology for decoding information encoded in this way. To carry out cryptography, an algorithm known as a *cryptosystem* (*cipher*) is first applied that transforms (*encrypts*) the information to be kept secret, referred to as a *plaintext message*, which requires the use of additional information, referred to as a *key*, resulting in the production of a *cryptogram* (*ciphertext*). Ideally, as in the case of one-time-pad encryption (using the *Vernam cipher*), the encryption algorithm and decryption algorithm are chosen such that, without possessing the key, an unintended recipient (or *eavesdropper*) can decrypt the message no more efficiently than by performing an exhaustive search over all possible cryptographic keys [434, 435].

---

[1] Quantum key distribution evolved from the idea of "quantum money" introduced by Charles Wiesner in the early 1970s but left unpublished until 1983 [455].

Cryptosystems are of one of two types, symmetrical or asymmetrical. *Symmetrical* systems are those that use the *same* (shared) key for encryption and decryption; *asymmetrical* systems are those that use *different* keys for encryption and decryption. *Public-key cryptosystems* are asymmetrical, and function as follows. The receiver of a private message, Bob, first creates or obtains a *private key* that he keeps secret. Bob then creates a corresponding *public key* using this private key and provides it to the eventual sender, Alice. Alice uses Bob's public key to encrypt her message and transmits this encrypted message to him. Finally, Bob decrypts the cryptogram produced by Alice using his private key. The security of a public-key cryptosystem is based on the difficulty—more precisely, the *computational complexity*—of decrypting encoded messages. Traditionally, mathematical one-way functions, which are functions that are *believed* to be hard to invert, have been used in public key cryptosystems. There is currently no publicly known *proof* that such functions are so; the deduction of x from a putative one-way function, $f(x)$, is believed to be hard in the sense that the time required to invert the function grows exponentially with the size of the input information, as opposed to growing at most polynomially.[2] Examples of such functions include exponentiation modulo p, the RSA function, and the Rabin function [296]. Multiplying integers is a convenient method of encryption because the inverse operation of *factoring* has a *trap door*, that is, it is easy to perform the required inversion with additional information.[3]

However, Shor's polynomial-complexity *quantum* algorithm enables the fast factorization of integers; see Chapter 14 below. If realized in practice for sufficiently large numbers, this algorithm would render insecure cryptography using such factoring-based public-key cryptosystems. Thus, quantum information processing poses a potentially great threat to this popular form of information security. The only cipher *proven* to be secure is a symmetrical cipher, the *one-time pad* (also known as the *Vernam cipher* [434]), which proceeds as follows.[4] Sender Alice encrypts her message, M, originally put into the form of a string of plaintext bits, m_i, using a random string (or keystream, K) of bits, k_i, by summing via binary addition ($\oplus$) each bit of the message to a corresponding key bit, resulting in an encrypted string of bits $c_1 c_2 \ldots c_n = (m_1 \oplus k_1)(m_2 \oplus k_2) \ldots (m_n \oplus k_n)$ that comprises the cryptogram, C. After transmission, receiver Bob decrypts the message using the *same shared key*, by the inverse operation of binary subtraction, returning the $m_i = c_i \oplus k_i$. With this cryptosystem, one has $H(M|C) = H(M)$ and

---

[2] For a discussion of computational complexity, see Chapter 13.1.

[3] Again, it is not presently publicly known whether factoring is indeed hard by traditional computational means, but it appears to be.

[4] The one-time pad cipher was first discovered in 1917 by Gilbert Vernam of AT&T and Joseph Mauborgne, who was then a captain in the U.S. Army and later was head of the U.S. Signal Corps lab in the U.S. Bureau of Standards and Chief of the Signal Corps, continuing until several months before the attack on Pearl Harbor.

$I(M : C) = 0$, so that the cryptogram provides *no* information about the plaintext message. This method is unconditionally secure no matter what the statistical properties of the plaintext message. The difficulty of this method, in addition to the property that the key must be as long as the plaintext message, lies in the problem of Alice and Bob coming to *share* the key, the problem of *key distribution*. The secret key must be transmitted by some *trusted* means, such as a courier who is incapable of being compromised.

Quantum key distribution (QKD) offers a solution to the cryptographic key distribution problem through the quantum encoding of binary information, potentially "secured by the laws of physics," in particular, the principles of quantum mechanics [40]. When combined with the Vernam cipher, quantum key distribution offers an unprecedented level of cryptographic security by *eliminating* the need for a trusted human courier. Thus, a quantum key distribution system is sometimes referred to as *untrusted* because no trust of the system is required beyond its proper physical operation. QKD operates as follows. The sender of the (random) cryptographic key material, Alice, encodes random bits of information in a set of *not all mutually orthogonal* quantum states, for example, of individual photons, that are then directly or indirectly provided to her intended recipient, Bob, through a quantum channel or by entanglement sharing. If Bob receives this information in an uninfluenced state, the transmitted photon states can be considered to be *unknown* to any potential eavesdropper, usually known as an *Eve* (*cf.* Fig. 12.1). In this case, any eavesdropping agent obtains *none* of the desired information about the bit-coding subspaces of the photon states due to the physical constraints imposed by the Heisenberg uncertainty relation and the no-cloning theorem, both of which follow from the quantum superposition principle. In practice, by communicating through a *classical* communication channel, Alice and Bob can check whether Eve may have been attempting to obtain key material from their shared quantum channel by compare a randomly chosen subset of their (now shared) bit material. If any perturbation *has* occurred, the associated (random) key material is considered compromised and so is disregarded. Traditional means of improving key quality can also be employed at this stage. A more sophisticated method for detecting the presence of an eavesdropper, described below, makes use of entanglement and Bell's theorem.

12.2 QKD systems

Current quantum-key-distribution methods for quantum cryptography use one of two physical means, either a single-photon qubit (or qu-d-it) or one of a number of such systems from a compound system such as a photon pair in an entangled state. Encoding states are chosen from a number of states, not all mutually orthogonal and hence imperfectly distinguishable without knowledge of the basis in which each lies. The first, and most common, approach uses quantum coherent states produced by a laser source that is attenuated so as to adequately approximate single-photon states. The second approach uses photon pairs in entangled states, such as those produced by spontaneous parametric down-conversion and spatial filtering, and exploits quantum nonlocality. This second approach directly provides cryptographic key information with just those statistical characteristics required by QKD without the need of *active* physical encoding or decoding of individual qubits.

To accomplish the local transmission of quantum keys (less than 10 km), optical-fiber-based systems can use wavelengths in the telecommunications wavelength window near 800 nm, for which efficient silicon-based avalanche photodiode (APD) detectors are available. Larger transmission distances have used telecommunications wavelength windows near 1300 nm and 1550 nm, for which undesirable effects such as attenuation and mode dispersion are less pronounced but for which photon detection has so far proven more difficult; signal attenuation in optical fibers imposes clear efficiency-over-distance limits on QKD.

For distances greater than 10 km, free-space links from ground to satellite have been developed after years of successful short-distance ground-to-ground tests using free-space as the medium of transmission. Quantum cryptographic-key generation over distances on the order of tens of kilometers have been realized in metro-area networks (MANs) and currently achieve megabit-per-second rates [159, 160]. Optical-fiber-based systems, however, can suffer from significant polarization-mode dispersion (PMD) effects. In turn, these can be mitigated by actively measuring this effect in real-time and compensating for their variation using feedback methods.

An elegant solution to the problem of quantum state dephasing due to PMD is the *plug-and-play* QKD line [308]. Such *autocompensating* systems are often favored for QKD using current telecommunications technology over existing optical-fiber-based networks. These systems have the receiver send relatively bright, orthogonally polarized pairs of optical pulses to the sender, who encodes her qubits in *relative phase* and reflects the light with a 90° polarization, attenuating it to the appropriately weak intensity as it is returned back to the receiver. The receiver then interferometrically obtains the phase-encoded information. The optical phase of the photon quantum amplitude contributed by polarization-mode dispersion in the fiber is, as a result, self-canceled due to the round-trip nature of its journey.

It is impossible to amplify single-photon quantum states in order to eavesdrop on QKD communication, because accurate such amplification amounts to quantum cloning of unknown quantum states, which is disallowed by the no-cloning theorem. However, undesirable multiple-photon pulses arise in current

QKD practice because genuinely single-photon sources are presently unavailable, making QKD vulnerable to undetectable monitoring—for example, by the use of a beam-splitter—because photons must be transmitted from sender to receiver in determinate signal states. True single-photon light sources are thus now being aggressively sought.

12.3 The BB84 (four-state) protocol

The most easily understood quantum-key-distribution method is based on attenuated coherent states and polarization coding of qubits according to the *BB84* protocol, which uses four pure quantum states from two conjugate Hilbert-space bases each capable of being encoding with two bit values [41]. Because all QKD implementations use methods and apparatus very similar to those for implementing this protocol, this section is pertinent to other protocols discussed later and, as a result, more briefly.

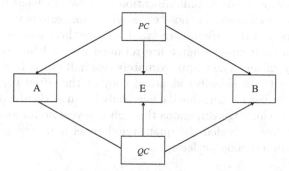

Fig. 12.1. Schematic of a quantum cryptography system including sender, A, and receiver, B. The quantum channel, QC, appearing at bottom is a private one-way channel. The two-way classical channel appearing at top, PC, is a public channel. An eavesdropper, E, with access to both channels is also shown.

To carry out QKD in this realization, ideally Alice prepares single-photon states; in current practice, this is done to a good approximation by attenuating pulsed laser light so as to provide a photon rate on the order of 0.1 photon per pulse, so that "nonempty" pulses with very high probability provide only one photon. Bit values 0 and 1 can, for example, be encoded in one basis as vertical and horizontal linear polarization states of single photons, $|H\rangle$ and $|V\rangle$, respectively; see Section 1.2 and Fig. 1.1 for details. The second basis of states must be conjugate to the first basis and can be chosen to be the diagonal basis in which, without loss of generality, $|\nearrow\rangle$ encodes the bit value

1 and $| \searrow \rangle$ encodes the bit value 0.[5] The required communications channels between Alice and Bob, and those of a possible Eve, are shown in Fig. 12.1.

Alice sends the individually encoded photons through a quantum channel to Bob using a random sequence of encoding-basis/bit-value pairs, each corresponding to one of the four state-vectors designated, making use of a random-number generator, one preferably based on a physical source of genuine randomness as opposed to a mathematical pseudo-random source. Bob then measures each of the incoming qubits in one of the two bases, similarly chosen at random, obtaining a bit value at the end of each measurement. After a sequence of such actions, Bob and Alice each possess random strings, their *raw keys*. Whenever the transmitter and receiver use the same basis for a given qubit, the corresponding bit values are perfectly correlated, provided the channel is noise-free and Eve has not intervened. For each bit, Bob tells Alice through a classical public channel the *basis* in which he measured, without having to mention the *bit-value* obtained (because Alice will already know it having encoded that value in the first place). Alice similarly reveals only whether the state in which she encoded that qubit was from same basis as his. This two-way classical communication is known as *basis reconciliation*.

In basis reconciliation, in those cases where measurement bases differ the results of Alice and Bob are *uncorrelated* and are thrown away, whereas those bits for which their choices agree are retained as good bits constituting the process of *key sifting*. Thus, approximately one half of the bits are eliminated leaving each with an (ideally) identical copy of the *sifted key*. The classical channel used for basis reconciliation need only be an *authentic* one rather than a secure one, because transmissions through it reveal no information useful to eavesdropper Eve, provided the quantum channel is indeed noiseless and all equipment is functioning perfectly.

If Eve were to intercept *every* photon used in this form of QKD, measure, and then replace the photon with one matching her measurement result, an error rate of 25 to 50 percent would be induced in an ideal system. The lower limit is reached only when Eve's measurements are made in the same bases chosen by both Alice and Bob, or in a basis lying in the same great circle on the Poincaré–Bloch sphere; the upper limit is reached when she measures in the basis corresponding to the Poincaré–Bloch-sphere poles relative to this great circle.

---

[5] The circular-polarization basis can equally well be used as one of the two encoding bases, replacing either the computational or diagonal basis, because all three are mutually conjugate.

In practice, the sifted key will actually contain some erroneous bit-values, occurring with a rate referred to as the *quantum bit-error rate* (QBER) that is typically a few percent of the total. Such errors arise because of imperfect equipment, noise in the quantum channel, or the interaction of Eve with the channel, which for security purposes are *all* attributed to Eve's intervention. These errors can then be eliminated using traditional classical error-correction methods, as is discussed in Section 12.7, below.

12.4 The E91 (Ekert) protocol

By contrast with the above single-photon method, the entangled-photon (EPR) pair scheme introduced by Artur Ekert, commonly known as *E91*, is based on measurements of two particles in a Bell state, such as $|\Psi^-\rangle$, shared by the two communicating parties [150]. It is hard to over-emphasize the remarkable elegance of this protocol for QKD, because its nature is precisely matched to its task: cryptographic key distribution is the provision of fully random bits to two parties, where these bits are fully correlated with each other. These are precisely the statistics of an n-fold tensor product of maximally entangled two-qubit states wherein one qubit of each pair is shared by the two parties.

In this QKD method, each photon that is created and measured is accompanied by exactly one other perfectly synchronized photon, preventing any attempts at undetected beam-splitting by Eve. Attenuation of the required photon pair beams is generally about 0.01, reducing the key transmission rate by an order of magnitude relative to the single-photon method. In the realization originally suggested by Ekert, both Alice and Bob receive one particle of each entangled pair and perform measurements along at least *three* different polarizer orientations during the course of measurement of the emitted ensemble. Measurements along parallel axes are used for key generation, whereas those along oblique angles are not thrown out as in BB84, but are used for security verification. Measurements are performed with just the sort of apparatus used in linear optical tests of Bell's theorem; see Fig. 3.1 of Section 3.5.

As in the BB84 protocol, Alice and Bob both randomly and independently measure their respective qubits. To take full advantage of the situation provided by entangled-photon pairs, one may use a "passive basis choice" method where a beam-splitter is placed in each of the detection suites in the laboratories of Alice and Bob, allowing each photon to "choose" its own coding basis, obviating the need of introducing (inevitably pseudo-random) basis-choice input by the experimenter. Alice and Bob, as in BB84, retain outcomes of measurements taken in the same basis, in which case their results are perfectly correlated, providing them with a shared key. The security of this protocol can be checked using the Bell inequality, as follows [150]. The oblique angle of the third, nonconjugate basis is chosen so as to effectively perform a test for violation of a Bell-type inequality (*cf.* Eq. 3.8); the probability that they choose the same basis is reduced, but incompatible choices can then be used to verify the quality of the source and line, and to provide information as to whether an eavesdropper may be present.

In single-photon BB84 implementations, when qubit states may be encoded using different sources for different bit values, the state of the qubit as encoded in one degree of freedom may be leaked to another, ostensibly nonencoding degree of freedom. This could allow Eve to detect the qubit state without inducing errors in the encoding subspace. By contrast, in a symmetric configuration of the entangled-photon E91 protocol in which neither Alice *nor* Bob possesses the source, such a leakage of information to other quantum degrees of freedom of the photon is detectable through an increased QBER. Even if Eve were to take control of the source of entangled states, she would be able to obtain no additional information because the source itself plays no role in encoding or decoding. Moreover, because the key does not exist, even in raw form, until the two photons are measured (if one has access to a quantum memory) then key material can be created even when the quantum channel is no longer available.

12.5 The B92 (two-state) protocol

The security of quantum cryptography depends on the indistinguishability of nonorthogonal quantum states and the inevitable perturbation arising from measurements of these states when the sequence of encoding states is unknown to an eavesdropper. In fact, just *two* nonorthogonal pure states are sufficient to implement QKD. A simple QKD scheme demonstrating this is the *two-state protocol*, also known as the B92 protocol [37]. In this protocol, Alice and Bob choose two nonorthogonal states of a single-photon, say $|\phi\rangle \equiv \cos\theta|0\rangle + \sin\theta|1\rangle$ and $|\phi'\rangle \equiv \sin\theta|0\rangle + \cos\theta|1\rangle$, and associate the bit value 1 with the former and 0 with the latter. Alice then uses this correspondence to send a random binary sequence to Bob. Bob performs measurements of qubits which, as in the case of BB84, sometimes fail, but now when a measurement succeeds it *always* provides the bit correctly, that is, he uses unambiguous state discrimination, as described in detail in Section 9.4.

In particular, Bob measures at random one of the POVM elements, E_1, E_2, and $E_3 = \mathbb{I} - E_1 - E_2$, where $\langle\phi|\phi'\rangle = \sin 2\theta$; see Section 2.7 and [152]. The probability of obtaining any given bit correctly is then $1 - \sin 2\theta$, in accordance with Eqs. 9.19–20. As in BB84, when Bob's measurement choice does not coincide with Alice's choice, these events are disregarded, leaving only good bits, with strings verified by parity checks of strings between parties. When implemented using interference between a macroscopic bright laser pulse and a dim laser pulse, with less than one photon per pulse on average, this protocol becomes more resistant to eavesdropping: Bob can monitor the bright pulses, ensuring that Eve is not removing pulses, in as much as removing a dim pulse noticeably alters the interference of the bright pulse with the resulting empty state.

12.6 The six-state protocol

A complex six-state QKD protocol has also been proposed, which has the advantage of greater symmetry within the Poincaré–Bloch sphere [87]. In this protocol, the six states used are the four used in the BB84 protocol plus the two remaining conjugate-basis states; see Fig. 1.1a. These states are sent with equal probabilities. The probability that, choosing bases randomly, Alice and Bob will use the same basis for a given bit is thereby reduced, but the optimal information gain for Eve at a given error rate is also lowered for her eavesdropping strategy in the single-qubit-based attack; by contrast to the BB84 protocol, the six-state protocol compels Eve to measure in an additional basis in a single-qubit-based attack, which increases her concomitant bit-error rate. If Eve measures every photon, the QBER is, in particular, one-third rather than one-fourth, as in the BB84 protocol, enhancing the detectability of eavesdropping. The enhancement of Eve's error rate under this protocol has been verified experimentally in an entanglement-based implementation [163]. In the range of low error rates, however, the efficiency of secret-key generation is found to be higher when using the four-state BB84 protocol.

12.7 Eavesdropping

The individual-bit eavesdropping strategies available to Eve may induce state projection and/or decoherence of transmitted qubits and so give rise to observable effects. From the point of view of Alice and Bob, any such effect that could have been produced by an eavesdropper must be assumed to be a result of successful eavesdropping, because in quantum key distribution the goal is to support absolute security. Alice and Bob must find the bounds of tolerable error within which a truly secure key can be recovered [291]. In the protocols discussed above, there is a region of safety in which such security exists.

To understand the effect of a given eavesdropping strategy, one can calculate the quantum bit-error rate (QBER), Q, and inconclusive transmission rate, R, and study the mutual information between Alice and Bob, between Alice and Eve, and between Bob and Eve [152]. Alice and Bob can remove and compare a randomly chosen substring of their sifted key in order to obtain its specific QBER. Although Eve may but does not necessarily alter R, again it is assumed that any change in R is due to the presence of Eve; if R is found to differ from its expected value R_0, this is taken to be an indication of eavesdropping. Let ϵ be the "raw" error probability characterizing a QKD channel before any inconclusive bits have been eliminated. The resulting portion of errors after these bits have been eliminated is

$$Q = \epsilon/(1 - R) \ . \tag{12.1}$$

Alice, Bob, and Eve may be attributed random variables A, B, and E, described by a joint probability distribution $P(A, B, E)$ describing their key bits. The information known to Alice and Bob is the marginal distribution $P(A, B)$. A lower bound on $P(A, B, E)$ is the larger of the differences between the (classical) mutual Shannon information $I(A : B)$ of Alice and Bob and the (classical) mutual information of Eve with either of them, that is,

$$H(A, B|E) \geq \max\{I(A : B) - I(A : E), I(A : B) - I(B : E)\} \ . \tag{12.2}$$

Eve's information can be made smaller by a privacy-amplification method, either classical [44, 77, 326] or quantum [131] in nature. If Alice and Bob share *more* information with each other than with Eve, they can use a classical protocol to eliminate all errors. For this purpose, Alice and Bob can use a classical cryptographic *privacy-amplification protocol* such as the following. One of the parties chooses pairs of bits at random, computes their XOR value, and discloses the *position* of the bits in the string, but not the value obtained. These bits are then replaced by their XOR value, shortening the key without introducing any additional errors and, with repeated implementation, reducing the eavesdropper's information [44]. The effect is to "smear out" the value of each initially shared bit across the remaining bits remaining in the shortened key. By sufficiently shortening the resulting key, the eavesdropper's information about the key can be made arbitrarily small.

When the eavesdropper carries out attacks by attaching independent probe systems to each qubit and measures them individually, the attack is referred to as an *individual* (or *incoherent*) *attack*. This sort of attack is the most realistic given current technology. Eve is usually taken to measure her probe systems as soon as the public-channel reconciliation of basis choice is performed by Alice and Bob. The most general sort of attack, known as the *coherent attack*, is based on the simultaneous manipulation of probes attached to *several* qubits. A coherent attack where no conditions are placed on how probes are attached to several qubits is known as a *joint attack*. In a coherent attack, if Eve constrains herself to attaching one probe per qubit but measures these probes collectively, then the resulting attack is referred to as a *collective attack*. In the case of coherent attacks, Eve is usually assumed to refrain from measuring probe systems until the full procedure of basis reconciliation, error correction, and privacy amplification has been completed, so as to be able to maximize the value of information available in the full set of probes by measurement of them as a *single composite system* rather than as a number of individual systems.

Opaque strategies involve the capture of signal states that are subsequently measured and then resent, and are also called *intercept–resend strategies*. Eve measures the quantum signal by a standard quantum measurement in such a way that her information about its state is maximized. This procedure will not affect R but increases Q, because Eve will *fail* some fraction of the time by performing measurements in a basis not used for encoding, causing Eve to misidentify the state she retransmits in an attempt to avoid detection. To understand such a procedure, information effectively transmitted from Alice to Eve can be viewed as a symmetric channel; the information effectively transmitted from Eve and Bob can be viewed as an erasure channel.[6] If the intercept-resend procedure is carried out as an individual attack, the receiver and eavesdropper can analyze the situation on the basis of the random variables obtained through the individual measurements as classical information, allowing them to make use of traditional cryptanalysis techniques.

Translucent eavesdropping strategies can instead be used, wherein the reading of signal photon states with small disturbance is followed by the reading out of the desired information. Such approaches are translucent in the sense that Eve may disturb the photon state but *does not intercept it*. In order to do this, she can use a POVM involving an ancillary probe system that becomes correlated with the transmitted particle under a unitary evolution. Furthermore, this probe can be stored indefinitely by Eve in a quantum memory, which allows her to make use of basis-choice information that is later communicated classically between Alice and Eve in the basis reconciliation stage of their quantum-key-distribution protocol [76, 163].

12.8 Security proofs

Ideally, proofs of the security of quantum key distribution demonstrate *unconditional security*, proving security in the face of *all conceivable attacks* within the bounds of quantum mechanical and other physical principles. However, currently available sources fail to produce, for example, single-photon quantum states, and current detectors have less than perfect efficiency. Recall also that the QKD agents, Alice and Bob, are assumed initially to share *no* information and are taken to share *identical* copies of a secret key (bit string) at the end of key distribution. Eve must be unable to obtain *any information whatsoever* about the resulting shared key. Furthermore, any QKD protocol will fail if Eve is capable of *impersonating* Alice or Bob during the execution of the protocol. In practice, Alice and Bob must execute classical authentication protocols on classical messages between each other to prevent such spoofing. For these required messages to be secure, Alice and Bob must share *at least* a small secret key for authentication purposes before executing the protocol, say by meeting each other before executing the protocol. Thus, in practice, any automated QKD protocol not initially using a trusted courier will, strictly speaking, ultimately implement a *key expansion* protocol, rather than a key distribution protocol. Moreover, Eve usually can with relative ease *jam* the

---

[6] The *quantum erasure channel* is one where an initial qubit state is replaced by a state "$|2\rangle$" orthogonal to *both* computational basis states $|0\rangle$ and $|1\rangle$ with some probability p.

communication protocol, requiring Alice and Bob to continually revalidate their communication security.

Security proofs for QKD are generally applicable only to particular protocols faced with a *limited set* of possible attack methods, but remain significant because the options of a real-world Eve are limited. One generally quantifies secrecy in practical quantum key distribution under a given realization of a given protocol. As an example, therefore, consider the implementation of the BB84 protocol with single-photon states and an eavesdropper who is restricted to carrying out individual translucent attacks on one qubit at a time using a probe particle that becomes entangled with the attacked photon. Take m to be the number of bits of raw key transmitted and n to be the number of bits of sifted key remaining after the discarding of $(m - n)$ inconclusive bits, ultimately leaving $(n - \bar{e})$ bits after the removal of $\bar{e}$ erroneous bits, constituting the error-corrected key. One can then implement privacy amplification to obtain a shorter key that is more secure, as discussed previously, removing an additional number of bits, corresponding to the privacy-amplification *compression level*

$$s = t(n, \bar{e}) + q + \nu + g \, , \qquad (12.3)$$

where $t(n, \bar{e})$ is the defense function, q is the number of bits estimated to have been "leaked" during error correction, and ν is the estimate of bits leaked due to multiple-photon transmission during putative single-photon transmission intervals, leaving $(n - \bar{e} - s)$ bits of key material, if g is a "safety margin;" see, for example [75].

The *defense function* is an estimate of the greatest possible leakage of information due to eavesdropping and depends in general on the number of sifted data bits and the number of errors and is found by maximizing the total Rényi information gained by the eavesdropper as determined by minimizing the quantum overlap, o, of measured eavesdropping ancilla states and receiver Bob's signal states for a fixed induced error rate, namely,

$$I^R_{\text{optimal}} = \log_2(2 - o^2) \qquad (12.4)$$

[396, 397]. The privacy-amplified key can be kept secret provided one takes into account the upper bound on the maximum Rényi information gained by eavesdropping on the corrected key material. The average *secrecy capacity* of the quantum cryptosystem under consideration is given by the number of secret bits obtained $(n - \bar{e} - s)$ over the number of bits m originally transmitted in the long transmission limit, $m \to \infty$.

13

Classical and quantum computing

The fundamental limitations of any form of computation can be expressed in terms of the resource requirements of standard computational tasks under it. Within traditional models of computation, such as the Turing machine model, many problems are found to be intractable due to the limited computational capabilities of classical physical systems. However, quantum systems allow the range of tractable computations to be extended beyond that achievable by classical computation because the superposition principle offers a radically different sort of computational parallelism. The *quantum circuit model* (or *gate array model*), in which networks composed of quantum logic gates act on sets of qubits, is the dominant model of quantum computation and has an equivalent (quantum) Turing machine model. Both of these models and the general principles of quantum computation are discussed in this chapter. A number of specific algorithms, which illustrate the novel character of quantum computation, are described in the following chapter.

Any good model of computation will allow arbitrary operations to be accurately performed through a set of basic operations, as the quantum circuit model does. A very simple picture of how quantum computation works in a qubit architecture, in which operations are performed on elements of a Hilbert space with a dimension that is a power of 2, is the following. One begins with classical input information, that is, bits encoded in qubits in the computational basis, and a number of ancillary qubits each generally assumed to be in the standard state $|0\rangle$. In the quantum circuit model, a sequence of logic operations described by unitary transformations then takes place on this full collection of qubits, in parallel, as a result of which their values may be changed. After some specific number of logical operations, one measures designated output qubits to obtain the result (*readout*) of computation.

In this chapter, we begin by discussing Turing machines of various sorts, quantum as well as classical. We then examine computational complexity in quantum computing. Finally, we briefly consider fault tolerance in quantum computing and a specific proposal for accomplishing quantum computation in linear optics.

13.1 Classical computing and computational complexity

Alan Turing was responsible for introducing what is now known as the *Turing machine*, an abstract machine he considered to be a universal computing device in the sense that it is assumed to be capable of simulating any other conceivable computing device. Specifically, he assumed that the class of functions that can be evaluated via algorithms is identical to the class of computable functions, an assumption that has been taken both as a fundamental principle and an empirical hypothesis and is now known as the *Church–Turing thesis* [107, 426].

The Turing machine is now understood to be universal in at least two respects. First, it provides a universal definition of computability in the sense of computability in Lambda calculus.[1] Secondly, all reasonable computing machines—by which is meant all machines that can in principle be physically realized—are polynomially equivalent to Turing machines: the minimal time required for any reasonable machine and for a Turing machine to produce an output given an input of a given size are polynomially related. That such a *single* universal device could actually exist still remains open to question. It is noteworthy, for example, that the running times of Turing machines are not predictive of those of digital computers that use floating point numbers to address scientific problems, as is commonly the case in practice. In particular, such computations have computational costs associated with operations that are *independent of* operand size, unlike Turing machines. Nonetheless, the Turing machine model *is* a central element of the dogma of current computer science and, as such, is a good model in which to consider quantum computation here. Furthermore, Turing machines have the very attractive feature of capturing the intuitive essence of computation in a simple schema.

Because quantum computing devices have other properties not likely foreseen by Turing, they provide a ground for scrutinizing the Church–Turing thesis. For example, the Turing machine is a finite machine not designed to address continuous mathematical models. Richard Feynman touched on the question of the adequacy of the Turing machine model from the physical point of view when he raised the question of the ability of quantum systems to perform efficient simulations of other physical systems, particularly *quantum systems* [170].

The question of the difficulty of performing a computational task, such as computing the value of a function for a given argument, can be addressed both qualitatively and quantitatively. A function is *computable* if there exists an algorithm for computing it. Computational complexity theory seeks *bounds* on the resources necessary for solving computational problems.[2] The *computational complexity* of obtaining the value of a function can be given in terms of the number of computational steps required to calculate it for a

---

[1] For information regarding Lambda calculus, see [25].

[2] It has, as a result, been referred to as the *thermodynamics of computation* [318].

generic argument. Those problems for which there exists *no algorithm at all*, as a matter of principle, are known as *uncomputable problems*. The following "halting problem" is an example of an uncomputable problem [426].[3] Consider the problem of determining whether any given Turing machine will halt given *any* input. Suppose there were an algorithm for determining this using a Turing machine. Then it would be possible to have a Turing machine that *halts* when provided the description of *any* Turing machine yet *does not halt* if the machine does not halt given its *own* description, which is absurd [123]. The situation is akin to that of the famous Liar Paradox.

The complexity of computable problems can also be assessed in terms of the time and space requirements for their solution. Computational problems can be classified as either "easy" or "hard." *Easy problems* are those for which the required computation time is a polynomial function of the size of the input, as measured in bits. *Hard problems* are those for which the required computation time is an exponential function of the size of the input measured in bits. The majority of problems of interest appear to be *hard* problems, though hardness is often difficult to prove for a given problem. As is often the case in general, the reduction of one problem to another the properties of which are already known is a good method for tackling assessments of complexity. Thus, when addressing computational complexity in the context of quantum computation, it is often helpful to make use of the results obtained for similar classical situations.

Classical algorithms can be classified as *effective* (or *polytime*) when the number of steps, and the *time* to execute them all, grows as a *polynomial function of the size of the input*, that is, $f \leq an^b$ for some constants a, b and for input sizes n sufficiently large.[4] The computational complexity class of effectively solvable problems is accordingly denoted **P**. The class of problems that are solvable *nondeterministically* in polynomial time is denoted **NP**, an example being the problem of factoring integers.[5] The subclass of **NP** consisting of the hardest problems of this class are the **NP-complete** problems, a large number of which have been identified, examples being the Traveling Salesman Problem and the determination of membership in a correlation polytope; see Section A.8 and [451]. The **NP-complete** problems are those such that if any one of them is solvable by an efficient algorithm then *all* **NP** problems are so solvable. The class of problems solvable with an *amount of memory*

---

[3] The halting problem for quantum computers has been explicitly addressed; see [311, 313].

[4] One must take care to note the encoding of input as well, because the number of steps may depend on the encoding used. Unless otherwise stated, we assume here n is given in *bits*.

[5] Note that this class is only defined for *predicates*; see, for example, Part 1.3 of [252] for several definitions of **NP**, the complement of which is designated **coNP**, corresponding to the class of languages decided by a probabilistic Turing machine always accepting members of the language and rejecting any string not in the language with finite probability.

polynomial in the input size is **PSPACE**. The class of problems that can be solved *with high probability* in polynomial time with the help of a random number generator is called **BPP**, for "bounded error probability in polynomial time." The class of problems that can be solved in polynomial time if sums of exponentially many addends (themselves computable in polynomial time) can is denoted $\mathbf{P}^{\#\mathbf{P}}$. These classes are related as

$$\mathbf{P} \subset \mathbf{BPP}, \mathbf{P} \subset \mathbf{NP} \subset \mathbf{P}^{\#\mathbf{P}} \subset \mathbf{PSPACE} .$$

The relation of **BPP** to **NP** is currently unknown, as is the question of whether **P** and **NP** are, in fact, identical.[6]

These classes can be defined precisely in terms of Turing machines. A special class of problems, the *decision problems*, which are equivalent to the evaluation of predicate functions, have yes/no answers and can be used to clarify further such an assessment. Decision problems can be cast as questions about formal languages, defined as the set of all finite strings constructible from a given alphabet of symbols. Turing machines also allow one to formalize the notion of language (decision problem) reduction: a given language may be *reduced* to another if there exists a Turing machine operating in polynomial time for which, given an input in the reduced language, it provides an output, and for which this input is in the reduced language if and only if the output is in the reducing language. There are several types of Turing machines, corresponding to different types of computation, some of which are briefly examined next.

13.2 Deterministic Turing machines

The deterministic Turing machine is attributed a finite set of fundamental information units (*letters*) known as its *alphabet*, A, a finite set of *control states*, Q, and a *transition function*,

$$\delta : Q \times A \to Q \times A \times D , \tag{13.1}$$

where $D = \{1, 0, -1\}$. If necessary, A may be replaced by the null set. Strings of letters in A are known as *words*.[7] A *deterministic Turing machine* may be defined as $(Q, A, \delta, q_0, q_\alpha, q_r)$; the *state* of the machine is specified by $q \in Q$ where, in particular, $q_0, q_\alpha, q_r \in Q$ act as the *initial machine state*, the *accepting state*, and the *rejecting state*, respectively. The *configuration* of the machine at any stage is given by $c = (q, x, y)$, where $x, y \in A^*$, A^* being the set

---

[6] A more detailed hierarchy of complexity classes is succinctly presented in Part 1.5 of [252]. For more on the **P-NP** question, see [395]. Quantum computational complexity is addressed in Sect. 13.6, below.

[7] Here, and in the cases of the other machines below, often one considers instead of A a subset, A', of A, of input/output strings not including the blank symbol, in which case A' is referred to as the *external alphabet* [216].

of all words from A. One also has use of the set A^0 that contains only the *empty word*, ϵ, consisting of *no* letters. The transition function δ describes transitions between configurations, the first of which *yields* the second, its *successor*. The Turing machine also has a *tape* that we consider here to contain a two-letter word xy, the first letter of which it is said to be *scanning* (or *reading*).

In each computational step, the symbol being scanned is replaced with another symbol, which is said to be *printed* in the process; the machine enters a new state, q', scanning either the same symbol, the symbol to the left, or the symbol to the right, as dictated by the given value $d \in D$ that represents the behavior of the *read-write head* of the machine at the computational step. Every Turing machine computes a partial function $f : A^* \to A^*$. A *computation* is a sequence of configurations beginning with an initial *input word* and an initial configuration, c_0, proceeding toward a *halting configuration*; a computation *halts* after t computational steps when *either* one of its configurations has no successor *or* if its state is either q_α or q_r in which it is said to be *accepting* or *rejecting*, respectively; a Turing machine *accepts* its input if it halts in the accepting state, or *rejects* its input if it halts in the rejecting state *or* fails to halt. Let $g_M(a)$ denote the output string for a given machine, M, and input string, a. The partial function f is *computable* if there exists a Turing machine M such that $g_M(a) = f(a)$ for all $a \in A^*$, in which case M is said to *compute* f.

A set of words over an alphabet forms a *recursively enumerable language* if there exists a Turing machine that accepts a word if and only if it is a member of the set. A set of words forms a *recursive language* if there exists a Turing machine such that every computation halts and accepts a word if and only if it is a member of the set. In this light, the decision problems can be seen as the set of problems in which an input word w must be found either *to belong* to a given language or *not to belong* to that language. In the case of recursive languages, a Turing machine accepting a given language provides an algorithm for solving a given decision problem. To each of these two sorts of language there corresponds a class of algorithmic decision problems: those constituting the class **R** are *recursively solvable* (or *decidable*), whereas those not in **R** are *recursively unsolvable* (or *undecidable*).

In addition to these broad classifications, one can examine the specific computational time required for solutions to be found in order to determine their computational complexity. For a Turing machine that halts for every input, the *time-complexity function*, $T(n)$, is the greatest computation time required for inputs of length n.

13.3 Probabilistic Turing machines

The *probabilistic Turing machine* is more general in character than the deterministic Turing machine. It has a transition function that is a mapping assigning *probabilities* to possible operations of the machine, of the form

$$\delta : Q \times A \times Q \times A \times D \to [0,1] \ . \tag{13.2}$$

For well-definedness, the sum of probabilities over all successor states for each configuration is required to be unity. As a result, the machine-state transitions may be described by a stochastic matrix. The deterministic machine can then be viewed as a *special case* of the probabilistic machine. A given configuration is said to *yield* a successor configuration with a probability given by δ. A given final configuration is computed from an initial configuration with a probability given by the product of the probabilities of those configurations leading to it by a particular *computation* defined by the sequence of states leading to it.[8] Unlike a deterministic Turing machine, a probabilistic Turing machine may accept an input in one execution but reject it in another.

The family of languages L that can be accepted in polynomial time by a probabilistic Turing machine, **RP** (for "randomized polynomial time"), is such that if a word is in L the probabilistic Turing machine accepts it with a probability at least $\frac{1}{2}$, and if not, the probabilistic Turing machine *rejects it with certainty*. **BPP** can be identified with the family of languages such that if a word is in L then the probabilistic Turing machine accepts it with probability at least $\frac{2}{3}$, and if the word is not in L, it is rejected with probability at least $\frac{2}{3}$. In addition to $\mathbf{P} \subset \mathbf{BPP}$, as mentioned in Section 13.1, it is the case that $\mathbf{P} \subset \mathbf{RP}$. Probabilistic Turing machines are helpful in defining the class of *interactive proof systems*, which are machines modeling computation as communication between two parties for the purpose of determining membership of strings in L, because they allow the verifier, which is attributed finite computation resources, to make use of randomness against the prover, which is attributed unlimited computational resources.

13.4 Multi-tape Turing machines

An *m-tape* (*deterministic*) *Turing machine* is one attributed m tapes, an alphabet, A, a finite set of control states, Q, and a transition function δ such that

$$\delta : Q \times A^m \to Q \times (A \times D)^{\times m} \ , \tag{13.3}$$

and is defined by $(Q, A, \delta, q_0, q_\alpha, q_r)$, where as usual $q_0, q_\alpha, q_r \in Q$ act as the initial machine state, accepting state, and rejecting state, respectively. A configuration for such a machine is given by $(q, x_1, y_1, \ldots, x_m, y_m)$, where the current machine state is q, $(x_i, y_i) \in A^* \times A^*$, and the *content of the ith tape* is $x_i y_i$. What is computable by a single-tape machine in time t is computable by a multi-tape machine in time $O(\sqrt{t})$. Such a machine is naturally suited to the study of situations involving computational parallelism. However, from

---

[8] Note that although quantum Turing machines also have transitions that are non-deterministic, the property of factorability of probabilities in particular *does not hold* for them.

the point of view of computational complexity, multi-tape Turing machines are no more powerful than single-tape machines.

13.5 Quantum Turing machines

Charles Bennett has shown that traditional multi-tape Turing machines can be simulated by reversible Turing machines, in particular, that arbitrary irreversible computations can be seen as reversible ones, often with little reduction in efficiency [36]. A reversible computation operates as a permutation of input bit-strings: Tommaso Toffoli has shown that any finite mapping can be reversibly computed by a process of padding strings with zeros, their permutation, and the projection of some bit strings onto other bit strings; see, for example, [423]. In order to represent permutations of bit-strings by logic circuits, one can use elementary reversible gates each having the same number of outputs as inputs; a logic-gate output leads to precisely one input of a succeeding gate. This picture of computing bears a strong resemblance to quantum computing in the quantum circuit model. Paul Benioff first showed that unitary quantum state evolution, which is inherently reversible, can be as powerful as a Turing machine, demonstrating that quantum mechanical systems can be at least as computationally powerful as classical computers are [34, 35]; the corresponding machine is known as the *quantum* Turing machine.

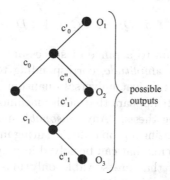

Fig. 13.1. The behavior of quantum probability amplitudes c_i in, for example, the apparatus illustrated in Figure 1.5. Interferometric background counts correspond to outputs O_1 and O_3; quantum interference arises in output O_2.

To distinguish quantum computing from classical computing, it is valuable to consider a probabilistic classical process, such as coin-tossing. Such a process has a state description that is an exponentially growing tree; in the case of coin-tossing there are 2^n possible outcomes. The quantum computing process is similar, with the important difference that the probabilities for each branch are given by *quantum probability amplitudes* for the quantum computer that are *capable of interfering* with each other, provided coherence within its quantum state is maintained, as discussed in Chapter 1; there is a *global* dependency of any given branch on the remainder of branches, which comes into play whenever there is quantum indistinguishability between paths through the tree to a given node; see output O_2 in Fig. 13.1.

> The extent to which classical systems can simulate quantum computation was addressed by Emanuel Knill and Daniel Gottesman, who showed that any quantum computation beginning with computational-basis state preparation and involving *only* the quantum gates of the (Clifford group) set $\{\frac{\pi}{2}\text{-phase}, H, \sigma_i, \text{C-NOT}\}$ can be efficiently simulated by a classical system [254]. This is so because such computations have only polynomial-size state descriptions rather than exponential-size descriptions, as discussed in Section 1.7. Not all quantum computations have polynomial state descriptions, however.

Let us now formally consider quantum computation, in order to better understand the relationship between it and classical computation. As with classical computation, quantum computation may be investigated via a type of Turing machine. The *quantum Turing machine* is a Turing machine with a transition function of the form

$$\delta : Q \times A \times Q \times A \times D \to \mathbb{C}, \tag{13.4}$$

from one configuration to a *range* of successors, each attributed a complex quantum probability *amplitude*, corresponding to a unitary transformation of a quantum state. Efficient universal quantum Turing machines have been shown to exist [54]. It appears that such machines violate the strong version of the Church–Turing thesis, "Any *reasonable* model of computation can be efficiently simulated using a probabilistic Turing machine," because there exist computational problems that can be solved in polynomial time by a quantum Turing machine but that are solvable only in *superpolynomial* time for a *probabilistic* Turing machine [56, 57].

Any computation that can be performed by a classical Turing machine can be performed by a reversible Turing machine viewable as a three-tape machine equipped with an input tape, a history tape, and an output tape, where the history tape records the history of the computation needed to reverse the computation.[9] Everything that can be computed in polynomial time on such a classical machine can be calculated on a quantum Turing machine [54].

---

[9] A traditional Turing machine is said to be *reversible* if each configuration has a unique predecessor.

13.6 Quantum computational complexity

There are several ways of describing computations on *qubit architectures*, which involve tensor products of spaces having dimensions that are powers of two; see Section 1.7. The dominant model of quantum computation is the *quantum circuit* (or *quantum-gate array) model*, in which one studies *quantum networks* composed of quantum gates. This model is polynomially equivalent the quantum Turing machine model just sketched. Because the quantum circuits have been shown to be polynomially equivalent to the quantum Turing machines [464], here we work within the quantum circuit model, which is helpful in understanding specific quantum algorithms in detail, as we show in the following chapter. The physical systems required to implement these models, the *quantum hardware*, must have several important properties for reliable computing: the ability to sustain the coherence of multi-qubit states for times sufficient for completing computations in the face of decoherence effects (described in Chapter 10), the ability to read out results via reliable measurements (described in Chapter 2), and the ability to perform controlled gating operations (described in Chapters 1,3, and 7) with precision [138].

The quantum-mechanical state-space of an n-qubit quantum computer is the 2^n-dimensional Hilbert space that is the tensor product of its individual qubit subspaces

$$\mathcal{H}^{(n)} = \mathbb{C}^2 \otimes \mathbb{C}^2 \otimes \cdots \otimes \mathbb{C}^2 \tag{13.5}$$

(see Sect. B.1). Every state $|\Psi\rangle \in \mathcal{H}^{(n)}$ is a superposition of n-qubit states, each representing a binary word $\mathbf{x}_i \in GF(2)^n$, that can be written

$$|\Psi\rangle = \sum_i c_i |\mathbf{x}_i\rangle , \tag{13.6}$$

where $\sum_i |c_i|^2 = 1$. In the quantum circuit model, quantum computation is realized by performing unitary transformations on n qubits describable as sequences (polynomial in n) of quantum gates from a universal family of gates, examples of which are given below. In the quantum circuit model, one commonly uses a graphical notation based on quantum wires, corresponding to qubits, which operate from left to right, as done in the discussions of quantum gates in previous chapters.[10] When a qubit is transformed, it is left unconnected at a given position in the sequence of operations, except when it controls another qubit in which case a dot is placed on its wire connected by vertical lines to any qubit it controls and it is referred to as a *control bit*.[11] When qubits may be acted upon nontrivially, a symbol describing the operation of the corresponding gate is placed over them (*cf.* Fig. 1.2). Qubits are typically but not necessarily ordered in location from top to bottom in a

---

[10] As an example, see the circuit for Bell-state synthesis shown in Fig. 6.1.

[11] See, for example, the two-qubit and three-qubit gates in Sect. 3.8, which are shown in Figs. 3.4-8

complex circuit according to their significance in it; for examples, see Fig. 7.1 and 8.2 and the following chapter.

The set of all permutations of computational-basis states corresponds to the set of all invertible Boolean transformations of $GF(2)^n$ and is a subgroup of the unitary group $U(2^n)$. Because, with a polynomial number of constant inputs and empty ports, any Boolean transformation can be embedded in a unitary one, a quantum computer is universal once the generic unitary operation, which is reversible by definition, can be performed with it. A given unitary transformation U is *implementable* if $U \approx U_{g_k} \cdots U_{g_1}$ for $k = O(\mathrm{poly}(n))$. A set, G, of quantum gates is *universal* if, for any $\epsilon > 0$ and any unitary transformation U on n qubits, there is a sequence of gates $g_1, \ldots, g_l \in$ G such that $\|U - (u_{g_l} \cdots u_{g_2} u_{g_1})\| \leq \epsilon$, where u_{g_j} is $V^{(k)} \otimes \mathbb{I}^{(n-k)}$, $V^{(k)}$ being the unitary transformation on k qubits operated on by the quantum gate g_j, and $\mathbb{I}^{(n-k)}$ is the identity acting on the remaining $n - k$ qubits, $\| \cdot \|$ being the spectral norm defined in Section A.3. Output information is finally extracted as classical information from the quantum computing system via measurement. The most well-known examples of *universal* sets of quantum gates are

(i) the C-NOT and all single-qubit gates;
(ii) the C-NOT, Hadamard, and suitable phase-flip gates;
(iii) the Toffoli and Hadamard gates.

[23, 24, 130, 137, 286].[12] Any particular gate in a given such family can be well approximated by a finite number of gates in another such family. By contrast, for *classical* reversible computation there exists a universal three-bit logic gate, the classical Toffoli gate, but *not* a universal two-bit gate; there is not even an adequate *set* of two-bit gates.

Any useful model of computation must describe processes including arbitrary operations carried out by the execution of elementary operations that can be counted, allowing one to describe the complexity of computational problems. Given a computation $U \in U(2^n)$, the *quantum computational complexity*, $\kappa(U)$, is the minimum number of operations in the chosen universal set of gates necessary for it to be carried out as a sequence of elementary quantum operations $U = u_1 u_2 \ldots u_k$. Due to the linearity of quantum mechanics, one finds that $\kappa(A \otimes B) \leq \kappa(A) + \kappa(B)$, for all $A \in U(2^{n_1})$ and $B \in U(2^{n_2})$. It is very often useful to consider unitary operations described by networks that approximate a given desired unitary transformation in order to discover whether it can be implemented. The complexity for such approximations is written as κ_ϵ, when

$$\left\| U - \prod_{i=1}^{n} u_i \right\| < \epsilon . \tag{13.7}$$

Specific algorithms have been discovered that demonstrate that the use of quantum systems for computation allows some speedup over classical compu-

---

[12] The Toffoli gate is a three-qubit gate that flips the third bit if and only if the first two, control bits both take the value 1; see Sect. 7.9.

tation. For example, the Grover search algorithm provides a quadratic speedup relative to any classical algorithm requiring a search component. It is important to note that there have been, to date, a limited number of specific examples of quantum algorithms that achieve *superpolynomial* speedups over classical algorithms. Most significant of these is the quantum Fourier transform, which plays a role in Shor's factoring algorithm. These two algorithms, and others, are discussed in detail in the following chapter.

Quantum-computational complexity theory focuses primarily on the class of problems known as **BQP**, for "bounded quantum polynomial" or "bounded error probability, quantum polynomial time," which is the quantum analogue of the **BPP** class and which can in principle be solved in polynomial time by a quantum computer. It can be seen that **BPP⊆BQP**, because a quantum computer can simulate a probabilistic classical computer by simply preparing the state $| \nearrow \rangle$ and projecting it to the single-bit computational basis to generate a random bit. It has also been shown that there is an oracle relative to which there are problems in **BQP** that *cannot* be solved with small error probability by probabilistic Turing machines running in $n^{\mathcal{O}(\log n)}$ steps, which is evidence that quantum computers are *fundamentally more powerful* than classical computers [56].

A quantum computer can be simulated on a classical computer with a memory of polynomial size, so that **BQP⊆PSPACE**, despite the intuitive sense one may have that exponential memory resources might be needed, simply due to the fact that simulation of an n-qubit quantum circuit involves matrices of size 2^n. In particular, it is known that **BQP⊆P$^{\#P}$**, which is contained in **PSPACE**. The difference between classical and quantum computation is rather one of *efficiency*. The relationships between **BQP**-class and **NP**-class problems and between the class **BQP** and the class **PH**, "polynomial hierarchy," are currently of great interest. It is known that, relative to a random oracle **NP⊄BQP**: a quantum computer is incapable of inverting black-box (oracle) one-way functions [38].[13]

The quantum analogue of the **NP** class for classical computation is the class **BQNP**, "bounded error probability, quantum nondeterministic polynomial." A problem is a member of **NP** if an offered answer can be verified within a time period growing no more rapidly than polynomially in the size of the input; a computational problem is in **BQNP** if its solution can be checked within a polynomial-time period on a *quantum* computer. Relative to an oracle, the class **BQP** is not contained in the class known as **MA** (the Merlin–Arthur class), the probabilistic generalization of **NP** [54].[14] Relative to a random oracle, quantum computers *also* cannot solve **NP**-complete

---

[13] Note, however, oracle results must be dealt with carefully, because they have often proven misleading in the past.

[14] Even were it found that **P=NP**, quantum computers may still be capable of being more efficient than their classical counterparts. In the context of an interactive proof system, Merlin plays the role of prover and Arthur, a probabilistic polynomial-time machine, plays the role of verifier; see Sect. 13.3.

problems. Finally, note that the class denoted **QSAT**, for "quantum analogue of satisfiability problem," is known to be complete for **BQNP** and thus
BQNP⊆PSPACE [251].[15]

13.7 Fault-tolerant quantum computing

Fault-tolerance is the ability of an information-processing system to operate
properly even when its elements function imperfectly. Before we consider practical proposals for realizing quantum computation, it is important to have an
idea of the nature of fault tolerance in quantum computing; as we have seen,
quantum computers are even more susceptible to errors than classical digital
computers, which themselves often include such a capability. Not only must
errors in quantum memories be corrected but also those arising from imperfect
quantum gates. This makes quantum computation an affair involving quantum networks significantly more complex than the simple ideal ones discussed
in relation to the quantum-communication tasks discussed so far and, indeed,
the basic descriptions of fundamental algorithms appearing in the following
chapter [140, 342, 389]. An extremely brief sketch of some basic elements of
fault-tolerant quantum computing in relation to methods described previously
is now given.[16]

Recall that quantum computational complexity theory focuses on **BQP**,
the class of problems that are efficiently solvable using quantum resources.
Efficient fault-tolerant quantum computing is possible provided the decoherence rate during computation is below a certain *threshold*, η.[17] A valid fault-
tolerant encoded operation can be performed by a combination of SWAP
operations within a block of an error-correcting code and transversal operations on the block, in such a way that the elements of the stabilizer are
permuted. The set operations of this kind can be identified with the automorphism group, $A(S)$, of the stabilizer S; see Section 10.6. Given an encoded
operation U, its effect on encoded states can be found by studying $N(S) \setminus S$
under its action: the action on $N(S) \setminus S$ describes the result of the operation
on the k encoded qubits, where $N(S)$ is the normalizer [192, 193].

Ancilla preparation and measurement can also be used for fault-tolerant
quantum computing. Measurements are equivalent to random applications of
a set of projection operators. One can apply an element of a set of operators conditioned on the quantum state that results. In the case of a binary
measurement—those with eigenvalues ±1—one eigenvalue can indicate that a
projection should be applied and the other eigenvalue indicates the contrary.

---

[15] For a more detailed discussion of these classes, see [38, 252].

[16] A detailed discussion of fault-tolerant computing, using the seven-qubit Steane
code as its central example, is given in [343].

[17] The hard bound on η is $\eta < \frac{1}{2}$ (unless **BQP=BQNC**); the best available methods
of fault-tolerant quantum computing suggest that $10^{-3} \geq \eta \geq 10^{-4}$, with $\eta \approx$
10^{-2} being highly desirable [347].

Preparing an ancilla in a known state and applying a known set of operations in the normalizer of the Pauli group can provide a state describable using a stabilizer S. One can measure elements of the Pauli group fault tolerantly when they anti-commute with an element of S in order to carry out the corresponding projection and to correct the result [192].

Rendering quantum gates fault-tolerant adds considerably to their size. For example, a three-qubit Toffoli gate is rendered fault-tolerant by constructing a quantum circuit involving 63 qubits and a number of measurement processes using the Steane code [343].[18] However, when quantum computations provide considerable speedup, as in the case of the quantum factoring algorithm discussed in the following chapter, the performance of a quantum computer will nonetheless greatly outperform any classical computer.

13.8 Linear optical quantum computation

Ideally, quantum gates perform their designated operations perfectly and the state transformations they induce are deterministic, even though quantum computations must end with quantum measurements. However, linear optics are insufficient for deterministically realizing, say, the two qubit gates C-NOT and controlled phase-flip, as one may wish to use for quantum computation on generic photonic qubits. The principal difficulty for deterministic implementations of these gates is that achieving nonlinear coupling of two optical modes containing only a few photons, which is the typical situation in optical quantum computing, is quite difficult. Nonetheless, *nondeterministic* linear optical realizations are possible.

It is easier to realize nondeterministic gates than deterministic ones, for example, by rendering computations conditional on appropriate measurement outcomes. It has been shown by Emanuel Knill, Raymond LaFlamme, and Gerald Milburn (KLM) that single-photon sources, passive linear optics, and photodetectors are collectively sufficient for the implementation of reliable quantum computation along these lines [257]. In such linear optical implementations of quantum computing (LOQC), one can consider optical modes as realizing "bosonic qubits." For example, qubits can be encoded using two spatial modes, one of which contains a single photon excitation while the other does not. Spontaneous parametric down-converting systems may be used as nondeterministic sources of individual photons by using the detection of one photon of a down-conversion pair to "herald" the presence of the other individual photon.[19]

The properties of a mode i are those corresponding to the application of a particular annihilation operator $\hat{a}_j^{(i)}$. The vacuum state, in which *all* modes are in the occupation-number operator $\hat{N} = \hat{a}_j^{(i)\dagger}\hat{a}_j^{(i)}$-eigenstate $|0\rangle$, is

---

[18] See Fig. 3.6 in Sect. 3.8.
[19] For details of the down-conversion process, see Sect. 6.16.

here written "$|vac\rangle$" to distinguish it from ground states of particular modes. Measurements of such modes are generally made destructively using particle detectors (assumed to have nearly 100% efficiency) that measure whether photons are present in the mode, producing outcomes that can be used for controlling linear-optical elements.

Probabilistic quantum computing can be performed fault-tolerantly provided the probability of error in the performance of gates is sufficiently small [259]. An exponential improvement in the probability of success for gates and state preparation can ultimately be obtained using quantum codes, as needed. Furthermore, as described in the examination of the process of partial Bell-state analysis in Section 9.8, for example, the Bose statistics of photons and the nonlinearity of measurements can be used as an effective substitute for interactions unavailable when one is limited to linear optics.

An essential part of the KLM proposal for LOQC is the use of the nonlinear sign-change operation to achieve a controlled phase-flip gate, which is now explicated in some detail;[20] the C-NOT can be realized nondeterministically once the controlled phase-flip has been [257]. Consider two (dual-rail) qubits encoded in *four* spatial modes corresponding to mode-pairs 1, 2 and 3, 4. One first applies 50–50 beam-splitters to the odd-numbered optical modes.[21] When the two dual-rail qubits are both $|1\rangle$, the resulting state for the modes 1 and 3 will be

$$|\Phi^+\rangle = \frac{1}{\sqrt{2}}(|20\rangle + |02\rangle) \ . \tag{13.8}$$

One then applies the nondeterministic *nonlinear sign-change* (NS) transformation

$$|\psi\rangle = c_0|0\rangle + c_1|1\rangle + c_2|2\rangle \tag{13.9}$$
$$\rightarrow c_0|0\rangle + c_1|1\rangle - c_2|2\rangle = |\psi'\rangle \tag{13.10}$$

to modes 1 and 3 using a sequence of beamsplitters and measuring the ancilla, accepting the result only when the ancilla is found to be in the desired state, which by design is also the desired state of the computational qubit. Each instance of this operation can be performed with a success probability $\frac{1}{4}$. Finally, one performs the inverse of the original beam-splitter operation. The desired *conditional phase-flip* (also known as the *controlled sign-change* operation) is thus achieved with probability $\frac{1}{4} \times \frac{1}{4} = \frac{1}{16}$.

Quantum teleportation can be implemented as a fundamental computational element in this approach as well, reducing the controlled phase-flip gate to what is essentially a *state preparation* task.[22] To accomplish teleportation in this context, an initial qubit state $|\psi\rangle$ in mode 1 is combined with modes

---

[20] For elements of the formal constitution of this gate, see Sects. 1.4 and 3.8.

[21] Care should be taken here not to confuse occupation eigenstates with mode *labels*. Here c's are used for amplitudes instead of a's to avoid confusion with annihilation operators.

[22] Quantum teleportation is described in detail in Sect. 9.9

2 and 3 in the state $|\Phi^+\rangle_{23}$, and the composite system of modes 1 and 2 is measured in the Bell basis: the parity of the number of photons in these modes and the sign of the two-mode superposition state is determined (*cf.* Section 9.9). When the parity is odd, if the sign is positive the state of mode 3 is the initial state of mode 1, accomplishing the task; if the sign is negative then mode 2 has an inverted single-mode superposition sign, which is then inverted to complete teleportation. Teleportation is achievable with a probability $\frac{1}{2}$ of success by sending modes 1 and 2 into a balanced beam-splitter and then measuring the number of photons in them. Two such teleportation operations are required to accomplished a controlled sign-change in this way, so that the probability of success is $\frac{1}{4}$. The probability of success of this sort of teleportation can be improved by increasing the number of modes to obtain quantum networks for state preparation of a size linear in the qubit number [257].

Photon losses and detector inefficiencies are treated as erasure events; they destroy all information about qubit states but can be counteracted after photon-loss detection. Such an approach is successful if the rate of failures due to loss is kept much smaller than unitary-gate failure rates. Phase-error corrections are handled by reducing unitary-gate failures using quantum repetition codes, allowing unknown phase-errors in half the qubits to be corrected. Using a two-qubit quantum code, arbitrarily high probabilities of gating success can be achieved in principle using concatenated stabilizer codes. In particular, one can use the two-qubit quantum code having the logical qubits

$$|0_L\rangle = |\Phi^+\rangle , \tag{13.11}$$
$$|1_L\rangle = |\Psi^+\rangle . \tag{13.12}$$

The full KLM proposal for quantum computation involves an array of techniques developed within quantum information science as it has matured. Only the basic elements have been very briefly sketched above. The implementation of quantum computing according to the methods of this proposal represents a significant step in moving quantum information processing forward toward a real-world technology, as has already been achieved in the case already with quantum cryptography. The following chapter will discuss a few of the earliest quantum-computing algorithms to be developed.

14

Quantum algorithms

There is no reason in principle why powerful quantum computers cannot be realized, despite the numerous technological challenges involved. As we have seen, the necessary conceptual basis now exists for real-world implementations of nontrivial algorithms that exploit the tremendous parallelism provided by quantum mechanics. Quantum algorithms are procedures for carrying out computations in quantum systems that are implementable as quantum circuits in those cases where finite numbers of gates are required. Fortuitously, the parallelism provided by quantum states grows exponentially with the size of the problem to be solved. However, one cannot simply read out the results from the output quantum states, because the required measurement set will *also* grow exponentially. Quantum algorithms are thus designed to map such large superposition states back to the computational basis in a way that allows them to be read out efficiently.[1] In some cases, quantum algorithms are exponentially faster than any corresponding classical algorithm.

Two broad classes of quantum algorithm have been identified so far. The first is that including the Grover search algorithm that make use of simple quantum superpositions but not necessarily entanglement. The second is that of those using the quantum Fourier transform, including the Shor factoring algorithm, which make essential use of entangled states. At this point in the history of quantum computing, a few extraordinary quantum algorithms have been produced that provide exponential speedup, as does the Shor algorithm. More algorithms providing exponential speedup are anticipated the discovery of which, however, appears to be just as challenging as the building of robust quantum hardware. In this chapter, the structures of the most well studied of the currently known algorithms are outlined. These examples exhibit the fundamental features and components characteristic of this novel form of computing and come from both of the above classes. A third class, not discussed in

---

[1] Fourier transforms, in particular, have proven valuable in assisting in achieving final states that can be read out efficiently.

any detail here, is sometimes added to this taxonomy, consisting of algorithms for simulating the behavior of quantum systems themselves.

14.1 The Deutsch–Jozsa algorithm

The first quantum algorithm that was found to provide a speedup relative to corresponding classical algorithms in accomplishing a computational task is the Deutsch algorithm in its extended form, known as the *Deutsch–Jozsa algorithm* [133, 110]. This algorithm is of the first of the above three classes of quantum algorithm and illustrates (subexponential) speedup within the quantum circuit model of quantum computation introduced by David Deutsch [129]. This algorithm serves primarily as an existence proof for the concept of the nontrivial quantum algorithm and can be realized using a very symmetrical circuit, an example of which is given below. It distinguishes two classes of binary function using n qubits. In particular, the Deutsch–Jozsa algorithm distinguishes members of the class of *constant functions*, which take all input values to a *single* output value, from the *balanced functions*, in which half of the input values are taken to *each* element of the range.

For clarity, let us consider here the $n = 3$ -qubit Deutsch–Jozsa algorithm, which classifies binary functions with a domain of just $n - 1 = 2$ bits and the usual range of one bit, $f : \{0, 1\} \times \{0, 1\} \to \{0, 1\}$ within the general context. In the Deutsch problem, the function is provided as an "oracle" unknown to the agent evaluating it, to whom it is therefore a black box. One sees directly that this domain and range together allow for a total of $N = 2^n = 8$ functions to be classified. The Deutsch–Jozsa algorithm allows one to decide, using only *one* query of the oracle function f, whether f is constant or whether it is balanced, based on the value of an output bit. The algorithm, in this simple case involving $f(i, j)$, proceeds as follows.[2]

(i) The $n = 3$–qubit system is prepared in the initial state $|0\rangle|0\rangle|1\rangle$;.

(ii) These $n = 3$ qubits are individually Hadamard transformed, so that

$$|0\rangle|0\rangle|1\rangle \to (|0\rangle + |1\rangle)(|0\rangle + |1\rangle)(|0\rangle - |1\rangle). \tag{14.1}$$

(iii) The oracle function f is then queried using the oracle operation U_f

$$|i\rangle|j\rangle|k\rangle \mapsto |i\rangle|j\rangle|f(i, j) \oplus k\rangle , \tag{14.2}$$

effectively producing phases with powers $f(i, j)$ in state components

$$|i\rangle|j\rangle|k\rangle \mapsto \sum_{j=0}^{1} \sum_{i=0}^{1} (-1)^{f(i,j)} |i\rangle|j\rangle(|0\rangle - |1\rangle) . \tag{14.3}$$

(iv) The inverses of the Hadamard transforms of step (ii) are performed.

---

[2] For clarity, state-vector normalization has been omitted here and later.

The phases do the work here. The result of the above operations is that the output quantum state-vector for the first $n - 1$ qubits in the case of the *constant* functions lies entirely in the direction $|0\rangle^{\otimes n-1}$; in the case of the *balanced* functions it lies entirely in the subspace *orthogonal* to this direction, that is, it has *no* component along the $|0\rangle^{\otimes n-1}$ direction. This allows the two classes to be distinguished by a projective measurement of the binary property of being an element of one subspace or not being so.

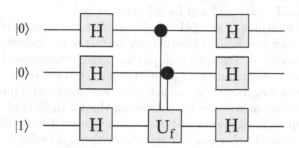

Fig. 14.1. A quantum circuit implementing the pre-measurement portion of the Deutsch–Jozsa algorithm for $f : \{0,1\}^{\times 2} \to \{0,1\}$. H indicates a Hadamard transformation and U_f the oracle. For an experimental three-qubit realization, see [413].

Any classical algorithm performing the same task requires the oracle f to be evaluated a total number of $2^{(n-1)} + 1$ times for the n-qubit version of the Deutsch–Jozsa algorithm to obtain a certainly correct classification; in the case of the above example, this number is five. The algorithm generalizes in the obvious way. This algorithm has already been carried out in several sorts of system. Let us now consider an algorithm of greater practical import.

14.2 The Grover search algorithm

Database searching is a fundamental task of information processing, having applications in cryptography, for example.[3] Given a database, one often desires to find a given one of its states among the set of its N possible states, provided as an unsorted list. This is equivalent to the problem of finding a particular "target" entry in the database that is "marked" in a situation where it is known that there exists one such entry in the database, because each possible state of the database corresponds to having a distinct entry marked. Lov Grover discovered a quantum algorithm for database searching having an efficiency that surpasses that of classical search algorithms, providing a *quadratic* speedup: a query taking $O(N)$ steps classically can be performed in

---

[3] For example, the 56-bit DES code can be broken with certainty by performing a complete search over the 2^{56} available possibilities [312].

$O(\sqrt{N})$ steps using this algorithm [199, 200]. Although less dramatic than the speedup provided by quantum algorithms involving period finding (discussed below), such an improvement *is* still significant, just as, for example, the classical Fast Fourier Transform is of practical importance as an improvement over the standard Fourier transform. The database involved in a search can be accessed, similarly to how the function in the Deutsch–Jozsa algorithm can be queried, using a representative binary oracle function f. The function in this case is defined over the pertinent set of states that takes the value 1 for a given "marked" state and is 0 for all other states.

Database searching is an "**NP**-oracle problem," a standard "oracle idealization" in which one has no internal access to the mechanism realizing the black box function f. The quantum circuit corresponding to the Grover operator, defined below, can "recognize" the target state but cannot directly provide its *location*. Rather, with each iteration, the central quantum circuit rotates the quantum register by increasing the magnitude of the quantum amplitude representing the actual database state relative to the amplitudes corresponding to all *other* possible database states, allowing it eventually to be distinguished from them. The oracle can be queried in this way as many times as necessary in the course of carrying out the algorithm which, however, involves only a prescribed number of queries. The best *classical* algorithm requires $\frac{N}{2}$ oracle evaluations, on average, to solve the database search problem for an N-element database, whereas this algorithm requires only $O(\sqrt{N})$ oracle evaluations. The full Grover algorithm takes only $O(\sqrt{N}\log_2 N)$ steps.

The Grover search is carried out primarily by repeatedly applying the unitary *Grover operator*, $U_G \equiv -I(|\Psi_{\mathrm{in}}\rangle)\mathrm{H}_n I(|m\rangle)\mathrm{H}_n$, involving one oracle query per iteration, beginning with an initial register state, $|\Psi_{\mathrm{in}}\rangle$, of which a given component $|m\rangle$ is sought; H_n is an n-qubit operation performing an individual Hadamard transformation on every qubit state within the total quantum system on which it acts; each operator of the Householder form $I(|\Phi\rangle) \equiv \mathbb{I} - 2P(|\Phi\rangle)$ inverts the register state by flipping the sign of its vector argument $|\Phi\rangle$ and leaving unchanged all the state-vectors orthogonal to it. The oracle "marks" the target amplitude using $I(|m\rangle) = \mathbb{I} - 2P(|m\rangle)$.

$I(|m\rangle)$ is a selective inversion about the hyperplane *orthogonal to* the target state $|m\rangle$ in a sense that is *away from it* within a plane $\mathcal{S}$ defined by $|\Psi_{\mathrm{in}}\rangle$ and $|m\rangle$); $-I(|\Psi_{\mathrm{in}}\rangle)$ is a selective inversion in the *opposite* sense by a *larger* angle in $\mathcal{S}$, back toward the target state. The planar subspace $\mathcal{S}$ is invariant under the actions of both $I(|m\rangle)$ and $I(|\Psi_{\mathrm{in}}\rangle)$. The net result of this pair of inversions is the pertinent rotation in $\mathcal{S}$, by twice the angle between the directions of $|\Psi_{\mathrm{in}}\rangle$ and $|m\rangle$, one step toward the target state: the effect of the (unitary) Grover operator U_G in each iteration is to provide a register superposition-state rotated by fixed-angle θ in this plane to a direction making a smaller angle with that of the target state.

The result of a single Grover transformation is a rotational step (see the box above). When this step is taken an appropriate number of times, a measurement in the computational basis finds the register in the target state $|m\rangle$ with very high probability, in essence completing the search. The effect of the $n + 1$ qubits repeatedly encountering the corresponding Grover circuit is a clock-like rotation of the register state by constant steps that progressively "amplify" the component along the target state vector, so that the register approaches the final state as the prescribed number of iterations is approached.

The initialized state for this algorithm is an $N = 2^n$–entry "unstructured database," where n is the size of the register R in qubits, that can be written

$$|\Psi_{\text{in}}\rangle = \frac{1}{\sqrt{N}} \sum_{i=0}^{N-1} |i\rangle_{\text{R}} = H^{\otimes n}|0\rangle^{\otimes n} , \tag{14.4}$$

where the index i counts over the 2^n elements of the computational basis, that is, all possible database states, of which *one* component, $|m\rangle_{\text{R}}$, is "marked" in the sense that the oracle picks it out. One adds a single ancillary qubit in the diagonal-basis state $|\searrow\rangle_{\text{A}}$, allowing the oracle output values $f(i)$ to influence the phases of the database amplitudes by giving $|m\rangle_{\text{R}}$ a minus sign; the ancilla A can be referred to as the *oracle's qubit* for this reason.

The initial state of the complete system of n-qubit register plus single ancillary qubit A in the combined Hilbert space $\mathcal{H}_{\text{R}} \otimes \mathcal{H}_{\text{A}}$ is the product

$$|\Psi\rangle = \frac{1}{\sqrt{N}} \sum_{i=0}^{N-1} |i\rangle_{\text{R}} |\searrow\rangle_{\text{A}} \tag{14.5}$$

$$= \frac{1}{\sqrt{N}} \sum_{i=0}^{N-1} |i\rangle_{\text{R}}|0\rangle_{\text{A}} - \frac{1}{\sqrt{N}} \sum_{i=0}^{N-1} |i\rangle_{\text{R}}|1\rangle_{\text{A}} . \tag{14.6}$$

Then, in *each* iteration, the following transformations constituting the Grover transformation take place, realizing the "amplitude amplification" procedure, beginning with the evaluation (or *query*) of f.

i) For the "marked" state, there is a state reflection $|m\rangle_{\text{R}} \rightarrow -|m\rangle_{\text{R}}$ via

$$I(|m\rangle_{\text{R}}) = \left[\mathbb{I} - 2P(|m\rangle)\right]_{\text{R}} \otimes \mathbb{I}_{\text{A}} . \tag{14.7}$$

ii) All qubits are Hadamard transformed;

iii) For $i \neq m$, there is a state reflection $|i\rangle_{\text{R}} \rightarrow -|i\rangle_{\text{R}}$ via

$$-I(|\Psi_{\text{in}}\rangle_{\text{R}}) = \left[\sum_{i\neq m} 2P(|i\rangle) - \mathbb{I}\right]_{\text{R}} \otimes \mathbb{I}_{\text{A}} . \tag{14.8}$$

iv) All qubits are again Hadamard transformed.

After each iteration, the overall system R+A is in a product state. The ancillary qubit in $\mathcal{H}_{\text{A}}$ is left in the same superposition state, $|\searrow\rangle_{\text{A}}$, as at the outset. By contrast, as a result of the above two inversions, the state in $\mathcal{H}_R$ is left rotated by a net angle θ with each iteration, where

$$\cos\theta = 1 - \frac{2}{N} \tag{14.9}$$

$$\sin\theta = 2\frac{\sqrt{N-1}}{N}, \tag{14.10}$$

and $\theta \approx \sin\theta = \frac{1}{\sqrt{N}}$, bringing it closer to the target state $|m\rangle_R$. Thus, the register state is sufficiently close to $|m\rangle_R$ after $\frac{\pi}{4}\sqrt{N}$ iterations, each inducing a rotation by the fixed angle θ, for a measurement to determine the state of the database with high probability. Hence only $O(\sqrt{N})$ queries are needed, an improvement over the $\mathcal{O}(N)$ queries required by a classical algorithm.

To summarize: the computational state is progressively rotated so as to allow for the identification of the marked state; the resulting state is measured after the required number of iterations, providing the the marked state among the possible database states.[4] If the problem is instead such that there is *more than one* marked state that is unknown, the situation is not greatly changed as long as the number M of marked states is much smaller than the size of the database, N. One can in that case choose, at random, the number of iterations to be performed as a number between 0 and $\frac{\pi\sqrt{N}}{4}$, because the probability of finding a marked state is close to 50 percent for every M. Any algorithm for this problem will require $\mathcal{O}(\sqrt{N/M})$ queries of the oracle.

The Grover search algorithm has been shown to provide an *optimal* search in the sense that it cannot be parallelized better than by assigning different parts of the searching space to a number of independent quantum processors [467].

14.3 The Shor factoring algorithm

The prime factoring of integers is of central importance to the security of classical public-key cryptosystems: the ability to efficiently factor allows one more easily to break cryptosystems that are dependent on the costliness of performing this operation using classical computers; see Section 12.1. Prime factoring was believed to be an **NP** problem because the difficulty of factoring increases *exponentially* with the size of the input number, in marked contrast to the "inverse" problem of multiplication which increases only polynomially with the size of the input. However, in 1993, Peter Shor discovered an efficient quantum algorithm for factoring integers that reduces the number of steps from the exponential time of $\exp(N^{1/3})$ steps classically required to the polynomial time of N^2 steps [388, 390]. This speedup can be understood by

---

[4] Note, however, that *after* the required number of iterations determined by the size of the database, if left unmeasured, the state will move past the desired state, overshooting it. Graphic illustrations of this rotation are often deceptive.

noting that the quantum version of the fast Fourier transform can be carried out efficiently, since the number of quantum-parallel computations increases exponentially with number with the number of bits involved.

Integer prime factoring is the discovery of the product of prime numbers, each potentially appearing repeatedly in the product, equal to the compound positive integer in question (see Eq. 14.16 below). Factoring a number N is easy, once finding the multiplicative order of an arbitrary element of $\mathbb{Z}_N^{\times}$ has been made easy, because integer factorization reduces to the problem of order finding [301].[5] Order finding is performed in the Shor algorithm by carrying out phase estimation, in particular, the estimation of the eigenvalues associated with eigenvectors of unitary operators. The phase-estimation algorithm, in turn, makes use of the quantum version of the discrete Fourier transform. We now examine these algorithmic elements individually before providing an explication of the (relatively complex) Shor algorithm.

The quantum Fourier transform is the quantum version of the discrete Fourier transform (DFT). Consider a function f that can be written as a normalized vector of the form

$$f = \sum_{i=0}^{n-1} c_i \beta_i , \qquad (14.11)$$

where $\{\beta_i\}$ is an orthonormal basis for the vector space formed by functions into $\mathbb{C}$, and its *discrete Fourier transform*, $\bar{f}$. In a quantum-state description, the β_i form an orthonormal basis for the Hilbert space of the system. The quantum Fourier transform corresponds to the operation

$$f = \sum_{i=0}^{n-1} c_i \beta_i \to \sum_{i=0}^{n-1} \bar{c}_i \beta_i = \bar{f} , \qquad (14.12)$$

which is linear and unitary.[6] The quantum Fourier transform accomplishes a reduction of the number of steps required to perform this operation by acting on the amplitudes of a quantum superposition-state. The *quantum Fourier transform* (QFT) is similar in character to the classical fast Fourier transform (FFT).[7]

---

[5] Here, $\mathbb{Z}_N^{\times}$ is the multiplicative group of units of the integers mod N, $\mathbb{Z}_N$, the cyclic group of order N; a *cyclic group* is a group that can be generated by a single element (the generator) by repeated multiplication.

[6] It is also the case that $||\bar{f}|| = 1$, as can be seen to be so by noting that a quantum Fourier transform pair must obey the Parseval identity $||f|| = ||\bar{f}||$, which can be thought of as the Pythagorean theorem for inner-product spaces, as does the traditional Fourier transform.

[7] The fast Fourier transform, originally used by Gauss and rediscovered by James Cooley and John Tukey in 1965, reduces the number of steps needed to compute the Fourier transform from $2N^2$ to $2N \log_2 N$ steps, by recursively reducing the transform for a number $N = 2^M$ to two transforms each of length $N/2$ [113, 178].

The QFT on the integers modulo n has the desired effect on basis states, namely, the mapping

$$|x\rangle \mapsto \frac{1}{\sqrt{n}} \sum_{k=0}^{n-1} e^{-2\pi ixy/n}|y\rangle , \qquad (14.13)$$

where x, y are representatives of cosets $x + n\mathbb{Z}$ and $y + n\mathbb{Z}$, respectively. However, the outcome of the quantum Fourier transform remains, in a sense, *hidden* in the complex amplitudes of the output state, so that only *global* properties such as the period of the function—which is fortunately what is of interest in this case—can be efficiently determined. As an example, the quantum circuit for performing a QFT on three qubits is shown in Fig. 14.2.

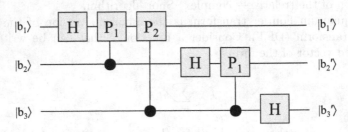

Fig. 14.2. A quantum circuit for carrying out the quantum Fourier transform on three qubits for input computational basis states specified by the b_i with outputs specified by the b_i'. The H indicate Hadamard gates and $P_1 = P(\frac{\pi}{2})$ and $P_2 = P(\frac{\pi}{4})$ phase-shift gates.

The *phase-estimation* component of the Shor algorithm determines the phase of a unit-magnitude complex eigenvalue from the solution of the eigenvalue problem of a unitary operator,

$$U|u\rangle = e^{2\pi i\phi}|u\rangle , \qquad (14.14)$$

making use of the QFT; the estimate is made via controlled-U operations involving a pair of registers in the product state

$$|0\rangle^{\otimes n}|u\rangle^{\otimes n} . \qquad (14.15)$$

The first register is acted on by an n-qubit Hadamard transformation and functions as an array of n control qubits for a sequence of controlled-U^m operations acting on collections of m qubits of the second register, where the power m is equal to the position of the control qubit in its register. An *inverse* discrete Fourier transform is then applied by the algorithm to the first register. This transform returns a product state that provides the solution to the phase-estimation problem upon measurement in the computational basis.

By the fundamental theorem of arithmetic, the prime-factored form of any positive integer N is

$$N = p_1^{\alpha_1} p_2^{\alpha_2} \cdots p_k^{\alpha_k} , \tag{14.16}$$

where p_i are the primes and α_i their powers, which are found using the (classical) Euclidean algorithm and a method of determining periods of (periodic) functions. The central component of the Shor algorithm, which is summarized briefly below, is the process of order finding, which makes use of the quantum phase-estimation procedure discussed above to assist in finding periods.

To factor N, consider an integer y between 1 and $N - 1$. One can use the (classical) Euclidean algorithm to find whether $\gcd(y, N) = 1$, in $\mathcal{O}((\log_2 N)^3)$ (classical) steps. If $\gcd(y, N) > 1$, this y is a desired nontrivial factor of N. Otherwise, i.e. if $\gcd(y, N) = 1$, one takes the (base N) order of y to be r, that is, $r = \text{ord}(y)$, $y^r = 1 \ (\text{mod} N)$ and N divides $y^r - 1$. If r is even, then $y^r - 1$ is readily factored as $(y^{r/2} - 1)(y^{r/2} + 1)$; N necessarily has a factor in common with at least one of the elements of this product, which is similarly found using the Euclidean algorithm. A function of the form

$$f_y(x) = y^x \text{mod } N, \tag{14.17}$$

is periodic because $f_y(x + r) = f_y(x)$; when f has period r, which is the case here, $a^r = a^0 = 1$.[8] To obtain the period, one takes a number of the form $M = 2^m$ as the length of the Fourier transform, where $N^2 \leq M \leq 2N^2$, allowing it to be efficiently performed, as below. With an efficient process for obtaining a nontrivial factor thus in hand, one applies it recursively to all factors until the full factorization of the form of Eq. 14.16 is obtained.

Factoring can therefore be carried out in the Shor algorithm as follows.[9]

(i) Randomly choose an integer y between 1 and $N - 1$, and find $\gcd(y, N)$. If $\gcd(y, N) > 1$, y is a nontrivial factor; otherwise, proceed.

(ii) Prepare two quantum registers in a tensor product state, one of m qubits for the arguments and a second one of $m' = \lceil \log_2 N \rceil$ qubits for the values of the periodic function, with each qubit taking the zero bit-value in the computational basis

$$|\Psi\rangle = |0\rangle_1 |0\rangle_2 , \tag{14.18}$$

where $|0\rangle_1 = |0\rangle^{\otimes m}$ and $|0\rangle_2 = |0\rangle^{\otimes m'}$.[10]

---

[8] A function $f(t)$ is periodic, with period T, if for all t, $f(t + T) = f(t)$, where T is a positive number.

[9] Here, again, most normalization factors have been omitted, cf. [60].

[10] Note that the number m' of qubits can be exponentially smaller than the period of the function $f_y(x)$.

(iii) Perform the Hadamard transform on every one of the m qubits of the first register, providing a superposition of all possible input number values, resulting in the state

$$|\Psi'\rangle = \sum_{x=0}^{M-1} |x\rangle_1 |0\rangle_2 . \tag{14.19}$$

(iv) Compute the function $f_y(x) = y^x \bmod N$ for each x using a unitary transformation taking $|x\rangle|0\rangle \mapsto |x\rangle|y^x \bmod N\rangle$, to obtain

$$|\Psi''\rangle = \xi \sum_{x=0}^{M-1} |x\rangle_1 |y^x \bmod N\rangle_2 , \tag{14.20}$$

where ξ provides normalization.

(v) *Measure the second register* in the computational basis, obtaining a value z_0, leaving the system in the quantum state

$$|\Psi'''\rangle = \sum_{k=0}^{K-1} |x_0 + kr\rangle_1 |z_0\rangle_2 , \tag{14.21}$$

where the first register is placed in a uniform superposition of all states $|x\rangle$ for which $f_y(x) = z_0$, where $y^{x_0} = z_0$ and $K = \lfloor \frac{M}{r} \rfloor$. The second register serves to *prepare* the first register; the particular value z_0 obtained in the process has in itself no significance for the operation of the algorithm.

(vi) Perform the (inverse) quantum Fourier transform on the first register, leaving the system in the state

$$|\Psi''''\rangle = \sum_{l=0}^{M-1} \sum_{k=0}^{K-1} e^{2\pi i(x_0+kr)l/M} |l\rangle_1 |z_0\rangle_2 , \tag{14.22}$$

with the effect that x_0 in the basis of the first register appears now in the *phases* of the superposition instead of the kets, "inverting" the periodicity of the input.

(vii) *Measure the first register* in the computational basis, providing the phase estimate from among the values of l.

(viii) Given the result, which has benefited from the speedup provided by quantum parallelism, the period r of the function is then determinable by *classical* computational methods.[11] The order, and hence a nontrivial factor, is obtained.

The above process is repeatedly applied until the full prime factorization, Eq. 14.16, is obtained.

---

[11] For example, see the discussion in Sect. 4.5 of [60].

14.4 The Simon algorithm

Factoring using the Shor algorithm involves finding the period of a function. Another algorithm that involves this is Simon's algorithm [60, 394], which is an instance of a more general problem, the *Abelian hidden-subgroup problem*.

Given a function f from $(\mathbb{Z}_2)^N$ to itself (*cf.* Section A.1), representable as a unitary map

$$|x\rangle|0\rangle \mapsto |x\rangle|f(x)\rangle , \qquad (14.23)$$

for all x, that can be evaluated on quantum states in the presence of a subgroup $\bar{H} \in (\mathbb{Z}_2)^N$ on which f takes a constant unique value on each right coset of $\bar{H}$, *Simon's problem* is to find the *generators* of this subgroup. Simon's algorithm uses a number, of order $\mathcal{O}(N)$, of evaluations of the oracle f, classical computation, and a polynomial number of quantum operations to solve the problem. This algorithm operates as follows.

(i) Prepare two N-qubit quantum registers in tensor product states wherein each qubit has the computational-basis-value zero,

$$|\Psi\rangle = |0\rangle_1^{\otimes N}|0\rangle_2^{\otimes N} . \qquad (14.24)$$

(ii) Perform a multiple-qubit Hadamard transformation on all qubits of the first register, providing a superposition of all possible inputs, namely,

$$|\Psi'\rangle = \frac{1}{\sqrt{2^N}} \sum_{x\in(\mathbb{Z}_2)^N} |x\rangle_1|0\rangle_2 . \qquad (14.25)$$

(iii) Act on this superposition with the unitary operation corresponding to f, yielding

$$|\Psi''\rangle = \frac{1}{\sqrt{2^N}} \sum_{x\in(\mathbb{Z}_2)^N} |x\rangle_1|f(x)\rangle_2 . \qquad (14.26)$$

(iv) *Measure the second register*, obtaining a value y from the second register and the coset of the hidden subgroup $\bar{H}$ in the *first* register:

$$|\Psi'''\rangle = \frac{1}{\sqrt{|\bar{H}|}} \sum_{f(x)=y} |x\rangle_1|y\rangle_2 . \qquad (14.27)$$

(v) Apply the multiple-qubit Hadamard transformation to the first register.

(vi) *Measure the first register* in the computational basis.

(vii) Repeat the above steps the anticipated number (that is, $\mathcal{O}(N)$) of times, generating with high probability the orthogonal complement of $\bar{H}$ with respect to the scalar product in $(\mathbb{Z}_2)^N$.

(viii) Solve linear equations, for example, via Gauss' classical algorithm for finding the kernel of a square matrix, to obtain the generators of $\bar{H}$.

A

Mathematical elements

The language of quantum mechanics is rooted in linear algebra and functional analysis. It often expands when the foundations of the theory are reassessed, which they periodically have been. As a result, additional mathematical structures not ordinarily covered in physics textbooks, such as the logic of linear subspaces and Liouville space, have come to play a role in quantum mechanics. Information theory makes use of statistical quantities, some of which are covered in the main text, as well as discrete mathematics, which are also not included in standard treatments of quantum theory. Here, before describing the fundamental mathematical structure of quantum mechanics, Hilbert space, some basic mathematics of binary arithmetic, finite fields and random variables is given. After a presentation of the basic elements of Hilbert space theory, the Dirac notation typically used in the literature of quantum information science, and operators and transformations related to it, some elements of quantum probability and quantum logic are also described.

A.1 Boolean algebra and Galois fields

A *Boolean algebra* B_n is an algebraic structure given by the collection of 2^n subsets of the set $I = \{1, 2, \ldots, n\}$ and three operations under which it is closed: the two binary operations of union ($\vee$) and intersection ($\wedge$), and a unary operation, complementation ($\neg$). In addition to there being complements (and hence the null set $\emptyset$ being an element), the following axioms hold.

   (i) Commutativity: $S \vee T = T \vee S$ and $S \wedge T = T \wedge S$;
   (ii) Associativity: $S \vee (T \vee U) = (S \vee T) \vee U$ and $S \wedge (T \wedge U) = (S \wedge T) \wedge U$;
   (iii) Distributivity: $S \wedge (T \vee U) = (S \wedge T) \vee (S \wedge U)$ and $S \vee (T \wedge U) = (S \vee T) \wedge (S \vee U)$;
   (iv) $\neg \emptyset = I$, $\neg I = \emptyset$, $S \wedge \neg S = \emptyset$, $S \vee \neg S = I$, $\neg(\neg S) = S$,
for all its elements S, T, U. The algebra B_1 is the *propositional calculus* arising from the set $I = \{1\}$, which is also used in digital circuit theory, where $\emptyset$

corresponds to FALSE, I to TRUE, $\vee$ to OR, $\wedge$ to AND, and $\neg$ to NOT. The "basic" logic operation XOR corresponds to $(S \vee T) \wedge (\neg S \vee \neg T)$.

A *field* is a set of elements, with two operations (called multiplication and addition) under which it is closed, that satisfies the axioms of associativity, commutativity and distributivity and in which there exist additive and multiplicative identity elements and inverses. The *order* of a field is the number of its elements. *Theorem* (Galois): There exists a field of order q if and only if q is a prime power, that is, $q = p^n$ where p is prime and n is a positive integer; furthermore, if q is a prime power, then there is (up to a relabeling), only *one* field of that order. A field of order q is known as a *Galois field* and is denoted $GF(q)$. If p is prime, then $GF(p)$ is simply the set $\{0, 1, ..., p-1\}$ with mod-p arithmetic. $GF(p^n)$ is a linear space of dimension n over $\mathbb{Z}_p$ (the set of integers 0 to $p-1$) containing a copy of it as a subfield. For $\mathbb{Z}_2$, $GF(2)$ is such a field over the set $\{0, 1\}$ with XOR as addition and AND as multiplication. In computing, one is primarily interested in functions $f : \{0,1\}^m \to \{0,1\}^n$; for example, two-bit logic gates $g : GF(2) \to GF(2)$.

A.2 Random variables

A *random variable* is a deterministic function from a given sample space S, that is, the set of all possible outcomes of a given experiment, the subsets of which are known as *events*, those containing only a single element being the *elementary events*, to the real numbers. The *expectation value*, $E[Y(X)]$, of a function $Y(X)$ of a random variable X is given by a linear operator such that

$$E[Y(X)] = \sum_{i=1}^{n} Y(x_i) P(X = x_i) , \qquad (A.1)$$

$$E[Y(X)] = \int_{-\infty}^{+\infty} Y(x) f(x) dx , \qquad (A.2)$$

in the cases of discrete and continuous variables, respectively, where in the former $P(X = x_i)$ is the *probability* and in the latter $f(x)$ is the *probability density* (p.d.f.), that is, a collection of numbers between 0 and 1 summing to unity. For a random variable with a (differentiable) cumulative distribution function $F(X)$, $f(x) \equiv dF/dx$; a random variable has the cumulative distribution function $F(X)$ if the probability of an experiment with sample space S to yield $x < X$ as an outcome is

$$F(x) = P(X < x) = \int_{-\infty}^{x} f(x') dx' . \qquad (A.3)$$

A *Markov chain* is a sequence of random variables $X_1, X_2, \ldots$ such that a given random variable X_i is independent of all the variables $X_1, X_2, \ldots X_{i-1}$, that is, is *memoryless*.

A.3 Vector Spaces and Hilbert space

The central mathematical structure of quantum mechanics is Hilbert space, which is a specific sort of vector space. The Dirac notation commonly used to describe Hilbert-space calculations in the physics literature is introduced below, after this structure is described in standard mathematical notation.

A *vector space* V over a field F is a set of vectors together with operations of scalar-multiplication and addition satisfying the following. For all elements $a, b \in F$ and $u, v, w \in V$ (with the zero vector denoted $\mathbf{0}$), scalar multiples and vector sums are elements of V such that:

(i) There is a unique scalar *zero element* $0 \in F$ such that $0u = \mathbf{0}$ for all u;
(ii) $\mathbf{0} + u = u$;
(iii) $a(u + v) = au + av$ and $(a + b)u = au + bu$; and
(iv) $u + v = v + u$, and $u + (v + w) = (u + v) + w$.

A (scalar) *inner product* (u, v) is assumed for all pairs $u, v \in V$ such that:

(i) $(u, v) = (v, u)^*$, where * indicates complex conjugation;
(ii) $(u, u) \geq 0$, with $(u, u) = 0$ if and only if $u = \mathbf{0}$;
(iii) $(u, v + w) = (u, v) + (u, w)$; and
(iv) $(u, av) = a(u, v)$.

The *norm* of a vector, $||u|| \equiv \sqrt{(u, u)}$, satisfies:

(i) $||u|| \geq 0$ for all $u \in V$, with $||u|| = 0$ if and only if $u = \mathbf{0}$;
(ii) $||u + v|| \leq ||u|| + ||v||$, for all $u, v \in V$; and
(iii) $||au|| = |a| \, ||u||$, for all $a \in F$ and all $u \in V$.

The field F used in quantum mechanics is usually taken to be that of the complex numbers, $\mathbb{C}$. The vectors for which $||u|| = 1$ are the *unit vectors*. Vectors u and v are *orthogonal* ($u \perp v$) if and only if $(u, v) = 0$.

A function of two vector arguments that is more general than the inner product, which serves similar purposes but is applicable in broader contexts, is the *Hermitian form*, $h(u, v)$, which satisfies:

(i) $h(u, v) = h(v, u)^*$;
(ii) $h(u, av) = ah(u, v)$, for all $a \in F$, $u, v \in V$; and
(iii) $h(u + v, w) = h(u, w) + h(v, w)$, for all $u, v, w \in V$.

A *positive Hermitian form* is an Hermitian form that also satisfies the condition $h(u, u) \geq 0$ for every $u \in V$. An Hermitian form is *positive-definite* if $h(u, u) = 0$ implies $u = \mathbf{0}$; a *positive-definite* Hermitian form is an inner product.

A *basis* for a vector space is a set of mutually orthogonal vectors such that every $u \in V$ can be written as a linear combination of its elements; a basis is *orthonormal* if it is composed entirely of unit vectors. A set S of vectors in V is a *subspace* of V if S is a vector space in the same sense as V itself is. The one-dimensional subspaces of V are called *rays*.

A *Hilbert space*, $\mathcal{H}$, is a complete complex vector space with an inner product for which $(u, u) \geq 0$ for all $u \in \mathcal{H}$. A map $A : \mathcal{H} \to \mathcal{H}$, $u \mapsto Au$ is a *linear operator* if, for all $u, v \in \mathcal{H}$ and scalars $a \in F$:

(i) $A(u + v) = Au + Av$; and

(ii) $A(au) = a(Au)$.

If, instead of (ii), one has $A(au) = a^*(Au)$, then A is an *anti-linear operator*. (Together, the linear and anti-linear operators are fundamental to quantum mechanics; see Wigner's theorem, below.)

A vector space with an inner product, such as a Hilbert space, can be attributed a norm $|| \cdot || = (v, v)^{1/2}$, which provides a distance via $d(v, w) = ||v - w||$. Such a vector space is *separable* if there exists a countable subset in the space that is everywhere dense, that is, for every vector there is an element of the space within a distance ϵ of it for every positive real ϵ; the space is *complete* if every Cauchy sequence—namely, every sequence such that for every $\epsilon > 0$ there is a number $N(\epsilon)$ such that $||v_m - v_n|| < \epsilon$ if $m, n > N(\epsilon)$— has a limit in the space. (For finite-dimensional spaces, one usually considers the norm topology, although weak topologies may be required to define needed limits and to give proper definitions of continuity.) A *subspace* of a Hilbert space $\mathcal{H}$ is a closed linear manifold, that is, a linear manifold containing its limit points; a linear manifold in $\mathcal{H}$ is a collection of vectors such that the scalar multiples and sums of all its vectors are in it.

A *bounded* linear operator is a linear transformation L between normed vector spaces V and W such that the ratio of the norm of $L(v)$ to the norm of v is bounded by the *same* number, for all non-zero vectors $v \in V$. The set of bounded linear operators on a Hilbert space $\mathcal{H}$ is designated $B(\mathcal{H})$. The sum of two operators A and B, $A + B$, is another operator defined by $(A + B)v = Av + Bv$, for all $v \in \mathcal{H}$; multiplication of an operator by a scalar a is defined by $(aA)v = a(Av)$, for all $v \in \mathcal{H}$; multiplication of two operators is defined by $(AB)v = A(Bv)$, for all $v \in \mathcal{H}$. The *zero operator*, $\mathbb{O}$, and the *unit operator*, $\mathbb{I}$, are defined by $\mathbb{O}v = \mathbf{0}$ and $\mathbb{I}v = v$, respectively. An operator B is the *inverse* of another operator A whenever $AB = BA = \mathbb{I}$; it can be written $B = A^{-1}$. Two operators A and B *commute* if the *commutator* $[A, B] \equiv AB - BA = 0$. A ordering relation $A \geq B$ for self-adjoint bounded linear operators is defined by $A - B \geq \mathbb{O}$.

A nonzero vector $v \in \mathcal{H}$ is an *eigenvector* of the linear operator A if $Av = \lambda v$, for any scalar λ, which is said to be the *eigenvalue* of A corresponding to v; one can then write

$$(A - \lambda \mathbb{I})v = \mathbf{0} . \tag{A.4}$$

By considering the linear operator A in its matrix representation, the solutions to this *eigenvalue problem* can be found by solving the *characteristic* equation $\det(A - \lambda \mathbb{I}) = 0$, the left-hand side of which, in cases where A has a *finite set* (*spectrum*) of eigenvalues, is an nth-degree polynomial in λ.

The *adjoint*, $A^\dagger$ of the operator A is defined by the property that $(A^\dagger v, w) = (v, Aw)$ for every $v, w \in \mathcal{H}$. Any operator for which $A^\dagger = A$ is said to be (Hermitian) *self-adjoint* and has the two following properties.

(i) All eigenvalues are real.

(ii) Any two eigenvectors v_1 and v_2 with corresponding eigenvalues λ_1 and λ_2, respectively, are orthogonal to each other when λ_1 and λ_2 are nonidentical.

A linear operator O is *unitary* if $OO^\dagger = O^\dagger O = \mathbb{I}$, in which case $O^\dagger = O^{-1}$. Unitary operators are usually designated by the symbol U and have the following properties.

(i) The rows of U form an orthonormal basis.

(ii) The columns of U form an orthonormal basis.

(iii) U preserves inner products, that is, $(v, w) = (Uv, Uw)$ for all $v, w \in \mathcal{H}$.

(iv) U preserves norms and angles.

(v) The eigenvalues of U are of the form $e^{i\theta}$.

The matrix representing any unitary transformation U on a Hilbert space of countable dimension d can be diagonalized as above, to take the form

$$
\begin{pmatrix}
e^{i\theta_1} & 0 & \cdots & 0 \\
0 & \ddots & \ddots & 0 \\
\vdots & \ddots & \ddots & \vdots \\
0 & \cdots & 0 & e^{i\theta_d}
\end{pmatrix}.
$$

A bounded linear operator $\mathcal{O} \in B(\mathcal{H})$ is *positive*, $\mathcal{O} \geq \mathbb{O}$, if $\langle \psi | \mathcal{O} | \psi \rangle \geq 0$ for all $|\psi\rangle \in \mathcal{H}$. The set of positive operators is convex. Positive operators are Hermitian, always admit a positive square root, and, if invertible, have a unique decomposition into *polar form*, by which is meant that such an operator $\mathcal{O}$ can be written $\mathcal{O} = |\mathcal{O}|\mathcal{U}$ with $\mathcal{U}$ a unitary operator and $|\mathcal{O}| = \sqrt{\mathcal{O}\mathcal{O}^\dagger}$, analogous to the polar form of a complex number.

The *singular-value decomposition theorem* for operators can be stated as follows. For any operator A representable as an $m \times n$ matrix of rank r, there exists an $m \times n$ matrix Σ of the form

$$
\Sigma = \begin{pmatrix} D & 0 \\ 0 & 0 \end{pmatrix}, \tag{A.5}
$$

where D is an $r \times r$ diagonal matrix having as diagonal entries the first r singular values of A, namely, σ_i such that $\sigma_1 \geq \sigma_2 \geq \cdots \geq \sigma_r$, and there exists an $m \times m$ matrix U and an $n \times n$ orthogonal matrix V such that

$$
A = U\Sigma V^{\mathrm{T}}, \tag{A.6}
$$

where ^{T} indicates matrix transpose, that is, $[A^{\mathrm{T}}]_{ij} = [A]_{ji}$. Any such factorization is a *singular-value decomposition* of A.

An Hermitian operator A is a *projection operator* (projector) if and only if $A^2 = A$, in which case it is usually denoted $P(S)$, where S is a subspace of $\mathcal{H}$, often a ray. A bounded linear operator is a *trace-class* operator if its *trace*,

$$\mathrm{tr}A \equiv \sum_k (Av_k, v_k) \, , \tag{A.7}$$

is absolutely convergent for any orthonormal basis $\{v_k\}$ of $\mathcal{H}$. The trace is a linear functional over the space of trace-class operators, that is, $\mathrm{tr}(aA + bB) = a\,\mathrm{tr}A + b\,\mathrm{tr}B$, for A, B trace-class, and is *cyclic*, *i.e.* $\mathrm{tr}(AB) = \mathrm{tr}(BA)$. The bilinear map $\langle A, B \rangle \equiv \mathrm{tr}(A^\dagger B)$ is an inner product on the trace-class operators, and provides the *Hilbert–Schmidt norm*. By contrast, the *spectral norm* of an operator described by an $n \times n$ complex matrix A is

$$\max\{|\lambda| \,|\, \lambda \in \mathrm{Spec}(A)\} \, , \tag{A.8}$$

where $\mathrm{Spec}(A)$ is the eigenvalue spectrum of A; it is the square root of the *spectral radius* of A, which is the largest eigenvalue of $A^\dagger A$.

KyFan's maximization principle provides a useful constraint on the eigenvalues of a sum of two Hermitian matrices. It can be stated as follows. Given an Hermitian operator A and a set of k-dimensional projectors $\{P_i\}$,

$$\sum_{j=1}^k \lambda_j = \max_{P_i}[\mathrm{tr}(AP_i)] \, , \tag{A.9}$$

where $\lambda_j \in \mathrm{Spec(A)}$.

Majorization is a method of ordering vectors that captures their relative degree of order. One writes $\mathbf{v}^\downarrow$ for $\mathbf{v}$ reordered so that $v_1 \geq v_2 \geq v_3 \geq \ldots \geq v_d$. Consider two d-dimensional vectors $\mathbf{u} = (u_1, u_2, u_3, \ldots, u_d)$ and $\mathbf{v} = (v_1, v_2, v_3, \ldots, v_d)$. The relation $\prec$ defined as follows.

$$\mathbf{u} \prec \mathbf{v} \quad \text{if} \quad \sum_{j=1}^k u_j^\downarrow \leq \sum_{j=1}^k v_j^\downarrow \, , \tag{A.10}$$

with $k \leq d$, equality occurring when $k = d$; one then says that $\mathbf{v}$ *majorizes* $\mathbf{u}$. One valuable result involving this ordering is that

$$\mathbf{u} \prec \mathbf{v} \quad \text{iff} \quad \mathbf{u} = \sum_j p_j \Pi_j \mathbf{v} \, , \tag{A.11}$$

for some probability distribution $\{p_j\}$ and permutation matrix Π_j, which is a binary matrix with exactly one entry 1 in each row and each column and zeros elsewhere that implements a permutation on a vector. With majorization, KyFan's maximization principle mentioned above gives rise to a helpful constraint on the sum of eigenvalue-vectors of two Hermitian matrices:

$$\boldsymbol{\lambda}(A + B) \prec \boldsymbol{\lambda}(A) + \boldsymbol{\lambda}(B) \, . \tag{A.12}$$

If one vector majorizes another, then the latter can be obtained from the former by multiplication by a doubly stochastic matrix—a matrix is said to be *doubly stochastic* when its entries are nonnegative entries and each row and each column sum to one. *Birkhoff's theorem* states that a $d \times d$ matrix D is doubly stochastic if and only if it can be written $D = \sum_j p_j \Pi_j$.

A.4 The standard quantum formalism

The *Hilbert-space formulation of quantum mechanics* has the following elements, which have associated postulates given in Appendix B.

(i) Every physical system is attributed a separable Hilbert space $\mathcal{H}$.

(ii) Every physical property is associated with a self-adjoint, not necessarily bounded, linear operator, O, called an *observable*, and vice versa.

(iii) Every state of a physical system is assigned a *statistical operator* ρ that is a linear bounded self-adjoint positive trace-class operator, and vice versa.

(See Appendix B for a statement of the quantum postulates, including its extension to composite systems via the tensor product, and the time-evolution of states.)

Wigner's theorem is a fundamental theorem of the Hilbert-space formulation of quantum mechanics: Let $\mathcal{I} : u \to v$ be a length-preserving transformation (that is, an *isometry*) with respect to the Hilbert-space norm; $\mathcal{I}$ is either unitary or anti-unitary.

The *expectation value*, $\langle O \rangle_\rho$, of an operator O for a quantum state ρ is given by

$$\langle O \rangle_\rho = \mathrm{tr}(\rho O) . \tag{A.13}$$

The inner product, (u, v) on a finite-dimensional Hilbert space $\mathcal{H}$ induces a natural geometry. Because global phases on the vectors have no effect on the probability for finding a system in a given eigenstate upon measurement, they are not a necessary part of the description of a state and all state vectors related to each other by such a phase can be identified.[1] Hence, *projective Hilbert space* is appropriate for the description of finite-dimensional quantum systems [431]. A natural metric for these projective Hilbert spaces is the *Fubini–Study metric*, which for finite values corresponds to the distance

$$d(u, v) = 1 - \min_v \left[\mathrm{Re}(u, e^{i\phi}v) \right]^2 . \tag{A.14}$$

A.5 The Dirac notation

The Hilbert-space structures described in the previous section can all be written in *Dirac notation*, which we now introduce. The state of a physical system

---

[1] Indeed, this phase naturally cancels out in the formation of the projector, P_u, corresponding to a given vector u.

described by a state-vector v is written as a *ket*, $|v\rangle$, and corresponds to a pure statistical operator; the corresponding Hermitian adjoint is given by a *bra*, $\langle v|$. The inner product (v, w) of two such vectors is written as the *braket* $\langle v|w\rangle$, and is a complex scalar (*cf.* Eq. A.16). Operators acting from the left on a ket yield a ket and acting from the right on a bra yield a bra. A *ketbra*, $|v\rangle\langle w|$, is an operator (*cf.* Eq. A.17) that, when acting on a ket $|u\rangle$, yields

$$|v\rangle\langle w|(|u\rangle) = \langle w|u\rangle|v\rangle = (w, u)|v\rangle . \tag{A.15}$$

Every statistical operator ρ can be written as a linear combination of (pure) *projector ketbras*, $P(|u_i\rangle) \equiv |u_i\rangle\langle u_i|$, having weights p_i. The inner product of vectors, $\langle v|w\rangle$ taking $|v\rangle = \sum_i \alpha_i|i\rangle$ and $|w\rangle = \sum_i \beta_i|i\rangle$, is therefore

$$\langle v|w\rangle \doteq (\alpha_1^* \; \alpha_2^* \cdots) \begin{pmatrix} \beta_1 \\ \beta_2 \\ \vdots \end{pmatrix} . \tag{A.16}$$

The row vector $(\alpha_1^*\alpha_2^* \cdots)$ represents $\langle v|$ and the column vector $(\beta_1\beta_2 \cdots)^{\mathrm{T}}$ represents $|w\rangle$. The general ketbra $|v\rangle\langle w|$ can be written as the outer product

$$|v\rangle\langle w| \doteq \begin{pmatrix} \alpha_1 \\ \alpha_2 \\ \vdots \end{pmatrix} (\beta_1^* \; \beta_2^* \cdots). \tag{A.17}$$

Recalling that the projector $P(|v\rangle) = |v\rangle\langle v|$, one thus has, for any $|w\rangle$, $P(|v\rangle)|w\rangle = (\langle v|w\rangle)|v\rangle$; see Fig. 2.1. Note that $P^2 = |v\rangle\langle v|v\rangle\langle v| = P$ because $|v\rangle$ has norm 1, that is, projectors are idempotent.

The matrix for an operator O and basis states $|i\rangle$ and $|j\rangle$ has elements $\langle j|O|i\rangle \in \mathbb{C}$. The representation of an operator by a collection of matrix elements is relative to the choice of eigenbasis; for example, see Section 1.4 for matrix representations of single-qubit gates. Hermitian operators (observables) correspond to physical properties and have matrices with real diagonal elements O_{ii} and complex off-diagonal elements such that $O_{ij} = O_{ji}^*$. The matrix representation of a statistical operator ρ, such as is necessary to describe mixed quantum states, is known as a *density matrix* and is designated by the same symbol. When the state is pure, the density matrix is of rank one. Recalling the spectral representation given in Eq. 2.15, one can define a function of an operator O by

$$f(O) = \sum_n f(o_n)P(|o_n\rangle) , \tag{A.18}$$

under appropriate conditions, as discussed in Footnote 9 of Chapter 2. The expectation value of an operator is given in Dirac notation by Eq. 2.16.

The *tensor product* of two vectors $|v\rangle$ and $|w\rangle$ is written $|v\rangle \otimes |w\rangle$. The tensor product space, written $V \otimes W$, is the linear space formed by such products of vectors: given bases $|v_1\rangle, \ldots, |v_k\rangle$ and $|w_1\rangle, \ldots, |w_l\rangle$ for two vector spaces V and W, respectively, a corresponding basis for $V \otimes W$ is given by

$$\{|v_i\rangle \otimes |w_j\rangle : 1 \leq i \leq k, 1 \leq j \leq l\},$$

and $\dim(V \otimes W) = kl$. Any vector $|\Psi\rangle \in V \otimes W$ can be written in the form

$$|\Psi\rangle = \sum_{ij} \alpha_{ij} |v_i\rangle |w_j\rangle, \tag{A.19}$$

where the α_{ij} are corresponding scalar components. Every linear operator O in such a tensor product space $V \otimes W$, where

$$O(v_i \otimes w_j) \equiv (O_1 v_i) \otimes (O_2 w_j), \tag{A.20}$$

is a linear combination of direct products of linear operators, namely,

$$O = \sum_i O_1^{(i)} \otimes O_2^{(i)}. \tag{A.21}$$

A.6 Groups of transformations

A set of elements G, together with a product map $G \times G \to G$, forms a *group* if it satisfies the following conditions.

(i) Multiplication of elements is *associative*: $a(bc) = (ab)c$ for all $a, b, c \in G$.

(ii) An *identity element* $e \in G$ exists, for which $eg = ge = g$ for all $g \in G$.

(iii) An *inverse* $g^{-1} \in G$ exists for every $g \in G$, such that $g^{-1}g = gg^{-1} = e$.

A map θ between two groups G and H is a *group homomorphism* if $\theta(g_1 g_2) = \theta(g_1)\theta(g_2)$ for all $g_1, g_2 \in G$. An *action* of a group G on another set S is given by a map $G \times S \to S$ such that $g_2(g_1 s) = (g_1 g_2)s$ and $es = s$ for any $s \in S$, that is, a homomorphism from the group into the group of one-to-one transformations of S.

A *unitary representation* of a group on a vector space assigns unitary operators U on the space such that $U(gh) = U(g)U(h)$ for all $g, h \in G$; a mere *projective representation* is a representation for which $U(gh) = \omega(g, h)U(g)U(h)$, where $\omega(g, h)$ is a phase term. A representation is *irreducible* if there is no vector subspace that is mapped to itself by every element of the representation. Representations are *equivalent* if there is an isomorphism M between them such that $MU(g) = U(g)M$.

The *orbit* $G \cdot S$ of an element m of a set S under the action of a group G is the subset of S given by $\{gm | g \in G\}$, g ranging over all elements of G. Orbits stratify the set of quantum states of a system. The orbit of a statistical operator ρ under the group $U(n)$ of unitary operators of dimension n is

determined by its spectrum. An orbit O can therefore be specified by a representative diagonal matrix, the eigenvalues of which are ordered from greatest
to smallest. The unitary group $U(n)$ partitions the set of density matrices
into an uncountably infinite family of orbits. Given two statistical operators
ρ_1 and ρ_2, the following statements are equivalent.

(i) ρ_1 and ρ_2 are *unitarily equivalent*: $\rho_2 = U\rho_1 U^\dagger$, for some unitarity U.

(ii) ρ_1 and ρ_2 have *identical eigenvalue spectra*.

(iii) $\mathrm{tr}\rho_1^r = \mathrm{tr}\rho_2^r$ for all $r = 1, 2, \ldots, n$, where $n = \dim\rho_1 = \dim\rho_2$.

Majorization can be used to provide a partial ordering of orbits, by taking
$O_1 \prec O_2$ if $\rho_1 \prec \rho_2$, where $O_i = O[\rho_i]$ designates the orbit of ρ_i.

A.7 Probability, lattices, and posets

Here we review the basic elements of structures appearing in traditional quantum logic and probability theory. Traditional quantum logic has a long history
stretching back to early work by Birkhoff and von Neumann [61]. More on the
quantum-logical approach to quantum mechanics can be found in Section A.9.

A *σ-algebra* is a nonempty collection S of subsets of a set X such that:

(i) The empty set $\emptyset$ is in S,

(ii) If A is in S, then the complement of A in X is in S,

(iii) If A_n is a sequence of elements of S, then the union of the elements
of the sequence is also in S.

A *partially ordered set* (*poset*) P is a set S together with a binary (partial
ordering) relation, $\leq$, that is

(i) Reflexive ($a \leq a$),

(ii) Antisymmetric ($a \leq b$ and $b \leq a$ implies that $a = b$, for all $a, b \in S$),

(iii) Transitive ($a \leq b$ and $b \leq c$ implies that $a \leq c$, for all $a, b, c \in S$).

The *least upper bound* (lub) of two elements, a and b, under $\leq$ is written
$a \vee b$, and the *greatest lower bound* (glb) is written $a \wedge b$.

An *orthomodular poset* is a poset, with a unary operation $'$, fulfilling:

(i) $0 \leq a \leq 1$ for all $a \in P$, 0 being the zero element and 1 the unit,

(ii) For all $a, b \in P$, $(a')' = a$, $a \leq b \Rightarrow b' \leq a'$, $a \vee a' = 1$,

(iii) If $a \leq b'$ then $a \vee b \in P$,

(iv) If $a \leq b$, then there is an element $c \in P$ such that $c \leq a'$ and $b = a \vee c$.

Condition (ii) ensures that the operation $' : P \to P$, corresponding to set-
theoretic complementation, is an orthocomplementation; (iv) is the orthomodular law. Two elements a and b of an orthomodular poset are *orthogonal*
($a \perp b$) if $a \leq b'$.

A *lattice* is a poset for which there exists both a lub and a glb for every
pair of elements. A lattice contains both a zero element, 0, and an identity
element, 1, if $0 \leq a$ and $a \leq 1$ for every one of its elements a. A lattice
is a *complemented lattice* if there exists a complement, a', for every one of

its elements, a—that is, if for every a there exists an element a', such that $a \vee a' = 1$ and $a \wedge a' = 0$. A lattice is a *distributive lattice* if for all triplets of elements a, b, c, $a \wedge (b \vee c) = (a \wedge b) \vee (a \wedge c)$ and $a \vee (b \wedge c) = (a \vee b) \wedge (a \vee c)$. An *orthomodular lattice* is an orthomodular poset that is a lattice. A *Hasse diagram* for a lattice is a figure in which elements are indicated by circles and are joined by a line in such a way that the "lesser" element is located below the "greater" element (*cf.*, for example, [411]).

A *Boolean lattice* (or *Boolean algebra*, see Section A.1 above) is a lattice that is both complemented and distributive. Every element of a Boolean lattice has a unique complement that is an orthocomplement. An *orthomodular lattice* is an orthomodular poset that is a lattice. Elements a and b of an orthomodular poset are *orthogonal* ($a \perp b$) if $a \leq b$. Given two orthomodular posets P_1 and P_2, P_1 is *orthorepresentable* in P_2 if there exists a mapping, the *orthoembedding*, $h\colon P_1 \to P_2$, such that, for every $a, b \in P_1$:

(i) $h(0) = 0$,
(ii) $h(a') = h(a)'$,
(iii) $a \leq b$ if and only if $h(a) \leq h(b)$, and
(iv) $h(a \vee b) = h(a) \vee h(b)$ whenever $a \perp b$.

An orthomodular poset P_1 is *representable* in another orthomodular poset P_2 if there exists a mapping $h : P_1 \to P_2$ such that h is an *orthoembedding* for which $h(a \vee b) = h(a) \vee h(b)$ for every $a, b \in P_1$. The set $h(P_1)$ is then an *orthorepresentation* of P_1 in P_2. A Boolean subalgebra of an orthomodular poset is a suborthoposet that is a Boolean algebra.

A *probability measure*, p, over the Hilbert space describing a quantum system is a mapping from the Hilbert-space projection operators onto the interval $[0, 1]$, that satisfies the *Kolmogorov probability axioms*, namely, given events $A, B, C, \ldots$ and the sample space S (the unit event being identified in quantum mechanics with the projector $\mathbb{I}$) defined as their union, the following conditions are satisfied by p.

(i) For any set of events $\{E_i\}$: $0 \leq p(E_i) \leq 1$, $p(E_i) \in \mathbb{R}$ being the *probability* of event E_i.
(ii) $p(S) = 1$.
(iii) For any countable sequence of mutually disjoint events $E_1, E_2, \ldots$, $p(E_1 \cup E_2 \cup \cdots) = \sum_i p(E_i)$ (σ-additivity).

In the quantum context, one requires for any countable set of mutually orthogonal projection operators $P(S_i)$ that $p\left(\sum_i P_i\right) = \sum_i p(P_i)$, $p(\mathbb{O}) = 0$ and $p(\mathbb{I}) = 1$. Writing the *conditional probability* of event B given event A as $p(B|A)$, one takes $p(AB) = p(A)p(B|A)$. Defining $\bar{A}$ as the *complement* of A in S, one has $p(\bar{A}) = 1 - p(A)$.[2]

A *generalized probability measure* on a non-Boolean lattice is taken to satisfy the Kolmogorov axioms on each Boolean sublattice of the lattice.

---

[2] The projection operators P_i are not to be confused with the posets above.

A.8 Projectors, correlations, and the Kochen–Specker theorem

There exists a geometrical representation of the bounds of probabilities in terms of correlation polytopes. Any classical probability distribution can be represented as a convex sum over all binary measures. It can, therefore, be represented by an element of the face of the *correlation polytope*, $C = \text{conv}(K)$, namely, the convex hull

$$\text{conv}(K) = \left\{ \sum_{i=1}^{2^n} \lambda_i \mathbf{x}_i \;\middle|\; \lambda_i \geq 0, \; \sum_{i=1}^{2^n} \lambda_i = 1 \right\} \tag{A.22}$$

of set

$$K = \{\mathbf{x}_1, \mathbf{x}_2, \ldots, \mathbf{x}_{2^n}\} = \big\{(t_1, t_2, \ldots, t_n, t_{i_1} t_{i_2}, \ldots),$$
$$n \geq k \geq 2, i_1 > \ldots > i_k \,\big|\, t_i \in \{0,1\}, \; i = 1, \ldots, n\big\},$$

where the terms $t_{i_1} t_{i_2}, \ldots$ are products associated with joint propositions. Every convex polytope has a equivalent description, as either

   (i) the convex hull of extreme points, or

   (ii) intersection of a finite number of half-spaces, each one given by a linear inequality.

The inequalities of (ii) correspond to Bell-type inequalities for a given physical situation [336, 424]. Consider an arbitrary number n of independent classical events $a_1, a_2, \ldots, a_n$. Probabilities $p_1, p_2, \ldots, p_n$ and joint probabilities $p_{12}, \ldots$ can be taken together to form a vector, $p = (p_1, p_2, \ldots, p_n, p_{12}, \ldots)$ in Euclidean space. Because the probabilities p_i, $i = 1, \ldots, n$ are assumed independent, each of them can in principle reach both 0 and 1. The combined values of $p_1, p_2, \ldots, p_n$ of these extreme cases $p_i = 0, 1$, and the associated joined probabilities $p_{ij} = p_i p_j$, can be interpreted as truth values; they correspond to a *two-valued* (or *dispersionless*) measure.

Consider the task of assigning real numbers to the operators of quantum mechanics applicable to a given system interpretable as the values of the corresponding properties of that system. Problems arise from placing conditions on such an assignment. In particular, the *Kochen–Specker paradox* arises when one attempts to assign values to all quantum properties, described by Hermitian self-adjoint operators, in all quantum states of a quantum system with a Hilbert space of dimension three or greater, under the natural constraint that the algebraic relations of these operators be reflected in the values assigned to them. The *Kochen–Specker theorem* provides a useful result for a finite sublattice of quantum propositions, specifically, that such a valuation cannot be found under this constraint by considering the associated coloring problem [260].[3]

---

[3] Because a detailed discussion of this theorem would be lengthy, requiring an independent chapter of its own, the reader is advised to look elsewhere for a

A.9 Traditional quantum logic

The original manner of associating logical states with quantum systems, introduced by von Neumann and Birkhoff, differs from that now primarily associated with quantum information processing [61]. Rather than being based on the computational basis and its relation to bit strings, it assigns binary values to closed linear subspaces of the Hilbert space of a quantum system. In this sense, it is a logic that is *more quantum* than the (standard) logic of quantum information processing. The field that resulted from this pioneering work is now known as that of *quantum logic* [168, 351, 408, 445].

This traditional quantum logic, $\bar{L}(\mathcal{H})$, sometimes simply called the "*logic of subspaces*," arises from the set of Hilbert subspaces of the complex Hilbert space, $\mathcal{H}$, describing the system. Each subspace, $\bar{h}$, is naturally identified with the projection operator $P_{\bar{h}}$ onto it. The lattice of closed linear subspaces of a Hilbert space $\mathcal{H}$ is equivalent to the lattice of projection operators on $\mathcal{H}$. One can thus define the two operations $\wedge$ and $\vee$ acting pairwise on any two projectors P_1 and P_2 by

$$P_1 \wedge P_2 = P_1 P_2 \;, \tag{A.23}$$

$$P_1 \vee P_2 = P_1 + P_2 - P_1 P_2 \;, \tag{A.24}$$

and identify the zero as the projector $\mathbb{O}$ onto the zero vector $\mathbf{0}$ and the identity as the projector $\mathbb{I}$ onto all of $\mathcal{H}$; $\vee$ corresponds to the linear span, $\wedge$ to set-theoretic intersection. The unit vectors of $\mathcal{H}$, the *rays*, are considered to be the atomic propositions of $\bar{L}(\mathcal{H})$.[4] Compound propositions formed from them correspond to higher-dimensional closed linear subspaces. The conjunction $\wedge$ is essentially the same as the conjunction of classical logic. However, the disjunction $\vee$ and negation $'$ behave much differently, being dissimilar to set-theoretic disjunction and negation (see Section A.7). The propositions of quantum logic refer to the state of the system at a given time; their semantical interpretation involves no reference to the preparation or measurement of the corresponding physical system.

Projection operators have an inherent algebraic structure, that of a partial Boolean algebra. A *partial Boolean algebra* is formed by a family of Boolean algebras, $B^{(i)}$ if the following conditions are satisfied.

(i) The set-theoretical intersection, $B^{(i)} \cap B^{(j)} = B^{(k)}$, of two members $B^{(i)}, B^{(j)}$ is a member of the family.

(ii) If three elements of the partial Boolean algebra are such that two of them belong to a given member of the family, then there exists a Boolean algebra of which all three are members.

---

thorough treatment of this result, such as found in Ch. 3 of [90] and Ch. 5 of [348]. Examples that simplify the Kochen–Specker theorem have been explored by N. David Mermin in [297].

[4] An excellent critical résumé of the quantum-logical approach to quantum theory can be found in [399]. For up-to-date reviews of quantum logic, see [117, 411].

The complement of any element of the partial Boolean algebra is its complement with respect to any of the members of the family to which it belongs. This complement is then unique and belongs to any member to which the element belongs. The complement of the projector P_i is the operator $\tilde{P}_i = \mathbb{I} - P_i$ such that $\tilde{P}_i \wedge P_i = \mathbb{O}$ and $\tilde{P}_i \vee P_i = \mathbb{I}$. The above-described matching of truth values with closed linear subspaces of Hilbert space does not require that the corresponding propositions are necessarily either true or false; this mathematical correspondence is compatible with the indefiniteness of truth values of physical properties inherent in quantum mechanics discussed in Chapter 1. Furthermore, the truth of an elementary quantum proposition is insufficient to determine the value of all other propositions.

B

The quantum postulates

As mentioned in the introduction to the first chapter, quantum information science has primarily been concerned with quantities having discrete eigenvalue spectra, or discretized version of those with continuous quantities. We now consider a set of postulates for standard quantum mechanics, similar to those set out by von Neumann, tailored to this context with occasional comments relating to other situations.[1] In the standard approach to quantum mechanics, the pure states of systems are elements of complex Hilbert spaces. They can also be viewed as statistical operators in these spaces, which are necessary in the case of open quantum systems and which can be described in the related construct of Liouville space that is sketched briefly here after a presentation of the standard quantum postulates.

B.1 The standard postulates

The first postulate is often referred to as the *superposition principle*.

Postulate I:
    Each physical system is represented by a Hilbert space and described by physical quantities and a state represented by linear operators in that space.

The Hilbert space in question is usually taken to be complex Hilbert space, described in Appendix A, which has proven adequate for all quantum mechanical situations encountered so far; states are discussed in Chapter 1.[2] This

---

[1] It is valuable to compare this formulation, for example, with the related formulation of Arno Bohm [64], where great care is taken to incorporate all situations in quantum mechanics through the use of variants of some postulates, each focusing on specific sorts of quantity, or with the classic textbook formulation of Albert Messiah, Ch. VIII [299].

[2] The quantum mechanics of particle motion in physical space requires the use of an infinite-dimensional separable Hilbert space. Note also that *quaternionic* Hilbert

postulate provides both quantum cryptography and quantum computation their unique properties, as discussed in Chapter 1. The second postulate is often referred to as the *Born rule*.

Postulate II:
Each physical quantity of a quantum system is represented by a positive Hermitian operator O, the expectation value of which is given by $\mathrm{tr}(\rho O)$, where ρ is the bounded positive Hermitian trace-class operator representing the state of the system.

This postulate and the following provide quantum mechanics with its essentially statistical nature, described in Appendix A and discussed in Chapter 1. The third postulate is commonly referred to as the *projection postulate*.

Postulate III:
When a physical quantity of a system initially prepared in a state represented by the statistical operator ρ is measured, the state of the system immediately after this measurement is represented by the statistical operator

$$\rho' = \frac{P_k \rho P_k}{\mathrm{tr}(\rho P_k)},\qquad\text{(B.1)}$$

where P_k is the projection operator onto the subspace corresponding to measurement outcome k, with a probability given by the expectation value of P_k for ρ.

This postulate and alternative versions of the projection postulate for different situations are addressed in Chapter 2. This postulate is essential for connecting the behavior of quantum systems with that of the classical systems used to measure them. In the context of quantum cryptography, it describes how random quantum-key material arises; in the context of quantum computation, it helps describe the readout process. The fourth postulate provides a prescription for describing composite systems in relation to descriptions of their parts.

Postulate IV:
Each physical system composed of two or more subsystems is represented by the Hilbert space that is the tensor product of the Hilbert spaces representing its subsystems; the operators representing its physical quantities act in this product space.

This postulate provides the basis for the ability of quantum computation to provide exponential speedup through the parallelism that, together with the first postulate, it enables. The tensor product is described in Appendix A.

---

space has yet to be ruled out as another possible vector space for formulating quantum mechanics [172].

The fifth postulate is commonly referred to as either the *Schrödinger evolution* or the *von-Neumann evolution*.

Postulate V:

The time-evolution of the state of each closed physical system, that is, each physical system not interacting with anything outside of itself, takes place according to

$$\rho(t) = U(t)\rho(0)U^\dagger(t) , \tag{B.2}$$

(in the "Schrödinger picture") where t is the time parameter and $U = e^{-itH/\hbar}$ is a *unitary operator*, H being the generator of time-translations.

This postulate provides the natural time-evolution of closed quantum systems, which corresponds to a linear transformation of associated state-vectors. The description of open quantum systems is treated in Section B.3, below.

B.2 The Heisenberg–Robertson uncertainty relation

The *dispersion* of an Hermitian operator A in a state ρ is given by

$$\mathrm{Disp}_\rho A = \langle (A - \langle A \rangle \mathbb{I})^2 \rangle \tag{B.3}$$

$$= \langle A^2 \rangle - \langle A \rangle^2 , \tag{B.4}$$

where the expectation values on the right-hand side (and below) are those for the statistical operator ρ.

The square root of the dispersion is the *uncertainty* of A in state ρ:

$$\Delta A \equiv \sqrt{\mathrm{Disp}_\rho A} . \tag{B.5}$$

An important implication of the postulates of quantum mechanics is the *Heisenberg–Robertson uncertainty relation* between quantities, which can be stated for two Hermitian operators A and B as

$$\langle (\Delta A)^2 \rangle \langle (\Delta B)^2 \rangle \geq \frac{1}{4} |\langle [A, B] \rangle|^2 . \tag{B.6}$$

Werner Heisenberg introduced a thought experiment involving the measurement of the position of a particle with a gamma-ray microscope as an illustration of the unusual nature of position and momentum and their interrelation in quantum mechanics in light of this relation [210, 354].

The uncertainty relation has also been viewed as expressing how the use of nonorthogonal qubit state-vectors in different encoding bases, such as those of conjugate bases, in quantum key distribution protocols provides security against potential eavesdroppers.[3]

---

[3] In this regard, it is worthwhile to consider the treatment of this issue by Chris Fuchs [175].

B.3 Liouville space and open quantum systems

The statistics of *open quantum systems*, that is, quantum systems that are in contact with an external environment the details of which are not well known, are often studied through the behavior of states in Liouville space [171]. *Liouville space* is the vector space formed by the set of all *linear operators* acting on the Hilbert space $\mathcal{H}$, with an inner product given by $(A|B) = tr(A^\dagger B)$. Its vectors, $|A)$, are in correspondence with the Hilbert-space operators A on $\mathcal{H}$. Consider a set of basis vectors $\{|u_1\rangle, |u_2\rangle, \ldots\}$ spanning $\mathcal{H}$. A basis in Liouville space is represented by the set of all operators $\{|uu')\}$ obtained from the elements of $\{|u_i\rangle\}$, via

$$|u'u) \doteq |u'\rangle\langle u| , \tag{B.7}$$

where $u, u' \in \{|u_i\rangle\}$ with the *orthogonality relation* expressed as

$$(u'u|v'v) = tr\big[|u\rangle\langle u'|v'\rangle\langle v|\big] = \delta_{u'v'}\delta_{uv} \tag{B.8}$$

and the *completeness relation* expressed as

$$\sum_{u'u} |u'u)(u'u| = 1 . \tag{B.9}$$

One sees, then, that the inner product of any Liouville vector $|A)$ with the basis vectors $|u'u)$ is the usual Hilbert-space matrix-element:

$$(u'u|A) = tr\big[|u'u\rangle\langle u'u|A\big] = \langle u'|A|u\rangle. \tag{B.10}$$

Superoperators, $\mathcal{O}$, in Liouville space are linear operators taking Liouville-space vectors $|A)$ to Liouville-space vectors in accordance with $\mathcal{O}|A) = |\mathcal{O}A)$. These operators represent the similarity transformations, $A \longrightarrow OAO^{-1}$, generated by operators O in Hilbert space. One has, in any basis $\{|uu'\rangle\}$,

$$(u'u|\mathcal{O}|A) = \sum_{v'v}(u'u|\mathcal{O}|v'v)(v'v|A) = \sum_{v'v}\mathcal{O}_{u'uv'v}A_{v'v} . \tag{B.11}$$

The *Liouville operator* $\mathcal{L}$, for a given Hamiltonian H on $\mathcal{H}$ is

$$\mathcal{L}|A) = \frac{1}{\hbar}|[H, A]) , \tag{B.12}$$

for the Hilbert-space operators A on $\mathcal{H}$. The von Neumann equation describing time evolution is

$$|\dot{\rho}) = -\frac{i}{\hbar}|[H, \rho]) = -i\mathcal{L}|\rho) , \tag{B.13}$$

having the solution

$$|\rho(t)) = \exp(-i\mathcal{L}t)|\rho(t_0)) = \mathcal{U}(t, t_0)|\rho(t_0)) , \tag{B.14}$$

where $\mathcal{U}(t, t_0) = e^{-i\mathcal{L}t}$ is the *time-evolution superoperator*.

References

1. Web-based versions of this bibliography with direct weblinks to most of its entries are available at the Springer website http://www.springeronline.com and the author's homepage http://math.bu.edu/people/jaeger/books.html
2. Abouraddy, A., M. B. Nasr, B. E. A. Saleh, A. V. Sergienko, and M. C. Teich, "Demonstration of the complementarity of one- and two-photon interference," Phys. Rev. A **63**, 063803 (2001).
3. Abouraddy, A., A. V. Sergienko, B. E. A. Saleh, and M. C. Teich, "Quantum entanglement and the two-photon Stokes parameters," Opt. Comm. **201**, 93 (2001).
4. Acín, A., D. Bruß, M. Lewenstein, and A. Sanpera, "Classification of mixed three-qubit states," Phys. Rev. Lett. **87**, 040401 (2001).
5. Aerts, D., M. Czachor, and M. Pawłowski, "Entangled-state cryptographic protocol that remains secure even if nonlocal hidden variables exist and can be measured with arbitrary precision," Phys. Rev. A **73** 034303 (2006); *ibid.* **73**, 059901(E) (2006).
6. Aharonov, Y., and T. Kaufherr, "Quantum frames of reference," Phys. Rev. D **30**, 368 (1984).
7. Alber, G., T. Beth, M. Horodecki, P. Horodecki, R. Horodecki, M. Röteller, H. Weinfurter, R. Werner, and A. Zeilinger (eds.) *Quantum information*, STMP **173** (Springer-Verlag; Berlin, 2001).
8. Alber, G., A. Delgado, and M. Mussinger, "Quantum error correction and quantum computation with detected-jump correcting quantum codes," Fortschr. Phys. **49**, 901 (2001).
9. Altafini, C., "Tensor of coherences parametrization of multiqubit density operators for entanglement characterization," Phys. Rev. A **69**, 012311 (2004).
10. Altafini, C., and T. F. Havel, "Reflection symmetries for multiqubit density operators," J. Math. Phys. **47**, 032104 (2006).
11. Altepeter, J. B., D. F. V. James, and P. G. Kwiat, "Qubit quantum state tomography," in M. Paris, J. Rehacek (eds.) *Quantum state estimation*, (Springer-Verlag; Berlin, 2004), Chapter 4.
12. Anscombe, F. R., "Quiet contributor: The civic career and times of John W. Tukey," (special Tukey issue) Stat. Sci. **18** (3), 287 (2003).

250   References

13. Armstrong, J. A., N. Bloembergen, J. Ducuing, and P. S. Pershan, "Interactions between light waves in nonlinear dielectric," Phys. Rev. **127**, 1918 (1962).

14. Arrighi, P., and C. Patricot, "A note on the correspondence between qubit quantum operations and special relativity," J. Phys. A **36**, L287 (2003).

15. Arrighi, P., and C. Patricot, "Conal representation of quantum states and non-trace-preserving quantum operations," Phys. Rev. A **68**, 042310 (2003).

16. Aspect, A., P. Grangier, and G. Roger, "Experimental test of realistic theories via Bell's inequality," Phys. Rev. Lett. **47**, 460 (1981).

17. Audenaert, K., M. B. Plenio, and J. Eisert, "Entanglement cost under positive-partial-transpose-preserving operations," Phys. Rev. Lett. **90**, 027901 (2003).

18. Audenaert, K., F. Verstraete, and B. De Moor, "Variational characteristics of separability and entanglement of formation," Phys. Rev. A **64**, 052304 (2000).

19. Audretsch, J. (ed.), *Entangled world: The fascination of quantum information and computation* (John Wiley and Sons; Hoboken NJ, 2006).

20. Averin, D. V., "Quantum nondemolition measurements of a qubit," Phys. Rev. Lett. **88**, 207901 (2002).

21. Ban, M., "Error-free quantum receiver for a binary pure quantum state signal," Phys. Lett. A **213**, 235 (1996).

22. Ban, M., K. Kurokawa, R. Momose, and O. Hirota, "Optimal measurement for discrimination among symmetric quantum states and parameter estimation," Int. J. Theor. Phys. **36**, 1269 (1997).

23. Barenco, A., "A universal two-bit gate for quantum computation," Proc. Roy. Soc. London **449**, 679 (1995).

24. Barenco, A., C. H. Bennett, R. Cleve, D. P. DiVincenzo, N. Margolus, P. Shor, T. Sleator, J. A. Smolin, and H. Weinfurter, "Elementary gates for quantum computation," Phys. Rev. A **52**, 3457 (1995).

25. Barendregt, H. P., *The Lambda Calculus: Its syntax and semantics* (North-Holland; Amsterdam, 1984).

26. Barnum, H., C. M. Caves, C. A. Fuchs, R. Jozsa and B. Schumacher, "Non-commuting mixed states cannot be broadcast," Phys. Rev. Lett. **76**, 2818 (1996).

27. Barnum, H., E. Knill, and N. Linden, "On quantum fidelities and channel capacities," IEEE Trans. Inform. Theory **46**, 1317 (2000).

28. Barnum, H., and N. Linden, "Monotones and invariants for multi-particle quantum states," J. Phys. A **34**, 6787 (2001).

29. Bartlett, S. D., D. A. Rice, B. C. Sanders, J. Daboul, and H. de Guise, "Unitary transformations for testing Bell inequalities," Phys. Rev. A **63**, 042310 (2001).

30. Bell, J. S., "On the Einstein–Podolsky–Rosen paradox," Physics **1**, 195 (1964).

31. Bell, J. S., "On the problem of hidden variables in quantum mechanics," Rev. Mod. Phys. **38**, 447 (1966).

32. Bell, J. S., *Speakable and unspeakable in quantum mechanics* (Cambridge University Press; Cambridge, 1987).

33. Benenti, G., G. Casati, and G. Strini, *Principles of quantum computation and information*, Volume 1 (World Scientific; Singapore, 2005).

34. Benioff, P., "The computer as a physical system: A microscopic quantum mechanical Hamiltonian model of computers as represented by Turing machines," J. Stat. Phys. **22**, 563 (1980).

35. Benioff, P., "Models of quantum Turing machines," Fortschr. Phys. **46**, 423 (1998).

36. Bennett, C. H., "Time/space trade-offs for reversible computation," SIAM J. Comput. **18**, 766 (1989).

37. Bennett, C. H., "Quantum cryptography using any two nonorthogonal states," Phys. Rev. Lett. **68**, 3121 (1992).

38. Bennett, C. H., E. Bernstein, G. Brassard, and U. Vazirani, "Strengths and weaknesses of quantum computing," SIAM J. Comp. **26**, 1510 (1997).

39. Bennett, C. H., H. Bernstein, S. Popescu, and B. Schumacher, "Concentrating partial entanglement by local operations," Phys. Rev. A **53**, 2046 (1996).

40. Bennett, C. H., F. Bessette, G. Brassard, L. Salvail, and J. Smolin, "Experimental quantum cryptography," J. Cryptology **5**, 3 (1992).

41. Bennett, C. H., and G. Brassard, "Quantum cryptography: Public key-distribution and cointossing," in *Proceedings of IEEE conference on computers, systems and signal processing*, Bangalore, India (IEEE; New York, 1984), p. 175.

42. Bennett, C. H., G. Brassard, C. Crépeau, R. Jozsa, A. Peres, and W. K. Wootters, "Teleporting an unknown quantum state via dual classical and Einstein–Podolsky–Rosen channels," Phys. Rev. Lett. **70**, 1895 (1993).

43. Bennett, C. H., G. Brassard, S. Popescu, B. Schumacher, J. Smolin, and W. Wootters, "Purification of noisy entanglement and faithful teleportation via noisy channels," Phys. Rev. Lett. **76**, 722 (1996).

44. Bennett, C. H., G. Brassard, and J.-M. Robert, "Privacy amplification by public discussion," SIAM J. Comput. **17**, 210 (1988).

45. Bennett, C. H., D. P. DiVincenzo, T. Mor, P. W. Shor, J. A. Smolin, and B. M. Terhal, "Unextendible product bases and entanglement," Phys. Rev. Lett. **82**, 5385 (1999).

46. Bennett, C. H., D. P. DiVincenzo, and J. A. Smolin, "Capacities of quantum erasure channels," Phys. Rev. Lett. **78**, 3217 (1997).

47. Bennett, C. H., D. P. DiVincenzo, J. A. Smolin, and W. K. Wootters, "Mixed-state entanglement and quantum error correction," Phys. Rev. A **54**, 3824 (1996).

48. Bennett, C. H., S. Popescu, D. Rohrlich, J. A. Smolin, and A. V. Thapliyal, "Exact and asymptotic measures of multipartite pure state entanglement," Phys. Rev. A **63**, 012307 (2001).

49. Bennett, C. H., and P. W. Shor, "Quantum information theory," IEEE Trans. Inform. Theory **44**, 2724 (1998).

50. Bennett, C. H., P. W. Shor, J. A. Smolin and A. V. Thapliyal, "Entanglement-assisted classical capacity of noisy quantum channels," Phys. Rev. Lett. **83**, 3081 (1999).

51. Bennett, C. H., P. W. Shor, J. A. Smolin, and A. V. Thapliyal, "Entanglement-assisted capacity of a quantum channel and reverse Shannon theorem," IEEE Trans. Inform. Theory **48**, 2637 (2002).

52. Bennett, C. H., and S. J. Wiesner, "Communication via one- and two-particle operations on Einstein–Podolsky–Rosen States," Phys. Rev. Lett. **69**, 2881 (1992).

53. Bergou, J., M. Hillery, and Y. Sun, "From unambiguous quantum state discrimination to quantum state filtering," Fortschr. Phys. **51**, 74 (2003).

54. Bernstein, E., and U. Vazirani, "Quantum complexity theory," in *Proceedings of 25th annual ACM symposium on theory of computing*, San Diego, CA (1993), p. 11; SIAM J. Comput. **26**, 1411 (1997).

55. Bernstein, H. J., D. M. Greenberger, M. A. Horne, and A. Zeilinger, "Bell's theorem without inequalities for two spinless particles," Phys. Rev. A **47**, 78 (1993).

56. Berthiaume, A., and G. Brassard, "The quantum challenge to structural complexity theory," in *Proceedings of 7th IEEE conference on structure in complexity theory*, Boston, MA (1992), p. 132.

57. Berthiaume, A., and G. Brassard, "Oracle quantum computing," J. Mod. Opt. **41**, 2521 (1994).

58. Beth, T., and M. Grassl, "Improved decoding of quantum error correcting codes from classical codes," in *Proceedings of PhysComp96* (Boston University; Boston, 1996), pp. 28.

59. Beth, T., and G. Leuchs (eds.), *Quantum information processing, Second Edition* (John Wiley and Sons; Hoboken NJ, 2005).

60. Beth, T., and M. Rötteler, in G. Alber, T. Beth, M. Horodečki, P. Horodečki, R. Horodečki, M. Rötteler, H. Weinfurter, R. Werner, and A. Zeilinger (eds.), *Quantum information*, STMP **173**, VII–X (Springer-Verlag; Berlin, 2001), p. 96.

61. Birkhoff, G., and J. von Neumann, "The logic of quantum mechanics," Ann. Math. **37**, 823 (1936).

62. Blum, K., *Density matrix theory and applications, 2nd Edition* (Kluwer Academic; Dordrecht, 1996).

63. Boeuf, N., D. Branning, I. Chaperot, E. Dauler, S. Guérin, G. Jaeger, A. Muller, and A. Migdall, "Calculating characteristics of noncollinear phase matching in uniaxial and biaxial crystals," Opt. Eng. **39**, 1016 (2000).

64. Bohm, A., *Quantum mechanics: Foundations and applications* (Springer-Verlag; Berlin, 1979).

65. Bohm, D., "A suggested interpretation of quantum theory in terms of 'hidden' variables," I. Phys. Rev. **85**, 166 (1952); II. Phys. Rev. **85**, 180 (1952).

66. Bohm, D., and Y. Aharonov, "Discussion of experimental proof for the paradox of Einstein, Rosen and Podolsky," Phys. Rev. **108**, 1070 (1957).

67. Bohr, N., in P. A. Schilpp (ed.) *Albert Einstein: Philosopher-scientist. The Library of Living Philosophers, Volume 7* (Open Court; Evanston, IL, 1949), p. 216.

68. Born, M., "Zur Quantenmechanik der Stoßvorgägne," Z. Phys. **37**, 863 (1926).

69. Boschi, D., S. Branca, F. De Martini, L. Hardy, and S. Popescu, "Experimental realization of teleporting an unknown pure quantum state via dual classical and Einstein–Podolsky–Rosen channels," Phys. Rev. Lett. **80**, 1121 (1998).

70. Bose, S., V. Vedral, and P. L. Knight, "Multiparticle generalization of entanglement swapping," Phys. Rev. A **57**, 822 (1998).

71. Bouwmeester, D., J. W. Pan, K. Mattle, M. Eibl, H. Weinfurter, and A. Zeilinger, "Experimental quantum teleportation," Nature **390**, 575 (1997).

72. Bouwmeester, D., A. Ekert, and A. Zeilinger (eds.) *The physics of quantum information* (Springer-Verlag; Berlin, 2000).

73. Bovino, F. A., G. Castagnoli, A. Ekert, P. Horodečki, C. Moura Alves, and A. V. Sergienko, "Direct measurement of nonlinear properties of bipartite quantum states," Phys. Rev. Lett. **95**, 240407 (2005).

74. Brandt, H. E., "Qubit devices and the issue of decoherence," Prog. Quant. Elect. **22**, 257 (1998).

75. Brandt, H. E., "Secrecy capacity in the four-state protocol of quantum key distribution," J. Math. Phys. **43**, 4526 (2002).

76. Brandt, H. E., J. M. Myers, and S. L. Lomonaco, "Aspects of entangled translucent eavesdropping in quantum cryptography," Phys. Rev. A **56**, 4456 (1997); *ibid.* **58**, 2617 (1998).

77. Brassard, G., and L. Salvail, "Secret key reconciliation by public discussion," *Advances in cryptology-EUROCRYPT'93* LNCS **765**, 410 (1994).

78. Braunstein, S. L., "Entanglement in quantum information processing," in Akulin, V. M., A. Sarfati, G. Kurizki, and S. Pellegrin (eds.), *Decoherence, entanglement and information protection in complex quantum systems,* (Springer; Dordrecht, 2005).

79. Braunstein, S. L., and Pati, A. K. (eds.) *Quantum information with continuous variables* (Springer-Verlag; Berlin, 2003).

80. Braunstein, S. L., C. A. Fuchs, D. Gottesman, and H.-K. Lo, "A quantum analogue of Huffman coding," IEEE Trans. Inform. Theory **46**, 1644 (2000).

81. Braunstein, S. L., and P. van Loock, "Quantum information with continuous variables," Rev. Mod. Phys. **77**, 513 (2005).

82. Raussendorf, R., and H. J. Briegel, "A one-way quantum computer," Phys. Rev. Lett. **86**, 5188 (2001).

83. Brown, J., *Quest for the quantum computer* (Simon & Schuster; New York, 2000).

84. Bruckner, Č., M. Żukowski, and A. Zeilinger, "Quantum communication complexity protocol with two entangled qubits," Phys. Rev. Lett. **89**, 197901 (2002).

85. Buhrman, H., R. Cleve, and W. van Dam, "Quantum entanglement and communication complexity," SIAM J. Comput. **30**, 1829 (2001).

86. Buhrman, H., W. van Dam, P. Høyer, and A. Tapp, "Multiparty quantum communication complexity," Phys. Rev. A **60**, 2737 (1999).

87. Bruß, D., "Optimal eavesdropping in quantum cryptography with six states," Phys. Rev. Lett. **81**, 3018 (1998).

88. Bruß, D., G. M. D'Ariano, M. Lewenstein, C. Macchiavello, A. Sen(De), and U. Sen, "Dense coding with multipartite quantum states," Int. J. Quant. Inf. (to appear).

89. Brylinski, R. K., and G. Chen (eds.), *Mathematics of quantum computation* (Chapman and Hall/CRC; Boca Raton, LA, 2000).

90. Bub, J., *Interpreting the quantum world* (Cambridge University Press; Cambridge, 1997).

91. Busch, P., "Quantum states and generalized observables: A simple proof of Gleason's theorem," Phys. Rev. Lett **91**, 120403 (2003).

92. Busch, P., M. Grabowski, and P. J. Lahti, *Operational quantum physics,* (Springer-Verlag; Berlin, 1995).

93. Busch, P., and P. J. Lahti, "The standard model of quantum measurement theory: History and applications," Found. Phys. **26**, 875 (1996).

94. Busch, P., P. J. Lahti, and P. Mittelstaedt, *The quantum theory of measurement, Second, Revised edition* (Springer-Verlag; Berlin, 1996).

95. Bužek, V., S. L. Braunstein, M. Hillery, and D. Bruß, "Quantum copying: A network," Phys. Rev. A **56**, 3446 (1997).

96. Bužek, V., and M. Hillery, "Universal optimal cloning of arbitrary quantum states: From qubits to quantum registers," Phys. Rev. Lett. **81**, 5003 (1998).

97. Cabello, A., "N-particle N-level singlets: Some properties and applications," Phys. Rev. Lett. **89**, 100402 (2002).

98. Calderbank, A. R., E. M. Rains, P. W. Shor, and N. J. A. Sloane, "Quantum error correction and orthogonal geometry," Phys. Rev. Lett. **78**, 405 (1997).

99. G. Cassinelli, E. De Vito, and A. Levrero, On the Decompositions of a Quantum State, J. Math. Anal. App. **210**, 472, (1997).

100. Caves, C., and C. A. Fuchs, "Quantum information: How much information is in a state vector?" in A. Mann and M. Revzen, *Sixty years of EPR*, Ann. Phys. Soc., Israel, 1996; also http://xxx.lanl.gov quant-ph/9601025 (1996).

101. Cerf, N. J., C. Adami, and P. G. Kwiat, "Optical simulation of quantum logic," Phys. Rev. A **57**, R1477 (1998).

102. Cerf, N. J., N. Gisin, and S. Popescu, "Simulating maximal quantum entanglement without communication," Phys. Rev. Lett. **94**, 220403 (2005).

103. Chefles, A., "Distinguishability measures and ensemble orderings," Phys. Rev. A **66**, 042325 (2002).

104. Chefles, A., "Unambiguous discrimination between linearly independent quantum states," Phys. Lett. A **239**, 339 (1998).

105. Chefles, A., C. Gilson, and S. M. Barnett, "Entanglement, information and multiparticle quantum operations," Phys. Rev. A **63**, 032314 (2001).

106. Choi, M.-D., "Positive semidefinite biquadratic forms," Lin. Alg. Appl. **12**, 95 (1975).

107. Church, A., "An unsolvable problem of elementary number theory," Amer. J. Math. **58**, 345 (1936).

108. Clauser, J. F., M. Horne, A. Shimony, and R. A. Holt, "Proposed experiments to test local hidden-variable theories," Phys. Rev. Lett. **23**, 880 (1973).

109. Cleve, R., and H. Buhrman, "Substituting quantum entanglement for communication," Phys. Rev. A **56**, 1201 (1997).

110. Cleve, R., A. Ekert, C. Macchiavello, and M. Mosca, "Quantum algorithms revisited," Proc. Roy. Soc. Lond. A **454**, 339 (1998).

111. Clifton, R., "The subtleties of entanglement and its role in quantum information theory," Phil. Sci. **69**, S150 (2002).

112. Collins, D., N. Gisin, N. Linden, S. Massar, and S. Popescu, "Bell inequalities for arbitrarily high-dimensional systems," Phys. Rev. Lett. **88**, 040404 (2002).

113. Cooley, J. W., "The re-discovery of the fast Fourier transform," Microchimica Acta (Historical Archive) **93**, 33 (1987).

114. Cooley, J. W., and J. W. Tukey, "An algorithm for the machine calculation of complex Fourier series," Math. Comput. **19**, 297 (1965).

115. Cubitt, T. S., F. Verstraete, W. Dür, and J. I. Cirac, "Separable states can be used to distribute entanglement," Phys. Rev. Lett. **91**, 037902 (2003).

116. Czachor, M., "Einstein–Podolsky–Rosen–Bohm experiment with relativistic massive particles," Phys. Rev. A **55**, 72 (1997).

117. Dalla Chiara, M., R. Giuntini, and R. Greechie, *Reasoning in quantum theory: Sharp and unsharp quantum logics*, (Springer-Verlag; Berlin, 2004).

118. D'Ariano, G. M., "Quantum tomography," Fortschr. Phys. **48**, 579 (2000).

119. D'Ariano, G. M., C. Macchiavello, and P. Perinotti, "Superbroadcasting of mixed states," Phys. Rev. Lett. **95**, 060503 (2005).

120. Datta, A., S. T. Flammia, and C. M. Caves, "Entanglement and the power of one qubit," Phys. Rev. A **72**, 042316 (2005).

121. Davies, P., and G. J. Milburn, *Schrödinger's machines* (W.H. Freeman; New York, 1997).

122. Davies, E. B. *Quantum theory of open systems* (Academic Press; London, 1976).

123. Davis, M., *The universal computer* (W. W. Norton and Co.; New York, 2000).
124. de Broglie, L., "La mécanique ondulatoire et la structure atomique de la matière et du rayonnement," J. Physique et du Radium **8**, 225 (1927).
125. de Broglie, L., "La nouvelle méchanique des quanta," in H. Lorentz (Ed.), *Rapports et discussions du cinquième conseil de physique Solvay* (Gauthier-Villars; Paris, 1928), p. 105.
126. De Muynck, W. M., "An alternative to the Lüders generalization of the von Neumann projection, and its interpretation," J. Phys. A **31**, 431 (1998).
127. De Muynck, W. M., *Foundations of quantum mechanics, an empiricist approach* (Kluwer Academic; Dordrecht, 2002).
128. D'Espagnat, B. *Conceptual foundations of quantum mechanics* (Benjamin; Menlo Park, CA, 1971).
129. Deutsch, D., "Quantum computational networks," Proc. Roy. Soc. London A **425**, 73 (1989).
130. Deutsch, D., A. Barenco, and A. Ekert, "Universality in quantum computation," Proc. Roy. Soc. London A **449**, 669 (1995).
131. Deutsch, D., A. Ekert, R. Jozsa, C. Macchiavello, S. Popescu, and A. Sanpera, "Quantum privacy amplification and the security of quantum cryptography over noisy channels," Phys. Rev. Lett. **77**, 2818 (1996).
132. Deutsch, D., and P. Hayden, "Information flow in entangled quantum systems," http://xxx.lanl.gov quant-ph/9906007 (1999).
133. Deutsch, D., and R. Jozsa, "Rapid solution of problems by quantum computation," Proc. Roy. Soc. London A **439**, 553 (1992).
134. DeWitt, B. S., and N. Graham, *The many-worlds interpretation of quantum mechanics* (Princeton University Press; Princeton, 1973).
135. Dieks, D., "Overlap and distinguishability of quantum states," Phys. Lett. A **126**, 303 (1988).
136. Dirac, P. A. M., *The Principles of Quantum mechanics* (Oxford University Press; Oxford, 1930).
137. DiVincenzo, D., "Two-bit gates are sufficient for quantum computation," Phys. Rev. A **51**, 1015 (1995).
138. DiVincenzo, D., "Quantum computation," Science **270**, 255 (1995).
139. DiVincenzo, D., and A. Peres, "Quantum code words contradict local realism," Phys. Rev. A **55**, 4089 (1997).
140. DiVincenzo, D., and P. W. Shor, "Fault-tolerant error correction with efficient quantum codes," Phys. Rev. Lett. **77**, 3260 (1996).
141. Duan, L.-M., G. Giedke, J. I. Cirac, and P. Zoller, "Inseparability criterion for continuous variable systems," Phys. Rev. Lett. **84**, 2722 (2000).
142. Duan, L.-M., and G.-C. Guo, "A probabilistic cloning machine for replicating two non-orthogonal quantum states," Phys. Lett. A **243**, 261 (1998).
143. Dür, W., and J. I. Cirac, "Classification of multiqubit states: Separability and distillability properties," Phys. Rev. A. **61**, 042314 (2000).
144. Dür, W., G. Vidal, and J. I. Cirac, "Three qubits can be entangled in two inequivalent ways," Phys. Rev. A. **62**, 062314 (2000).
145. Duwell, A., "How to teach an old dog new tricks: Quantum information, quantum computation, and the philosophy of physics," Ph.D. Thesis, University of Pittsburgh, 2004.
146. Dužek, M., M. Jahma, and N. Lütkenhaus, "Unambiguous state discrimination in quantum cryptography with weak coherent states," Phys. Rev. A **62**, 022306 (2000).

256   References

147. Einstein, A., B. Podolsky, and N. Rosen, "Can quantum-mechanical description of physical reality be considered complete?", Phys. Rev. **47**, 777 (1935).
148. Eisert, J. "Entanglement in quantum theory," Ph.D. Thesis, University of Potsdam, Postdam, Germany, 2001.
149. Eisert, J., and H. J. Briegel, "Schmidt measure as a tool for quantifying entanglement," Phys. Rev. A **64**, 022306 (2001).
150. Ekert, A. K., "Quantum cryptography based on Bell's theorem," Phys. Rev. Lett. **67**, 661 (1992).
151. Ekert, A. K., "Quantum interferometers as quantum computers," Physica Scripta **T76**, 218 (1998).
152. Ekert, A. K., B. Huttner, G. M. Palma, and A. Peres, "Eavesdropping on quantum-cryptographical systems," Phys. Rev. A **50**, 1047 (1997).
153. Ekert, A. K., and Jozsa, R., "Quantum computation and Shor's algorithm," Rev. Mod. Phys. **68**, 733 (1996).
154. Ekert, A. K., and P. L. Knight, "Entangled quantum systems and the Schmidt decomposition," Am. J. Phys. **63**, 415 (1995).
155. Ekert, A., and C. Macchiavello, "Error correction in quantum communication," Phys. Rev. Lett. **77**, 2585 (1996).
156. Ekert, A. K., C. Moura Alves, D. K. L. Oi, M. Horodecki, P. Horodecki, and L. C. Kwek, "Direct estimation of linear and non-linear fuctionals of a quantum state," Phys. Rev. Lett. **88**, 217901 (2002).
157. Eldar, Y. C., "A semidefinite programming approach to optimal unambiguous discrimination of quantum states," IEEE Trans. Inform. Theory **49**, 446 (2003).
158. Eldar, Y. C., A. Megretski, and G. C. Verghese, "Designing optimal quantum detectors via semidefinite programming," IEEE Trans. Inform. Theory **49**, 1007 (2003).
159. Elliott, C., A. Colvin, D. Pearson, O. Pikalo, J. Schlafer, and H. Yeh, "Current status of the DARPA quantum network," in Donkor, E., A. R. Pirich, and H. Brandt (eds.), *Quantum information and computation III*, Proc. SPIE **5815**, 138 (2005).
160. Elliott, C. *et al.* in [375].
161. Englert, B.-G., C. Kurtsiefer, and H. Weinfurter, "Universal unitary gate for single-photon two-qubit states," Phys. Rev. A **63**, 032303 (2001).
162. Englert, B.-G., J. Schwinger, and M. Scully, "Is spin coherence like Humpty-Dumpty? I. Simplified treatment," Found. Phys. **18**, 1045 (1988).
163. Enzer, D. G., P. G. Hadley, R. J. Hughes, C. G. Peterson, and P. G. Kwiat, "Entangled-photon six-state quantum cryptography," New J. Physics **4**, 45.1 (2002).
164. Ettinger, M., and P. Høyer, "A quantum observable for the graph isomorphism problem," http://xxx.lanl.gov quant-ph/9901029 (1999).
165. Everitt, H. O., *Experimental aspects of quantum computing* (Springer; Berlin, 2005).
166. Fano, U., "Remarks on the classical and quantum-mechanical treatment of partial polarization," J. Opt. Soc. Am. **39**, 859 (1949).
167. Fejer, M. M., G. A. Megel, D. H. Jundt, and R. L. Byer, "Quasi-phase-matched second harmonic generation: Tuning and tolerances," IEEE J. Quant. Electron. **28**, 2631 (1992).
168. Février, P., "Les relations d'incertitude de Heisenberg et la logique," Académie des Sciences (Paris) Comptes Rendus **204**, 481 (1937).

169. Feynman, R. P., *The Feynman lectures on physics* [A. J. G. Hey and R. W. Allen (eds.)] (Addison-Wesley; Reading, MA, 1965).

170. Feynman, R. P., *Feynman lectures on computation* (Addison-Wesley; Reading, MA, 1996).

171. Fick, E., G. Sauermann, and W. D. Brewer, *The quantum statistics of dynamic processes* (Springer-Verlag: Berlin, 1990).

172. Finkelstein, D., J. M. Jauch, and D. Speiser, "Notes on quaternionic quantum mechanics," CERN Report 59-7 (1959) [Reprinted in Hooker, C. A. (ed.), *Logico-algebraic approach to quantum mechanics. II* (D. Reidel; Dordrecht, 1979), p. 367].

173. Franken, P. A., and J. F. Ward, "Optical harmonics and nonlinear phenomena," Rev. Mod. Phys. **35**, 23 (1963).

174. Franson, J. D., "Two-photon interferometry over large distances," Phys. Rev. A **44**, 4552 (1991).

175. Fuchs, C., "Information gain vs. state disturbance in quantum theory," Fortschr. Phys. **46**, 535 (1998).

176. Fuchs, C., "Quantum foundations in the light of quantum information," http://xxx.lanl.gov quant-ph/0106166 (2001).

177. Furusawa, F., J. L. Sorensen, S. L. Braunstein, C. A. Fuchs, H. J. Kimble, and E. S. Polzik, "Unconditional quantum teleportation," Science **282**, 706 (1998).

178. Gauss, C. F., "Nachlass: Theoria interpolationis methodo nova tractata," in *Carl Friedrich Gauss, Werke*, Band 3 (Königlichen Gesellschaft der Wissenschaften; Göttingen, 1866), p. 265.

179. Gershenfeld, N., *The physics of information technology* (Cambridge Univ. Press; Cambridge, 2000).

180. Gibbons, K. S., W. J. Hoffman, and W. K. Wootters, "Discrete phase space based on finite fields," Phys. Rev. A **70**, 062101 (2004).

181. Giedke, G., B. Kraus, M. Lewenstein, and J. I. Cirac, "Entanglement criterion for all bipartite quantum states," Phys. Rev. Lett. **87**, 167904 (2001).

182. Gibbons, K. S., M. J. Hoffman, and W. K. Wootters, "Discrete phase space based on finite fields," Phys. Rev. A **70**, 062101 (2004).

183. Giles, R., *Mathematical foundations of thermodynamics. International series of monographs on pure and applied mathematics, Volume 53* (Pergammon; Oxford, 1964).

184. Gingrich, R. M., A. J. Bergou, and C. Adami, "Entangled light in moving frames," Phys. Rev. A **68**, 042102 (2003).

185. Gisin, N., "Quantum measurements and stochastic processes," Phys. Rev. Lett. **52**, 1657 (1984).

186. Gisin, N., "Bell's inequality holds for all non-product states," Phys. Lett. A **154**, 201 (1991).

187. Gisin, N., "Stochastic quantum dynamics and relativity," Helvetica Physica Acta **62**, 363 (1989).

188. Gisin, N., G. Ribordy, W. Tittel, and H. Zbinden, "Quantum cryptography," Rev. Mod. Phys. **74**, 145 (2002).

189. Gleason, A. M., "Measures on the closed subspaces of a Hilbert space," J. Math. Mech. **6**, 885 (1957).

190. Good, I. J., "Studies in the history of probability and statistics XXXVII. A. M. Turing's statistical work in World War II," Biometrika **66**, 393 (1979).

191. Gordon, J. P., "Noise at optical frequencies; information theory," in P. A. Miles (ed.), *Quantum electronics and coherent light. Proceedings of international school of physics 'Enrico Fermi' XXXI* (Academic Press; New York, 1964).

192. Gottesman, D., "Stabilizer codes and quantum error correction," Ph.D. thesis, California Institute of Technology, Pasadena, CA, 1997; also, http://xxx.lanl.gov quant-ph/9705052.

193. Gottesman, D., "Theory of fault-tolerant quantum computation," Phys. Rev. A **57**, 127 (1998).

194. Grassl, M., M. Rötteler, and T. Beth, "Computing local invariants of quantum-bit systems," Phys. Rev. A **58**, 1833 (1998).

195. Greenberger, D. M., M. A. Horne, A. Shimony, and A. Zeilinger, "Bell's theorem without inequalities," Am. J. Phys. **58**, 1131 (1990).

196. Greenberger, D. M., M. A. Horne, and A. Zeilinger, "Going beyond Bell's theorem," in M. Kafatos (ed.), *Bell's theorem, quantum theory and conceptions of the universe* (Kluwer Academic; Dordrecht, 1989), p. 79.

197. Greenberger, D., M. A. Horne, and A. Zeilinger, "Multiparticle interferometry and the superposition principle," Physics Today, August 1993, p. 22.

198. Groisman, B., S. Popescu, and A. Winter, "On the quantum, classical and total amount of correlations in a quantum state," Phys Rev A **72**, 032317 (2005).

199. Grover, L. K., "A fast quantum mechanical algorithm for database search," in *Proceedings of 28th Annual ACM Symp. Theory Comp.*, p. 212 (1996).

200. Grover, L. K., "Quantum mechanics helps in searching for a needle in a haystack," Phys. Rev. Lett. **79**, 325 (1997).

201. Gruska, J., *Quantum computing* (McGraw-Hill; London, 1999).

202. Gudder, S., "On hidden-variables theories," J. Math. Phys. **11**, 431 (1970).

203. Gühne, O., P. Hyllus, D. Bruß, A. Ekert, M. Lewenstein, C. Macchiavello, and A. Sanpera, "Detection of entanglement with few local measurements," Phys. Rev. A **66**, 062305 (2002).

204. Hall, M. J. W., "Universal geometric approach to uncertainty, entropy, and information," Phys. Rev. A **59**, 2602 (1999).

205. Hamming, R. W., "Error detecting and error correcting codes," Bell Syst. Tech. J. **29**, 147 (1950).

206. Hanna, D. C., "Introduction to $\chi^{(2)}$ processes," Quantum Semiclass. Opt. **9**, 131 (1997).

207. Hardy, L., and D. Song, "Universal manipulation of a single qubit," Phys. Rev. A **63**, 032304 (2001).

208. Hayashi, M., *Quantum information: An introduction* (Springer; Berlin, 2006).

209. Hayden, P. M., M. Horodecki, and B. M. Terhal, "The asymptotic entanglement cost of preparing a quantum state," J. Phys. A **34**, 6891 (2001).

210. Heisenberg, W. "Über den anschaulichen Inhalt der quantentheoretischen Kinematik und Mechanik," Z. Physik **43**, 172 (1927).

211. Helstrom, C. W., *Quantum detection and estimation theory* (Academic Press; New York, 1976).

212. Henderson, L., N. Linden, and S. Popescu, "Are all noisy quantum states obtained from pure ones?" Phys. Rev. Lett. **87**, 237901 (2001).

213. Herzog, T. J., P. G. Kwiat, H. Weinfurter, and A. Zeilinger, "Complementarity and the quantum eraser," Phys. Rev. Lett. **75**, 3034 (1995).

214. Hill, S., and W. Wootters, "Entanglement of a pair of quantum bits," Phys. Rev. Lett. **78**, 5022 (1997).

215. Hioe, F. T., and J. H. Eberly, "N-level coherence vector and higher conservation laws in quantum optics," Phys. Rev. Lett. **47**, 838 (1981).

216. Hirvensalo, M., *Quantum computing, 2nd Edition* (Springer-Verlag; Berlin, 2004).

217. Holevo, A. S., "Somes estimates for the information content transmitted by a quantum communication channel," Probl. Pered. Inform. **9**, 3 (1973) [Probl. Inf. Transm. (USSR) **9**, 177 (1973)].

218. Holevo, A. S., "Statistical decisions in quantum theory," J. Multivar. Anal. **3**, 337 (1973).

219. Holevo, A. S., *Statistical structure of quantum theory* (Springer-Verlag; Berlin, 2001).

220. Horne, M., A. Shimony, and A. Zeilinger, "Two-photon interferometry," Phys. Rev. Lett. **62**, 2209 (1989).

221. Horodecki, M., P. Horodecki, and R. Horodecki, "Limits for entanglement measures," Phys. Rev. Lett. **84**, 2014 (2000).

222. Horodecki, M., P. Horodecki, and J. Oppenheim, "Reversible transformations from pure to mixed states, and the unique measure of information," http://xxx.lanl.gov quant-ph/0212019 (2002).

223. Horodecki, P., "Separability criterion and inseparable mixed states with positive partial transposition," Phys. Lett. A **232**, 333 (1997).

224. Horodecki, P., R. Horodecki and M. Horodecki, "Entanglement and thermodynamical analogies," http://xxx.lanl.gov quant-ph/9805072 (1998).

225. Horodecki, M., J. Oppenheim and R. Horodecki, "Are the laws of entanglement thermodynamical?", http://xxx.lanl.gov quant-ph/0207177 (2002).

226. Hradil, Z., J. Summhammer, G. Badurek, and H. Rauch, "Reconstruction of the spin state," Phys. Rev. A **62**, 014101 (2000).

227. Huffman, D. "A method for the construction of minimum-redundancy codes," Proc. IRE **40**, 1098 (1952).

228. Hughes, R. I. G., *The structure and interpretation of quantum mechanics* (Harvard University Press; Cambridge, MA, 1989).

229. Hughston, L. P., R. Jozsa, and W. K. Wootters, "A complete classification of quantum ensembles having a given density matrix," Phys. Lett. A **183**, 14 (1993).

230. Ivanovic, D., "How to differentiate between non-orthogonal states," Phys. Lett. A **123**, 257 (1987).

231. Jackiw, R., and A. Shimony, "The depth and breadth of John Bell's physics," Phys. Perspect. **4**, 78 (2002).

232. Jaeger, G., "Bell gems: The Bell basis generalized," Phys. Lett. A **329**, 425 (2004).

233. Jaeger, G., "The Ehrenfest classification of phase transitions," Archives of History of Exact Sciences **53**, 51 (1997).

234. Jaeger, G., M. A. Horne, and A. Shimony, "Complementarity of one-particle and two-particle interference," Phys. Rev. A **48**, 1023 (1995).

235. Jaeger, G., and A. V. Sergienko, "Entangled states in quantum key distribution," in A. Khrennikov *et al.* (eds.), *Proceedings of the conference on quantum theory, Reconsideration of foundations - 3*, AIP Conf. Proc. **810**, 161 (2006).

236. Jaeger, G., and A. V. Sergienko, "Multi-photon interferometry," in E. Wolf (ed.), *Progress in optics*, Volume 42 (Kluwer Academic; Dordrecht, 2001), Chapter 5.

237. Jaeger, G., A. V. Sergienko, B. E. A. Saleh and M. C. Teich, "Entanglement, mixedness and spin-flip symmetry in multiple-qubit states," Phys. Rev. A **68**, 022318 (2003).

238. Jaeger, G., and A. Shimony, "Optimal distinction between two non-orthogonal quantum states," Phys. Lett. A **197**, 83 (1995).

239. Jaeger, G., A. Shimony, and L. Vaidman, "Two interferometric complementarities," Phys. Rev. A **51**, 54 (1995).

240. Jaeger, G., M. Teodorescu-Frumosu, A. V. Sergienko, B. E. A. Saleh, and M. C. Teich, "Multiphoton Stokes-parameter invariant for entangled states," Phys. Rev. A **67**, 032307 (2003).

241. James, D. F. V., P. G. Kwiat, W. J. Munro, and A. G. White, "Measurement of qubits," Phys. Rev. A **64**, 052312 (2001).

242. Jammer, M., *The philosophy of quantum mechanics* (John Wiley and Sons; New York, 1974).

243. Jarrett, J., "Bell's inequality: A guide to the implications," in Cushing, J. and E. McMullin (eds.) *Philosophical consequences of quantum theory* (Univ. Notre Dame Press; Notre Dame, IN, 1989).

244. Jozsa, R., "Fidelity for mixed quantum states," J. Mod. Opt. **41**, 2315 (1994).

245. Jozsa, R., and B. Schumacher, "A new proof of the quantum noiseless coding theorem," J. Mod. Opt. **41**, 2343 (1994).

246. Jozsa, R., and N. Linden, "On the role of entanglement in quantum computational speed-up," http://xxx.lanl.gov quant-ph/0201143 (2002).

247. Kempe, J., "Multiparticle entanglement and its applications to quantum cryptography," Phys. Rev. A **60**, 910 (1999).

248. Keyl, M., and R. F. Werner, "Estimating the spectrum of a density operator," Phys. Rev. A **64**, 052311 (2001).

249. Kirkpatrick, K. A., "The Schrödinger-HJW theorem," Found. Phys. Lett. **19**, 95 (2006).

250. Kirkpatrick, K. A., "Translation of G. Lüders' *Über die Zustandsänderung durch den Meßprozeß*," Ann. Phys. (Leipzig) **15**, 322 (2006).

251. Kitaev, A. Yu., "Quantum computation: Algorithms and error correction," Uspekhi Mat. Nauk **52**, 53 (1997) (also Russian Mathematical Surveys **52**, 1191 (1997)).

252. Kitaev, A. Yu. A. H. Shen, and M. N. Vyalyi, *Classical and quantum computation*, GSM **47** (American Mathematical Society; Providence, RI, 2002).

253. Klein, O., "Quantum-mechanical foundations of the second law of thermodynamics," Z. Phys. **72**, 767 (1931).

254. Knill, E., reported in [193].

255. Knill, E., and R. Laflamme, "Concatenated quantum codes," http://xxx.lanl.gov quant-ph/9608012 (1996).

256. Knill, E., and R. Laflamme, "Theory of quantum error-correcting codes," Phys. Rev. A **55**, 900 (1997).

257. Knill, E., R. Laflamme, and G. J. Milburn, "A scheme for efficient quantum computation with linear optics," Nature **409**, 46 (2001).

258. Knill, E., R. Laflamme, and W. H. Zurek, "Resilient quantum computation," Science **279**, 46 (1998).

259. Knill, E., R. Laflamme, and W. H. Zurek, "Resilient quantum computation: error models and thresholds," Proc. Roy. Soc. Lond. A **454**, 365 (1998).

260. Kochen, S., and E. Specker, "The problem of hidden variables in quantum mechanics," J. Math. Mech. **17**, 59 (1967).

261. Kraus, K., *et al.*, *States, effects, and operations. Springer Lecture Notes in Physics. SLNP 190* (Springer-Verlag; Berlin, 1983).
262. Kremer, I., "Quantum communication," M.Sc. thesis, Computer Science Department, Hebrew University, 1995.
263. Kullback, S., and R. A. Leibler, "On information and sufficiency," Ann. Math. Stat. **22**, 79 (1951).
264. Kwiat, P. G., A. J. Berglund, J. B. Altepeter, and A. G. White, "Experimental verification of decoherence-free subspaces," Science **290**, 498 (2000).
265. Kwiat, P. G., E. Waks, A. G. White, I. Appelbaum, and P. H. Eberhard, "Ultrabright source of polarization-entangled photons," Phys. Rev. A **58**, R2623 (1999).
266. Laflamme, R., C. Miquel, J. P. Paz, and W. H. Żurek, "Perfect quantum error correcting code," Phys. Rev. Lett. **77**, 198 (1996).
267. Lahti, P., P. Busch, and P. Mittelstaedt, "Some important classes of quantum measurements and their information gain," J. Math. Phys. **32**, 2770 (1991).
268. Landau, L., "Das Dämpfungsproblem in der Wellenmechanik," Z. Phys. **45**, 430 (1927).
269. Landauer, R., "Information is physical," Physics Today, May 1991, p. 23 (1991).
270. Landauer, R., "Irreversibility and heat generation in the computing process," IBM J. Res. Develop. **5**, 183 (1961).
271. Landauer, R., "The physical nature of information," Phys. Lett. A **217**, 188 (1996).
272. Landauer, R., in E. D. Haidemenakis (ed.), *Proceedings of the conference on fluctuation phenomena in classical and quantum systems*, Chiana, Crete, Greece, August 1969 (Gordon and Breach, Science Publishers Inc.; New York, 1970).
273. Law, C. K., I. A. Walmsley, and J. H. Eberly, "Continuous frequency entanglement: Effective finite Hilbert space and entropy control," Phys. Rev. Lett. **84**, 5304 (2000).
274. Leggett, A., B. Ruggiero, and P. Silvestrini (eds.), *Quantum computing and quantum bits in mesoscopic systems* (Springer-Verlag; Berlin, 2003).
275. Leifer, M. S., N. Linden, and A. Winter, "Measuring polynomial invariants of multiparty quantum states," Phys. Rev. A **69**, 052304 (2004).
276. Lendi, K., "Evolution matrix in a coherence vector formulation for quantum Markovian master equations of N-level systems," J. Phys. A: Math. Gen. **20**, 15 (1987).
277. Leonhardt, U., and H. Paul, "Realistic optical homodyne measurements and quasiprobability distributions," Phys. Rev. A **48**, 4598 (1993).
278. Leonhardt, U., "Quantum-state tomography and discrete Wigner function," Phys. Rev. Lett. **74**, 4101 (1995).
279. Levitin, L. B., in *Proceedings of all-union conference on information complexity and control in quantum physics* (Moscow-Tashkent, Tashkent, 1969), Sect. II (in Russian).
280. Levitin, L. B., in A. Blaquieve, S. Diner, and G. Lochak (eds.), *Information complexity and control in quantum physics* (Springer-Verlag; Berlin, 1987), pp. 15.
281. Levitin, L. B., "Physical information theory Part II: Quantum systems," in D. Matzke (ed.), *Workshop on physics and computation: PhysComp '92* (IEEE Computer Society; Los Alamitos, CA 1993), p. 215.

282. Lewenstein, M., and A. Sanpera, "Separability and entanglement of composite quantum systems," Phys. Rev. Lett. **80**, 2261 (1998).

283. Linblad, G. "Completely positive maps and entropy inequalities," Commun. Math. Phys. **40**, 147 (1975).

284. Linden, N., and S. Popescu, "On multiparticle entanglement," Fortschr. Phys. **46**, 567 (1998).

285. Linden, N., S. Popescu, and W. K. Wootters, "The power of reduced quantum states," http://xxx.lanl.gov quant-ph/0207109 (2002).

286. Lloyd, S., "Almost any quantum logic gate is universal," Phys. Rev. Lett. **75**, 346 (1995).

287. Lomonaco, S. J., (ed.) *Quantum Computation* (Proceedings of Symposia in Applied Mathematics), (American Mathematical Society; 2002).

288. Lo, H-K., and S. Popescu, "Concentrating entanglement by local actions— beyond mean values," http://xxx.lanl.gov quant-ph/9707038 (1999).

289. Lo, H.-K., S. Popescu, and T. Spiller (eds.), *Introduction to quantum computation and information* (World Scientific; Singapore, 1998).

290. Lüders, G., "Über die Zustandsänderung durch den Messprozess," Annalen der Physik **8**, 322 (1951).

291. Lütkenhaus, N., "Security against eavesdropping in quantum cryptography," Phys. Rev. A **54**, 97 (1996).

292. MacKay, D. J. C., *Information theory, inference, and learning algorithms* (Cambridge University Press; Cambridge, 2003).

293. Mahler, G., and V. A. Weberruß, *Quantum networks* (Springer-Verlag; Berlin, 1995).

294. Mandel, L., "Coherence and indistinguishability," Opt. Lett. **16**, 1882 (1991).

295. Margenau, H., "Philosophical problems concerning the meaning of measurement in quantum physics," Phil. Sci. **25**, 23 (1958).

296. Menezes, A. J., P. C. van Oorschot, and S. A. Vanstone, *Handbook of applied cryptography* (CRC Press; Boca Raton, 1997).

297. Mermin, N. D., "Simple unified form for the major no-hidden-variables theorems," Phys. Rev. Lett. **65**, 3373 (1990).

298. Mermin, N. D., "What do these corrlations know about reality," Found. Phys. **29**, 571 (1999).

299. Messiah, A., *Quantum mechanics*, Volume I (North-Holland: Amsterdam, 1961).

300. Mielnik, B., "Generalized quantum mechanics," Commun. Math. Phys. **37**, 221 (1974).

301. Miller, G. L., "Riemann's hypothesis and tests for primality," J. Comput. System Sci. **13**, 300 (1976).

302. Mitchell, M. W., C. W. Ellenor, S. Schneider, and A. M. Steinberg, "Diagnosis, prescription, and prognosis of a Bell-state filter by quantum process tomography," Phys. Rev. Lett. **91**, 120402 (2003).

303. Mittelstaedt, P., *The interpretation of quantum mechanics and the measurement process* (Cambridge University Press; Cambridge, 1998).

304. Mittelstaedt, P., *Sprache und Realität in der modernen Physik* (B. I. Wissenschaftsverlag; Mannheim, 1986).

305. Miyake, A., and F. Verstraete, "Multipartite entanglement in $2 \times 2 \times n$ quantum systems," Phys. Rev. A **69**, 012101 (2004).

306. Morikoshi, F., M. Santos, and V. Vedral, "Accessibility of physical states and non-uniqueness of entanglement measure," J. Phys. A: Math. Gen. **37**, 5887 (2004).

307. Moura Alves, C., P. Horodecki, D. K. L. Oi, L. C. Kwek, and A. Ekert, "Direct estimation of functionals of density operators by local operations and classical communication," Phys. Rev. A **68**, 032306 (2003).

308. Muller, A., T. Herzog, B. Huttner, W. Tittel, H. Zbinden, and N. Gisin, " 'Plug and play' systems for quantum cryptography," App. Phys. Lett. **70**, 793 (1997).

309. Munro, W. J., D. F. V. James, A. G. White, and P. G. Kwiat, "Maximizing the entanglement of two mixed qubits," Phys. Rev. A **64**, 030302 (2001).

310. Murdoch, D., *Niels Bohr's philosophy of physics* (Cambridge University Press; Cambridge, 1989).

311. Myers, J. M., "Can a universal quantum computer be fully quantum?" Phys. Rev. Lett. **78**, 1823 (1997).

312. National Institute of Standards and Technology, "Data Encryption Standard (DES)," FIPS Publication 46 (1977).

313. Nielsen, M. A., "Computable functions, quantum measurements, and quantum dynamics," Phys. Rev. Lett. **79**, 2915 (1997).

314. Nielsen, M. A., "On the units of bipartite entanglement: Is sixteen ounces of entanglement always equal to one pound?" J. Phys. A: Math. Gen. **34**, 6987 (2001).

315. Nielsen, M. A., and I. L. Chuang, *Quantum computing and quantum information* (Cambridge University Press; Cambridge, 2000).

316. Nielsen, M. A., and J. Kempe, "Separable states are more disordered globally than locally," Phys. Rev. Lett. **86**, 5184 (2000).

317. Omnès, R., "General theory of the decoherence effect in quantum mechanics," Phys. Rev. A **56**, 3383 (1997).

318. Packel, E. W., and J. F. Traub, "Information-based complexity," Nature **328**, 29 (1987).

319. Palma, G. M., K. Suominen, and A. K. Ekert, "Quantum computers and dissipation," Proc. Roy. Soc. London A **452**, 567 (1996).

320. Pan, J. W., D. Bouwmeester, H. Weinfurter, and A. Zeilinger, "Experimental entanglement swapping: Entangling photons that never interacted," Phys. Rev. Lett. **80**, 3891 (1998).

321. Parker, S., S. Bose, and M. B. Plenio, "Entanglement quantification and purification in continuous-variable systems," Phys. Rev. A **61**, 032305 (2000).

322. Pati, A. K., "Quantum superposition of multiple clones and the novel cloning machine," Phys. Rev. Lett. **83**, 2849 (1999).

323. Pati, A. K., and B. C. Sanders, "No partial erasure of quantum information," http://xxx.lanl.gov quant-ph/0503138 (2005).

324. Pauli, W., "Wellenmechanik," in *Handbuch der Physik, Band 24, I*, 120 (1933).

325. Pavičić, M., *Quantum information and quantum communication* (Springer; Berlin, 2006).

326. Pearson, D., "High-speed QKD reconciliation using forward error correction," Proc. 7th international conference on quantum communication, measurement and computing (QCMC), 299 (2004).

327. Peres, A., "Higher-order Schmidt decompositions," http://xxx.lanl.gov quant-ph/9504006 (1995).

328. Peres, A., "How to differentiate between non-orthogonal quantum states," Phys. Lett. A **128**, 19 (1988).

329. Peres, A., "Separability criterion for density matrices," Phys. Rev. Lett. **77**, 1413 (1996).

330. Peres, A., "Clasical interventions in quantum systems," Phys. Rev. A **61**, 022116 (2000); *ibid.* 022117 (2000).

331. Peres, A., and D. R. Terno, "Quantum information and relativity theory," Rev. Mod. Phys. **76**, 93 (2004).

332. Peres, A., and W. K. Wootters, "Optimal detection of quantum information," Phys. Rev. Lett. **66**, 1119 (1991).

333. Peters, N., J. Altepeter, E. Jeffrey, D. Branning, and P. Kwiat, "Precise creation, characterization and manipulation of single optical qubits," Q. Info. Comp. **3**, 503 (2003).

334. Peters, N., T.-C. Wei, and P. G. Kwiat, "Mixed state sensitivity of several quantum information benchmarks," Phys. Rev. A **70**, 052309 (2004).

335. Pittenger, A. O., *An introduction to quantum computing algorithms. Progress in computer science and applied logic* (Birkhäuser; Boston, 2001).

336. Pitowsky, I., *Quantum probability–quantum logic* (Springer-Verlag; Berlin, 1989).

337. Poincaré, H., *Théorie mathématique de la lumière, II* (Gauthier-Villars; Paris, 1892).

338. Popescu, S., and D. Rohrlich, "Generic quantum nonlocality," Phys. Lett. A **166**, 293 (1992).

339. Popescu, S., and D. Rohrlich, "Thermodynamics and the measure of entanglement," Phys. Rev. A **56**, R3319 (1997).

340. Poyatos, J. F., J. I. Cirac, and P. Zoller, "Complete characterization of a quantum process: the two-bit quantum gate," Phys. Rev. Lett. **78**, 390 (1997).

341. Preskill, J., *Physics 229: Advanced mathematical methods of physics— Quantum computation and information*, California Institute of Technology, 1998. http://www.theory.caltech.edu/people/preskill/ph229/

342. Preskill, J., "Reliable quantum computation," Proc. Roy. Soc. London A **454**, 385 (1998).

343. Preskill, J., "Fault-tolerant quantum computing," in H.-K. Lo, S. Popescu and T. Spiller (eds.), *Introduction to quantum computation and information* (World Scientific; Singapore, 1998), p. 213.

344. Proakis, J. G., and M. Salehi, *Communication systems engineering* (Prentice-Hall; Harlow, 2001).

345. Rahn, B., A. C. Doherty, and H. Mabuchi, "Exact and approximate performance of concatenated quantum codes," http://xxx.lanl.gov quant-ph/0111003 (2001).

346. Raymer, M. G., D. F. McAlister, and U. Leonhardt, "Two-mode quantum-optical state measurement: Sampling the joint density matrix," Phys. Rev. A **54**, 2397 (1996).

347. Razborov, A. A., "Quantum communication complexity of symmetric predicates," Izvestiya: Mathematics **67**, 145 (2003).

348. Redhead, M., *Incompleteness, nonlocality and realism* (Oxford University Press; Oxford, 1987).

349. Řeháček, J., B.-G. Englert, and D. Kaszlikowski, "Minimal qubit tomography," Phys. Rev. A **70**, 052321 (2004).

350. Řeháček, J., Z. Hradil, and M. Ježek, "Iterative algorithm for reconstruction of entangled states," Phys. Rev. A **63**, 040303(R).
351. Reichenbach, H., *Philosophic foundations of quantum mechanics* (University of California; Berkeley, 1994).
352. Rényi, A., *Valószínüségszámítás* (Tankönyvkiadó; Budapest, 1966) [English translation: *Foundations of probability* (North-Holland; Amsterdam, 1970)].
353. Richter, Th., "Interference and non-classical spatial intensity correlations," Quantum Opt. **3**, 115 (1991).
354. Robertson, H. P., "The uncertainty principle," Phys. Rev. **34**, 163 (1929).
355. Rodgers, P., "The double-slit experiment," Physics World editorial, Sept. 2002.
356. Rudolph, T., R. W. Spekkens, and P. S. Turner, "Unambiguous discrimination of mixed states," Phys. Rev. A **68**, 010301(R) (2003).
357. Rungta, P., V. Bužek, C. M. Caves, M. Hillery, and G. J. Milburn, "Universal state inversion and concurrence in arbitrary dimensions," Phys. Rev. A **64**, 042315 (2001).
358. Rungta, P., and C. M. Caves, "Concurrence-based entanglement measures for isotropic states," Phys. Rev. A **67**, 012307 (2003).
359. Sakurai, J. J., *Modern quantum mechanics, Revised edition* (Addison-Wesley; Reading MA, 1994).
360. Saleh, B. E. A., A. F. Abouraddy, A. V. Sergienko, and M. C. Teich, "Duality between partial coherence and partial entanglement," Phys. Rev. A **62**, 043816 (2000).
361. Scarani, V., A. Acín, G. Ribordy, and N. Gisin, "Quantum cryptography robust against photon number splitting attacks for weak laser pulse implementations," Phys. Rev. Lett. **92**, 057901 (2004).
362. Scarani, V., S. Iblisdir, N. Gisin, and A. Acín, "Quantum cloning," Rev. Mod. Phys. **77**, 1225 (2005).
363. Schmidt, E. "Zur Theorie der linearen und nichtlinearen Integralgleichungen," Math. Annalen **63**, 433 (1906).
364. Schrödinger, E. "Quantisierung als Eigenwertproblem," Annalen der Physik **81**, 109 (1926).
365. Schrödinger, E., "Die gegenwaertige Situation in der Quantenmechanik," Die Naturwissenschaften **23**, 807 (1935).
366. Schrödinger, E., "Discussion of probability relations between separated systems," Proc. Cambridge Philos. Soc. **32**, 446 (1935).
367. Schumacher, B. W., "Quantum coding," Phys. Rev. A **51**, 2738 (1995).
368. Schumacher, B. W., "Sending entanglement through noisy quantum channels," Phys. Rev. A **54**, 2614 (1996).
369. Schumacher, B. W., and M. A. Nielsen, "Quantum data processing and error correction," Phys. Rev. A **54**, 2629 (1996).
370. Schumacher, B. W., and M. D. Westmoreland, "Quantum mutual information and the one-time pad," http://xxx.lanl.gov quant-ph/0604207 (2006).
371. Schumacher, B. W., M. D. Westmoreland, and W. K. Wootters, "Limitation on the amount of accessible information in a quantum channel," Phys. Rev. Lett. **76**, 3452 (1996).
372. Scully, M. O., B.-G. Englert, and J. Schwinger, "Spin coherence and Humpty-Dumpty. III. The effects of observation," Phys. Rev. A **40**, 1775 (1989).
373. Scully, M. O., B.-G. Englert, and H. Walther, "Quantum optical tests of complementarity," Nature **351**, 111 (1991).

374. Scully, M. O., and M. S. Zubairy, *Quantum optics* (Cambridge University Press; Cambridge 1997).

375. Sergienko, A. V. (ed.), *Quantum communications and cryptography* (CRC Press; Boca Raton, 2005).

376. Sergienko, A. V., and G. S. Jaeger, "Quantum information processing and precise optical measurement with entangled-photon pairs," Contemp. Phys. **44**, 341 (2003).

377. Sergienko, A. V., M. Atatüre, Z. Walton, G. Jaeger, B. E. A. Saleh, and M. C. Teich, "Quantum cryptography using femto-second pulsed parametric down-conversion," Phys. Rev. A **60**, R2622 (1999).

378. Shannon, C. E., "A mathematical theory of communication," Bell System Technical Journal **27**, 379 (1948); *ibid.*, 623 (1948).

379. Shannon, C. E., and W. Weaver, *The mathematical theory of communication* (University of Illinois; Urbana, IL, 1949).

380. Shimony, A., "Approximate measurement in quantum mechanics II," Phys. Rev. D **9**, 2317; *ibid.*, 2321 (1974).

381. Shimony, A., "Conceptual foundations of quantum mechanics," in Paul Davies (ed.), *The new physics* (Cambridge University Press; Cambridge, 1989), Ch. 13.

382. Shimony, A., "Controllable and uncontrollable non-locality," in S. Kamefuchi *et al.* (eds), *Proceedings of the international symposium on foundations of quantum mechanics in the light of new technology* (Physical Society of Japan; Tokyo, 1984), p. 25.

383. Shimony, A., "Degree of entanglement," Ann. N.Y. Acad. Sci. **755**, 675 (1995).

384. Shimony, A., *Search for a naturalistic worldview, Volume II* (Cambridge University Press; Cambridge, 1993).

385. Shimony, A., "The logic of EPR," Annales de la Fondation Louis de Broglie **26**, 399 (2001).

386. Shimony, A. "Bell's theorem," *Standford encyclopedia of philosophy* (Summer 2005 edition), E. N. Zalka (ed.), http://plato.stanford.edu/entries/bell-theorem

387. Shimony, A., M. A. Horne, and J. F. Clauser, "An exchange on local beables," Dialectica **39**, 97; *ibid.*, 107 (1985).

388. Shor, P. W., "Algorithms for quantum computation: Discrete logarithm and factoring," *Proceedings of 35th annual symposium on foundations of computer science* (IEEE Press, 1994), p. 124.

389. Shor, P. W., "Fault-tolerant quantum computation," in *Proceedings of 37th Annual symposium on foundations of computer science, Los Alamitos, CA,* (IEEE Computer Science Press; Silver Spring MD, 1996), p. 56.

390. Shor, P. W., "Polynomial-time algorithms for prime number factorization and discrete logarithms on a quantum computer," SIAM J. Comp. **26**, 1484 (1997).

391. Shor, P. W., "Scheme for reducing decoherence in quantum computer memory," Phys. Rev. A **52**, R2493 (1995).

392. Shor, P. W., J. A. Smolin, and B. M. Terhal, "Nonadditivity of bipartite distillable entanglement follows from conjecture on bound entangled Werner states," Phys. Rev. Lett. **86**, 2681 (2001).

393. Simon, C., G. Weihs, and A. Zeilinger, "Optimal quantum cloning and universal NOT without quantum gates," J. Mod. Opt. **47**, 233 (2000).

394. Simon, D. R., "On the power of quantum computation," in S. Goldwasser (ed.), *Proceedings of 35th annual symposium on the foundations of computer science* (IEEE Society Press; Los Alamitos CA, 1994), p. 116.

395. Sipser, M., "The history and status of the P versus NP question," *Proceedings of 24th ACM symposium on the theory of computing* (1992), p. 603.

396. Slutsky, B. A., R. Rao, L. Tancevski, and S. Fainman, "Defense frontier analysis of quantum cryptographic systems," Applied Optics **37**, 2869 (1998).

397. Slutsky, B. A., R. Rao, P. C. Sun, and Y. Fainman, "Security of quantum cryptography against individual attacks," Phys. Rev. A **57**, 2383 (1998).

398. Srinivas, M. D.. "Collapse postulate for observables with continuous spectra," Comm. Math. Phys. **71**, 131 (1980).

399. Stachel, J., "Do quanta need a new logic?" in R. G. Colodny (ed.), *From quarks to quasars* (Univ. Pittsburgh Press; Pittsburgh, 1986), p. 229.

400. Stairs, A., "Quantum logic and the Lüders rule," Phil. Sci. **49**, 422 (1982).

401. Steane, A., "Multiple particle interference and quantum error correction," Proc. Roy. Soc. London A **452**, 2551 (1996).

402. Steane, A., "Error correcting codes in quantum theory," Phys. Rev. Lett. **77**, 793 (1996).

403. Steane, A., "Quantum computing," Rep. Prog. Phys. **61**, 117 (1998).

404. Stenholm, S., and K.-A. Suominen, *Quantum approaches to informatics* (Wiley–Interscience; Hoboken, NJ, 2005).

405. Sternberg, S., *Group theory and physics* (Cambridge University Press; Cambridge, 1994), Sect. 1.2.

406. Stokes, G. G., "On the composition and resolution of streams of polarized light from different sources," Trans. Camb. Philos. Soc. **9**, 399 (1852).

407. Stolze, J., and Suter, D., *Quantum computing: A short course from theory to experiment* (John Wiley and Sons; Hoboken NJ, 2004).

408. Strauss, M., "Zur Begründung der statistischen Transformationstheorie der Quantenphysik," Berliner Berichte 1936, 382 (1936). [English translation, "The Logic of complementarity and the foundation of quantum theory," in Strauss, M., *Modern physics and its philosophy*, (Reidel; Dordrecht 1972), p. 186.]

409. Sun, Y., J. Bergou, and M. Hillery, "Optimal unambiguous discrimination between subsets of nonorthogonal quantum states," Phys. Rev. A **66**, 032315 (2002).

410. Sun, X., S. Zhang, Y. Feng, and M. Ying, "Mathematical nature of and a family of lower bounds for the success of unambiguous discrimination," Phys. Rev. A **65**, 044306 (2002).

411. Svozil, K., *Quantum logic* (Springer-Verlag; Berlin, 1998).

412. Svozil, K., "Quantum logic. A brief outline," http://xxx.lanl.gov quant-ph/9902042 (1999).

413. Takeuchi, S. "Experimental demonstration of a three-qubit quantum computation algorithm using a single photon and linear optics," Phys. Rev. A **62**, 032301 (2000).

414. Teodorescu-Frumosu, M., and G. S. Jaeger, "Quantum Lorentz-group invariants of n-qubit systems," Phys. Rev. A **67**, 032307 (2003).

415. Terhal, B. M., "Detecting quantum entanglement," Theor. Comp. Sci. **287**, 313 (2002).

416. Ter Haar, D., "Theory and applications of the density matrix," Rep. Prog. Phys. **24**, 304 (1961).

417. Terhal, B. M., "A family of indecomposable positive linear maps based on entangled quantum states," Lin. Algebr. Appl. **323**, 61 (2001).
418. Thapliyal, A. V., "Multipartite pure-state entanglement," Phys. Rev. A **59**, 3336 (1999).
419. Thapliyal, A. V., reported in [48].
420. Thew, R. T., K. Nemoto, A. G. White, W. J. White, and W. J. Munro, "Qudit quantum-state tomography," Phys. Rev. A **66**, 012303 (2002).
421. Timpson, G. T., "Quantum information theory and the foundations of quantum mechanics," Ph.D. Thesis, Queen's College, The University of Oxford, 2004; also http://xxx.lanl.gov quant-ph/0412063.
422. Tittel, W., J. Brendel, N. Gisin, and H. Zbinden, "Long-distance Bell-type tests using energy-time entangled photons," Phys. Rev. A **59**, 4150 (1999).
423. Toffoli, T., "Reversible computing," in G. Goos and J. Hartmanis (eds.), *Automata, languages and programming*, Lecture Notes in Computer Science **85** (Springer-Verlag; Berlin, 1980) p. 632.
424. Tsirel'son, B. S., "Quantum generalizations of Bell's inequalities," Lett. Math. Phys. **4**, 93 (1980).
425. Tsirel'son, B. S., "Quantum analogues of the Bell inequalities," J. of Sov. Math. **36**, 557 (1987).
426. Turing, A. M., "On computable numbers, with an application to the Entscheidungsproblem," Proc. London Math. Soc. **42**, 230 (1936); *ibid.*, **43**, 544 (1937).
427. Uffink, J., "Measures of uncertainty and the uncertainty principle," Ph.D. Dissertation, University of Utrecht, 1990.
428. Umegaki, H., "Conditional expectation in an operator algebra, IV (entropy and information)," Ködei Math. Sem. Rep. **14**, 59 (1962).
429. Unruh, W. G., "Maintaining coherence in quantum computers," Phys. Rev. A **51**, 992 (1995).
430. Van Fraassen, B., *Quantum mechanics* (Clarendon; Oxford, 1991).
431. Varadarajan, V. S., *Geometry of quantum theory, 2nd Edition* (Springer-Verlag; Berlin, 1985).
432. Vedral, V., "The role of relative entropy in quantum information processing," Rev. Mod. Phys. **74**, 197 (2002).
433. Vedral, V., and E. Kashefi, "Uniqueness of the entanglement measure for bipartite pure states and thermodynamics," Phys. Rev. Lett. **89**, 037903 (2002).
434. Vernam, G. S., "Secret signaling system," U.S. Patent No. 1,310,719 (22 July 1919).
435. Vernam, G. S., "Cipher printing telegraph systems for secret wire and radio telegraphic communications," J. American Inst. Elec. Eng. **55**, 109 (1926).
436. Verstraete, F., J. Dehaene, and B. De Moor, "Lorentz singular-value decomposition and its application to pure states of three qubits," Phys. Rev. A **65**, 032308 (2002).
437. Vidal, G., "Entanglement monotones," J. Mod. Opt. **47**, 355 (2000).
438. Vidal, G., and J. I. Cirac, "Irreversibility in asymptotic manipulations of entanglement," Phys. Rev. Lett. **86**, 5803 (2001).
439. Vidal, G., D. Jonathan, and M. A. Nielsen, "Approximate transformations and robust manipulation of bipartite pure-state entanglement," Phys. Rev. A **62**, 012304 (2000).
440. Vidal, G., and R. F. Werner, "Computable measures of entanglement," Phys. Rev. A **65**, 032314 (2002).

441. Vogel, K., and Risken, H., "Determination of quasiprobability distributions in terms of probability distributions for the rotated quadrature phase," Phys. Rev. A **40**, 2847 (1989).

442. Vollbrecht, K.-G. H., and M. M. Wolf, "Conditional entropies and their relation to entanglement criteria," J. Math. Phys. **43**, 4299 (2002).

443. Von Neumann, J., "Mathematische Begründung der Quantenmechanik" Gött. Nach., Session of May 20, 1 (1927).

444. Von Neumann, J., *Mathematische Grundlagen der Quantenmechanik* (Julius Springer; Berlin, 1932) [English translation: *Mathematical foundations of quantum mechanics* (Princeton University Press; Princeton, NJ, 1955)].

445. Von Weizsäcker, C., "Die Quantentheorie der einfachen Alternative," Zeitschrift für Naturforschung **13a**, 245 (1958).

446. Wehrl, A., "General properties of entropy," Rev. Mod. Phys. **50**, 221 (1978).

447. Wei, T.-C., J. B. Altepeter, P. M. Goldbart, and W. J. Munro, "Measures of entanglement in multipartite bound entangled states," Phys. Rev. A **70**, 022322 (2004).

448. Wei, T.-C., and P. M. Goldbart, "Geometric measure of entanglement and applications to bipartite and multipartite States," Phys. Rev. A **68**, 042307 (2003).

449. Weihs, G., T. Jennewein, C. Simon, H. Weinfurter, and A. Zeilinger, "Violation of Bell's inequality under strict Einstein locality conditions," Phys. Rev. Lett. **81**, 5039 (1998).

450. Werner, R. F., "Quantum states with Einstein–Podolsky–Rosen correlations admitting a hidden-variable model," Phys. Rev. A **40**, 4277 (1989).

451. Werner, R. F., and M. M. Wolf, "Bell inequalities and entanglement," http://xxx.lanl.gov quant-ph/0107093.

452. Wheeler, J. A., "Information, physics, quantum: The search for links," in [476].

453. Wheeler, J. A., and W. H. Żurek (eds.), *Quantum theory and measurement* (Princeton University Press; Princeton, NJ, 1983).

454. White, A. G., D. F. V. James, P. H. Eberhard, and P. G. Kwiat, "Nonmaximally entangled states: Production, characterization, and utilization," Phys. Rev. Lett. **83**, 3103 (1999).

455. Wiesner, S., "Conjugate coding," SIGACT News **15**, 78 (1983).

456. Wigner, E., "On hidden variables and quantum mechanical probabilities," Am. J. Phys. **38**, 1005 (1970).

457. Wong, A., and N. Christensen, "Potential multiparticle entanglement measure," Phys. Rev. A **63**, 044301 (2001).

458. Wootters, W. K., "Statistical distance and Hilbert space," Phys. Rev. D **23**, 357 (1981).

459. Wootters, W. K., "A Wigner-function formulation of finite-state quantum mechanics," Ann. Phys. (N.Y.) **176**, 1 (1987).

460. Wootters, W. K., "Entanglement of formation of an arbitrary state of two qubits," Phys. Rev. Lett. **80**, 2245 (1998).

461. Wootters, W. K., and W. H. Żurek, "A single quantum cannot be cloned," Nature **299**, 802 (1982).

462. Wootters, W. K., and W. H. Żurek, "Complementarity in the double-slit experiment: Quantum nonseparability and a quantitative statement of Bohr's principle," Phys. Rev. D **19**, 473 (1979).

463. Wu, C. S., and I. Shaknov, "The angular correlation of scattered annihilation radiation," Phys. Rev. **77**, 136 (1950).

464. Yao, A. C.,"Quantum circuit complexity," in *Proceedings of 34th annual IEEE symposium on foundations of computer science* (IEEE Press; New York, 1993), p. 352.

465. Yu, T., and J. H. Eberly, "Qubit disentanglement and decoherence via dephasing," Phys. Rev. B **68**, 165322 (2003).

466. Yuen, H. P., R. S. Kennedy, and M. Lax, "Optimal testing of multiple hypotheses in quantum detection theory," IEEE Trans. Inform. Theory **IT-21**, 125 (1975).

467. Zalka, C., "Grover's quantum searching algorithm is optimal," Phys. Rev. A **60**, 2746 (1999).

468. Zanardi, P., and M. Rasetti, "Error avoiding codes," Mod. Phys. Lett. B **11**, 1085 (1997).

469. Zanardi, P., and M. Rasetti, "Noiseless quantum codes," Phys. Rev. Lett. **79**, 3306 (1997).

470. Zeh, H. D., "On the interpretation of measurement in quantum theory," Found. Phys. **1**, 69 (1970).

471. Zeilinger, A., "A foundational principle for quantum mechanics," Found. Phys. **29**, 631 (1999).

472. Zeilinger, A., "Quantum entanglement: A fundamental concept finding its applications," Physica Scripta **T-76**, 203 (1998).

473. Zeilinger, A., H. J. Bernstein, and M. A. Horne, "Information transfer with two-state two-particle quantum systems," J. Mod. Opt. **41**, 2375 (1994)

474. Żukowski, M., A. Zeilinger, M. A. Horne, and A. K. Ekert, " 'Event-ready-detectors' Bell experiment via entanglement swapping," Phys. Rev. Lett. **71**, 4287 (1993).

475. Żurek, W., in J. J. Halliwell *et al.* (eds.), *Physical origins of time asymmetry* (Cambridge Univ. Press; Cambridge, 1994), p. 175.

476. Żurek, W. H. (ed.), *Complexity, entropy and the physics of information* (Addison-Wesley, Redwood City, CA, 1990).

477. Żurek, W., "Decoherence, einselection, and the origins of the classical," Rev. Mod. Phys. **75**, 715 (2003).

Index